AF411407

# Bacillus subtilis as a Model Organism to Study Basic Cell Processes

# Bacillus subtilis as a Model Organism to Study Basic Cell Processes

Editor

**Imrich Barák**

MDPI • Basel • Beijing • Wuhan • Barcelona • Belgrade • Manchester • Tokyo • Cluj • Tianjin

*Editor*
Imrich Barák
Department of Microbial
Genetics
Institute of Molecular Biology,
SAS
Bratislava
Slovakia

*Editorial Office*
MDPI
St. Alban-Anlage 66
4052 Basel, Switzerland

This is a reprint of articles from the Special Issue published online in the open access journal *Microorganisms* (ISSN 2076-2607) (available at: www.mdpi.com/journal/microorganisms/special_issues/Bacillus_Subtilis).

For citation purposes, cite each article independently as indicated on the article page online and as indicated below:

LastName, A.A.; LastName, B.B.; LastName, C.C. Article Title. *Journal Name* **Year**, *Volume Number*, Page Range.

**ISBN 978-3-0365-3744-3 (Hbk)**
**ISBN 978-3-0365-3743-6 (PDF)**

# Contents

 *microorganisms*

*Editorial*

# Special Issue "*Bacillus subtilis* as a Model Organism to Study Basic Cell Processes"

Imrich Barák 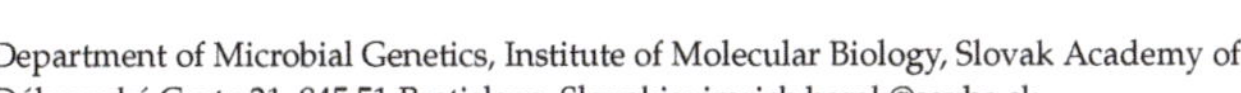

Department of Microbial Genetics, Institute of Molecular Biology, Slovak Academy of Sciences, Dúbravská Cesta 21, 845 51 Bratislava, Slovakia; imrich.barak@savba.sk

**Citation:** Barák, I. Special Issue "*Bacillus subtilis* as a Model Organism to Study Basic Cell Processes". *Microorganisms* **2021**, *9*, 2459. https://doi.org/10.3390/microorganisms9122459

Received: 31 October 2021
Accepted: 15 November 2021
Published: 28 November 2021

**Publisher's Note:** MDPI stays neutral with regard to jurisdictional claims in published maps and institutional affiliations.



*Bacillus subtilis* has served as a model microorganism for many decades. There are a few important reasons why this Gram-positive bacterium became a model organism to study basic cell processes. Firstly, *B. subtilis* has proven highly amenable to genetic manipulation and has become widely adopted as a model organism for laboratory studies. Secondly, it is considered as the Gram-positive equivalent of *Escherichia coli*, an extensively studied Gram-negative bacterium, and both microorganisms have served as examples for bacterial cell division studies in the last decades. Thirdly, unlike *E. coli*, it can form endospores and the sporulation represents the simplest cellular differentiation process. In addition, this bacterium became an exceptional specimen to study motility, chromosome segregation, competence, host system for bacteriophages, transcription regulation, biofilm formation, and *B. subtilis* is also widely used to produce various enzymes, such as proteases, amylase, etc.

The objective of this Special Issue on "*Bacillus subtilis* as a Model Organism to Study Basic Cell Processes" was to provide a platform to researchers for sharing their new studies on advances in basic cell processes studies in the model organism *B. subtilis*. The Special Issue includes papers from several leading scientists in the field studying sporulation, biofilm formation, bacteriophages, and transcription in this bacterium and its close relatives. The present Special Issue comprises nine research articles and two reviews.

Two articles were focused on studies of transcription and gene expression regulation in *B. subtilis*. Transcription of specific genes and production of physiologically relevant proteins is crucial for the adaptation of bacteria to changing environmental conditions. Sudzinová et al. [1] determined the role of DNA topology on transcription from rRNA promoters. Interestingly, they showed that the more negative DNA supercoiling in the exponential growth phase increases transcription from rRNA promoters. On the other hand, DNA relaxation in the stationary phase contributes to a decrease in their activity. This study points out the importance of DNA topology for the expression of rRNA and this is directly linked to nutrient availability. Vohradsky et al. [2] used an in silico approach with literature-mined data, gene expression modelling, and promoter sequence analysis to identify the regulon of *B. subtilis* sigma factor $\sigma^B$ as a subunit of RNA polymerase. They concentrated on $\sigma^B$-dependent genes expressed specifically during germination and the outgrowth of spores. They determined the binding motif of a subset of $\sigma^B$ regulated genes during these parts of *B. subtilis* life cycle. Importantly, they also experimentally verified sixteen selected SigB dependant promoters.

Ermi et al. [3] studied the influence of Non-B DNA structures on stationary-phase mutagenesis in *B. subtilis*. They concentrated on G4 DNA and hairpin-forming motifs and they showed that these non-B DNA-forming structures promote genetic instability and may have spatial and temporal mutagenic effects. Faßhauer et al. [4] analyzed the role of two *B. subtilis* cold shock proteins, CspB and CspD proteins. These proteins belong to the most abundant proteins in the cell and thus it suggesting their possible crucial function. Deletion of either of the genes has no clear effect on phenotype. However, the simultaneous loss of both proteins results in severe growth effects and the appearance of suppressor mutations. Interestingly, the global RNA profile of the double mutant suggests that these proteins are important for transcription elongation and termination.

The next three articles are focused on different aspects of biofilm formation. *B. subtilis* and related bacteria belong to bacteria that can form a surface-associated multicellular assemblage, so-called biofilms. Dergham et al. [5] concentrated on the multi-culturing comparison between macrocolony, swarming, pellicle, and submerged biofilm of undomesticated *B. subtilis* NDmed strain. By using different 15 mutant strains they identified genes important for all biofilm phenotypes. Špacapan et al. [6] studied the role of quorum sensing protein ComX on biofilm formation. Their results indicate that ComX can mediate early commitment to sporulation and slow down biofilm formation. This quorum-sensing system can be important for modulating the coexistence of multiple biological states, biofilm formation, and sporulation. Stoll et al. [7] studied the surfactin production of two different *Bacillus velezensis* strains and analyzed its impacts on biofilm formation and stable colonization on different plant surfaces which finally enables its activity as an elicitor of induced systemic resistance.

*B. subtilis* has been especially well known for many decades of studies on sporulation, an important mechanism by which bacteria can survive harsh environmental conditions, and it also represents the simplest cell differentiation process. The spore resistance arises from several protective layers that surround the spore core. In addition to the cortex, a peptidoglycan layer, the coat is the main defense system against the challenges of the environment. More than 80 coat proteins are localized into four distinct morphological layers. Krajčíková et al. [8] investigated the interactions between SpoVM and SpoIVA and the proteins essential for cortex synthesis. They found and characterized their protein partners localized in the basement spore coat layer. Their results are linking the processes of cortex and coat formation. Tu et al. [9] analyzed the wet heat resistance of spores of *Bacillus subtilis* A163 in comparison with laboratory *B. subtilis*, PY79 strain. They showed extremely high resistance to wet heat compared to spores of laboratory strain. They determined the proteome of vegetative and sporulating cells of both strains A163 and PY79 and the results revealed proteomic differences of the two strains what should help to explain the high heat resistance of *B. subtilis* A163 spores.

This Special Issue also includes two review articles. Rismondo and Schulz [10] provided an overview of the alternative roles of ABC transporters affecting antibiotic resistance, cell wall biosynthesis, cell division, and sporulation in *B. subtilis* and *Listeria monocytogenes*. The authors focused on several ABC transporters, which are not only required to inactivate or export drugs but are essential for drug sensing, and on ABC transporters, which affect cell wall biosynthesis and remodeling. The second review by Łubkowska et al. [11] provides a reference compilation of 56 bacteriophages, which infect the species of thermophilic 'Bacillus group' bacteria. This review of bacteriophages of thermophiles is an important overview not only of their previously described characteristics but also their roles in many biogeochemical, ecological processes, and biotechnology applications, including emerging bionanotechnology.

**Funding:** The work in author's laboratory is supported by VEGA—Grant No. 2/0001/21 from the Slovak Academy of Sciences, and Grant from the Slovak Research and Development Agency under contract APVV-18-0104.

**Informed Consent Statement:** Not applicable.

**Acknowledgments:** I would like to thank all the authors who contributed their papers to this Special Issue and the reviewers for their invaluable revisions. I am also grateful to all the members of the Editorial Office of *Microorganisms*, and especially to Athena Wang, for their continuous support.

**Conflicts of Interest:** The author declares no conflict of interest.

# References

1. Sudzinová, P.; Kambová, M.; Ramaniuk, O.; Benda, M.; Šanderová, H.; Krásný, L. Effects of DNA Topology on Transcription from rRNA Promoters in *Bacillus subtilis*. *Microorganisms* **2021**, *9*, 87. [CrossRef] [PubMed]
2. Vohradsky, J.; Schwarz, M.; Ramaniuk, O.; Ruiz-Larrabeiti, O.; Vaňková Hausnerová, V.; Šanderová, H.; Krásný, L. Kinetic Modeling and Meta-Analysis of the *Bacillus subtilis* SigB Regulon during Spore Germination and Outgrowth. *Microorganisms* **2021**, *9*, 112. [CrossRef] [PubMed]
3. Ermi, T.; Vallin, C.; García, A.; Bravo, M.; Cordero, I.; Martin, H.; Pedraza-Reyes, M.; Robleto, E. Non-B DNA-Forming Motifs Promote Mfd-Dependent Stationary-Phase Mutagenesis in *Bacillus subtilis*. *Microorganisms* **2021**, *9*, 1284. [CrossRef] [PubMed]
4. Faßhauer, P.; Busche, T.; Kalinowski, J.; Mäder, U.; Poehlein, A.; Daniel, R.; Stülke, J. Functional Redundancy and Specialization of the Conserved Cold Shock Proteins in *Bacillus subtilis*. *Microorganisms* **2021**, *9*, 1434. [CrossRef] [PubMed]
5. Dergham, Y.; Sanchez-Vizuete, P.; Le Coq, D.; Deschamps, J.; Bridier, A.; Hamze, K.; Briandet, R. Comparison of the Genetic Features Involved in *Bacillus subtilis* Biofilm Formation Using Multi-Culturing Approaches. *Microorganisms* **2021**, *9*, 633. [CrossRef] [PubMed]
6. Špacapan, M.; Danevčič, T.; Štefanic, P.; Porter, M.; Stanley-Wall, N.; Mandic-Mulec, I. The ComX Quorum Sensing Peptide of *Bacillus subtilis* Affects Biofilm Formation Negatively and Sporulation Positively. *Microorganisms* **2020**, *8*, 1131. [CrossRef] [PubMed]
7. Stoll, A.; Salvatierra-Martinez, R.; González, M.; Araya, M. Implication of surfactin in the biofilm formation of *Bacillus velezensis* on plant surfaces and its ISR effect. *Microorganisms* **2021**, *9*, 2251. [CrossRef]
8. Krajčíková, D.; Bugárová, V.; Barák, I. Interactions of *Bacillus subtilis* Basement Spore Coat Layer Proteins. *Microorganisms* **2021**, *9*, 285. [CrossRef] [PubMed]
9. Tu, Z.; Setlow, P.; Brul, S.; Kramer, G. Molecular Physiological Characterization of a High Heat Resistant Spore Forming *Bacillus subtilis* Food Isolate. *Microorganisms* **2021**, *9*, 667. [CrossRef] [PubMed]
10. Rismondo, J.; Schulz, L. Not Just Transporters: Alternative Functions of ABC Transporters in *Bacillus subtilis* and *Listeria monocytogenes*. *Microorganisms* **2021**, *9*, 163. [CrossRef] [PubMed]
11. Łubkowska, B.; Jeżewska-Frąckowiak, J.; Sobolewski, I.; Skowron, P. Bacteriophages of Thermophilic '*Bacillus* Group' Bacteria—A Review. *Microorganisms* **2021**, *9*, 1522. [CrossRef] [PubMed]

 *microorganisms*

*Article*

# Effects of DNA Topology on Transcription from rRNA Promoters in *Bacillus subtilis*

Petra Sudzinová, Milada Kambová, Olga Ramaniuk, Martin Benda, Hana Šanderová and Libor Krásný *

Institute of Microbiology of the Czech Academy of Sciences, 142 00 Prague, Czech Republic;
petra.sudzinova@biomed.cas.cz (P.S.); mkambova@gmail.com (M.K.); ramaniuk@biomed.cas.cz (O.R.);
mb.martin.benda@gmail.com (M.B.); sanderova@biomed.cas.cz (H.Š.)
* Correspondence: krasny@biomed.cas.cz

**Abstract:** The expression of rRNA is one of the most energetically demanding cellular processes and, as such, it must be stringently controlled. Here, we report that DNA topology, i.e., the level of DNA supercoiling, plays a role in the regulation of *Bacillus subtilis* $\sigma^A$-dependent rRNA promoters in a growth phase-dependent manner. The more negative DNA supercoiling in exponential phase stimulates transcription from rRNA promoters, and DNA relaxation in stationary phase contributes to cessation of their activity. Novobiocin treatment of *B. subtilis* cells relaxes DNA and decreases rRNA promoter activity despite an increase in the GTP level, a known positive regulator of *B. subtilis* rRNA promoters. Comparative analyses of steps during transcription initiation then reveal differences between rRNA promoters and a control promoter, P*veg*, whose activity is less affected by changes in supercoiling. Additional data then show that DNA relaxation decreases transcription also from promoters dependent on alternative sigma factors $\sigma^B$, $\sigma^D$, $\sigma^E$, $\sigma^F$, and $\sigma^H$ with the exception of $\sigma^N$ where the trend is the opposite. To summarize, this study identifies DNA topology as a factor important (i) for the expression of rRNA in *B. subtilis* in response to nutrient availability in the environment, and (ii) for transcription activities of *B. subtilis* RNAP holoenzymes containing alternative sigma factors.

**Keywords:** *Bacillus subtilis*; transcription; ribosomal RNA; DNA topology

**Citation:** Sudzinová, P.; Kambová, M.; Ramaniuk, O.; Benda, M.; Šanderová, H.; Krásný, L. Effects of DNA Topology on Transcription from rRNA Promoters in *Bacillus subtilis*. *Microorganisms* **2021**, *9*, 87. https://doi.org/10.3390/microorganisms9010087

Received: 2 December 2020
Accepted: 17 December 2020
Published: 1 January 2021

**Publisher's Note:** MDPI stays neutral with regard to jurisdictional claims in published maps and institutional affiliations.



## 1. Introduction

Bacterial cells need to adapt to environmental changes. In nutrient-rich environments, cells grow and divide rapidly, and this requires a large number of ribosomes to satisfy the need for new proteins. In nutritionally poor environments, the synthesis of new ribosomes stops. As the production of new ribosomes is energetically costly for the cell, it must be tightly regulated. The number of ribosomes in the cell is regulated mainly on the level of transcription initiation of ribosomal RNA (rRNA) [1].

Transcription initiation can be divided into several steps. First, when the RNA polymerase (RNAP) holoenzyme (the core RNAP subunits [$\alpha 2\beta\beta'\omega$] in complex with a $\sigma$ factor) binds to specific DNA sequences, promoters, it forms the closed complex where DNA is still in the double-helical form [2]. The specificity of RNAP for promoter sequences is provided by the $\sigma$ factor [3–6]. Subsequently, this complex isomerizes and forms the open complex where the two DNA strands are unwound, and the transcription bubble is formed. At this stage, initiating nucleoside triphosphates NTPs (iNTPs) can enter the active site and transcription can begin. RNAP then leaves the promoter and enters the elongation phase of transcription [7].

In bacteria, the concentrations of iNTPs act as key regulators of transcription and directly affect RNAP at some promoters. These promoters form relatively unstable open complexes where the time window available to iNTPs to penetrate into the active site and initiate transcription is relatively short. The higher the concentration of the respective iNTP, the higher the chance that it penetrates into the active site while the transcription bubble is still open. Hence, increases in intracellular concentrations of iNTPs stimulate transcription whereas low levels of iNTPs result in inefficient transcription initiation [8–10].

rRNA promoters are prime examples of where transcription initiation is regulated by the concentration of the iNTP. In *Bacillus subtilis*, a model soil-dwelling, spore-forming Gram-positive bacterium, the iNTP of the tandem rRNA promoters of all 10 rRNA operons is exclusively GTP [11]. The GTP level in *B. subtilis* is affected by (p)ppGpp, an alarmone that is produced at times of stress, such as amino acid starvation or heat shock. (p)ppGpp inhibits GuaB, the first enzyme in the de novo GTP biosynthesis pathway, which results in decreased GTP levels and increased ATP levels as more of the last common intermediate for the synthesis of both GTP and ATP, inosine monophosphate (IMP), is now available for ATP synthesis only [12,13]. By affecting the GTP level (p)ppGpp indirectly affects the activity of rRNA promoters in *B. subtilis* [14–16]. This might be similar in other Gram-positive microorganisms, such as *Staphylococcus aureus*, where the GTP concentration ([GTP]) affects rRNA promoter activity under stringent conditions [17].

Another important factor for transcription initiation in bacteria is the topological state of DNA, i.e., the levels of supercoiling. DNA in the cells is usually underwound and this results in negative supercoiling [18]. Negative supercoiling then helps RNAP to melt DNA in promoter regions. In general, bacterial cells display more pronounced negative supercoiling in exponential than in stationary phase of growth and initiation from a number of promoters is sensitive to this parameter [19–23].

Here, we investigated how the activity of rRNA promoters in *B. subtilis* changes when the cells transition from exponential to stationary phase. These promoters depend on the primary σ factor, $\sigma^A$. We show that their activity decreases during the transition and this correlates with a decrease in the GTP concentration. Nevertheless, there is a point in the process where the level of GTP does not decrease any further but the activity of rRNA promoters does. We show that besides [GTP], *B. subtilis* rRNA promoters are regulated by the level of their supercoiling, and we dissect the effects of supercoiling on the formation of closed and open complexes, thereby providing mechanistic insights into the process. Finally, we show that supercoiled (SC) DNA is a more efficient template for transcription for all alternative σ factors tested with the exception of $\sigma^N$, a recently discovered sigma factor encoded on the pBS32 plasmid of the NCIB 3610 strain [24,25]. In summary, a newly updated model of *B. subtilis* promoter regulation is presented here.

## 2. Materials and Methods

### 2.1. Media and Growth Conditions

Cells were grown at 37 °C, either in LB or in rich MOPS supplemented with 20 amino acids: 50 mM MOPS (pH 7.0), 1 mM $(NH_4)_2SO_4$, 0.5 mM $KH_2PO_4$, 2 mM $MgCl_2$, 2 mM $CaCl_2$, 50 μM $MnCl_2$, 5 μM $FeCl_3$, amino acids (50 μg/mL each), and 0.4% glucose. Antibiotics used: ampicillin 100 μg/mL, chloramphenicol 5 μg/mL, novobiocin 5 μg/mL, and rifampicin 2 μg/mL. Strains used are listed in Table 1.

**Table 1.** Bacterial strains used in a study.

| Name | Original code | Construct | Description | Reference |
|---|---|---|---|---|
| | | | *B. subtilis* | |
| **LK134** | RLG7554 | *rrn*B P1-*lacZ* | MO1099 *amyE*::Cm *rrn*B P1 (−39/+1)-*lacZ* | [11] |
| **LK135** | RLG7555 | P*veg*-*lacZ* | MO1099 *amyE*::Cm P*veg* (−38/−1, +1G)-*lacZ* | [11] |
| **LK41** | RLG6943 | RM-*lacZ* | MO1099 *amyE*::Cm *rrn*O P2 (−77/+50)-*lacZ* | [11] |
| **LK1723** | RLG7024 | *wt* RNAP | β′ with C-ter. His10x; MH5636 | [26] |
| **LK22** | | SigA | SigA; BL21(DE3) | [27] |
| **LK1207** | | SigB | SigB with C-ter. His6x; BL21(DE3) | This work |
| **LK1187** | | SigD | SigD; BL21(DE3) | [28] |

**Table 1.** *Cont.*

| Name | Original code | Construct | Description | Reference |
|---|---|---|---|---|
| | | | *E. coli* | |
| **LK2580** | | SigE | SigE with C-ter. His6x; BL21(DE3) | This work |
| **LK1425** | | SigF | SigF with C-ter. His6x; BL21(DE3) | This work |
| **LK1208** | | SigH | SigH with C-ter. His6x; BL21(DE3) | This work |
| **LK2531** | | SigN | His-SUMO-SigN in pBM05; BL21(DE3) | This work |
| **LK1177** | RLG7558 | P*veg* | pRLG770 with P*veg* ($-38/+1$) $+1G$; DH5$\alpha$ | [11] |
| **LK1522** | RLG7596 | *rrn*B P1core | pRLG770 with *rrn*B P1 ($-39/+1$); DH5$\alpha$ | [11] |
| **LK28** | RLG6927 | *rrn*B P1+P2 | pRLG770 with *rrn*B P1+P2 ($-248/+8$); DH5$\alpha$ | [15] |
| **LK17** | RLG6916 | *rrn*O P1+P2 | pRLG770 with *rrn*O P1+P2 ($-314/+9$); DH5$\alpha$ | This work |
| **LK1231** | | P*trxA* | pRLG770 with P*trxA* ($-249/+11$); DH5$\alpha$ | This work |
| **LK1233** | | P*motA* | pRLG770 with P*motA* ($-249/+11$); DH5$\alpha$ | This work |
| **LK2594** | | P*spoIIID* | pRLG770 with P*spoIIID* ($-150/+10$); DH5$\alpha$ | This work |
| **LK1495** | | P*spoIIQ* | pRLG770 with P*spoIIQ* ($-251/+9$); DH5$\alpha$ | This work |
| **LK1235** | | P*spoVG* | pRLG770 with P*spoVG* ($-94/+11$); DH5$\alpha$ | This work |
| **LK2672** | | *sigN* P2+P3 | pRLG770 with *sigN* P2+P3 ($-247/+159$); DH5$\alpha$ | This work |
| **LK2673** | | P*zpaB* | pRLG770 with P*zpaB* ($-266/+175$); DH5$\alpha$ | This work |
| **LK2608** | | P*zpbY* | pRLG770 with P*zpbY* ($-304/+155$); DH5$\alpha$ | This work |
| **LK2609** | | P*zpdG* | pRLG770 with P*zpdG* ($-244/+170$); DH5$\alpha$ | This work |

### 2.2. Bacterial Strains

### 2.3. Determination of ATP, GTP, and ppGpp concentrations

Strains of *B. subtilis* (LK134, for *rrn*B P1 and LK135 for P*veg*) were grown in the MOPS 20 AA medium supplemented with [$^{32}$P] $KH_2PO_4$ (100 µCi/mL) until early exponential phase ($OD_{600}$ ~ 0.3). Samples were taken until 250 min after $OD_{600}$ ~0.3 (time 0). Samples (100 µL) were pipetted into 100 µL 11.5 M formic acid, vortexed, left on ice for 20 min, and stored overnight at $-80$ °C [29]. After microcentrifugation (5 min, 4 °C) to remove cell debris, the samples (5 µL) were spotted on TLC (thin-layer chromatography) plates (Polygram® CEL 300 PEI, purchased from Macherey-Nagel), developed in 0.85 M (for ATP and GTP) or 1.5 M (for ppGpp) $KH_2PO_4$ (pH 3.4) and quantified by phosphorimaging. The identities of ATP, GTP, and ppGpp were verified by comparison with commercial preparations of these compounds run in parallel and visualized by UV shadowing [8].

To determine the relative ATP/GTP concentrations after novobiocin treatment, LK134 was grown to $OD_{600}$ ~0.3 (time 0) in medium supplemented with [$^{32}$P] $H_3PO_4$ (100 µCi/mL), and at time 5 min treated with novobiocin (5 µg/mL). Samples were taken at time points 0, 5, 10, 20, and 30 min and processed in the same way as above.

### 2.4. Promoter Activity Monitored by Quantitative Primer Extension (qPE)

Promoter constructs were fused to *lacZ* and activities were assayed by primer extension of the short-lived *lacZ* mRNA that allows to observe rapid decreases in promoter activity in time. The experiments were conducted as described in [15]. Typically, 1 mL of cells was pipetted directly into 2 mL phenol/chloroform (1:1) and 0.25 mL lysis buffer (50 mM Tris–HCl pH 8.0, 500 mM LiCl, 50 mM EDTA pH 8.0, 5% SDS). After brief vortexing, the recovery marker (RM) was added. The RM RNA was made from *B. subtilis* strain LK41 as described in [15]. This was followed by immediate sonication. Water was then added to increase the aqueous volume to 6 mL to prevent precipitation of salts, followed by two extractions with phenol/chloroform, two precipitations with ethanol, and suspension of the pellet in 20–50 µL 10 mM Tris–HCl, pH 8.0.

Primer extension was performed with M-MLV reverse transcriptase as recommended by the manufacturer (Promega) with 1–10 µL purified RNA. The $^{32}$P 5′-labeled primer (#2973) hybridized 89 nt downstream from the junction of the promoter fragment used for the creation of the *lacZ* fusion. Samples were electrophoresed on 7 M urea 5.5% or 9% polyacrylamide gels. The gels were exposed to Fuji Imaging Screens. The screens were scanned with Molecular Imager FX (Bio-Rad, Berkeley, CA, USA) and were visualized and analysed using the Quantity One software (Bio-Rad), and normalized to cell number ($OD_{600}$) and RM.

### 2.5. Promoter Activity Monitored by RT–qPCR

*rrn*B P1 and P*veg* promoters were fused to the marker *lacZ* gene (LK134 and LK135), yielding identical transcripts. The strains were grown to exponential phase ($OD_{600}$ ~0.5)—time point 0. Each culture was then divided into two flasks. Cells in one flask were treated with novobiocin (5 µg/mL) and cells in the other flask were left non-treated. At time points 0, 10, 20, and 30 min, 2 mL of cells were withdrawn and treated with RNAprotect Bacteria reagent (QIAGEN, Hilden, Germany), pelleted and immediately frozen. RNA was isolated with RNeasy Mini Kit (QIAGEN) and recovery marker RNA (RM RNA) was added at the time of extraction to control for differences in degradation and pipetting errors during extraction. The RM RNA was prepared from *B. subtilis* strain LK41 as for qPE. Finally, RNA was DNase treated according to manufacturers' instructions (TURBO DNA-free Kit, Ambion). Total RNA was then reverse transcribed to cDNA with reverse transcriptase (SuperScript™ III Reverse Transcriptase, Invitrogen, Waltham, MA, USA) using primer #2973 that targets *lacZ* (both in the test mRNA and RM). This was followed by qPCR in a LightCycler 480 System (Roche Applied Science, Penzberg, Germany) containing LightCycler® 480 SYBR Green I Master and 0.5 µM primers (each). RM cDNA was amplified with primers #2974 and #2973, and the test lacZ cDNA with primers #2975 and #2973. Sequences of primers were originally published in [15]. The final data were normalized to RM and the amount of cells ($OD_{600}$).

### 2.6. $^3$H-Incorporation in Total RNA

This experiment was conducted as described previously [30]. Briefly, strain LK134 was grown in LB medium to $OD_{600}$ ~0.3 (early exponential phase). Newly synthesized RNA in the cells was labeled with $^3$H-uridine (1 µCi/mL) (cold [non-radioactive] uridine was added to a final concentration of 100 µM); time point 0. The bacterial culture was divided into three flasks—non-treated, treated with novobiocin (5 µg/mL), and treated with rifampicin (2 µg/mL), respectively (time point 5). At 0, 5, 10, 20 and 30 min, 100 µL and 250 µL of cells were withdrawn to measure cell density and determine $^3$H-incorporation, respectively. The 250 µL cell sample was mixed with 1 mL of 10% trichloroacetic acid (TCA) and kept on ice for at least 1 h. Thereafter, each sample was vacuum filtered, using Glass Microfiber Filters (Whatman, Little Chalfont, UK), washed twice with 1 ml of 10% TCA and three times with 1 mL of ethanol. The filters were dried, scintillation liquid was added, and the radioactivity was measured. The signal was normalized to cell density ($OD_{600}$).

### 2.7. RNAP Levels in Time

Cells (strain LK134) were grown in LB rich medium to $OD_{600}$ 0.3 (time point 0). Subsequently, every 30 min 10 mL of cells were pelleted and $OD_{600}$ was measured. Pellets were washed with Lysis Buffer (20 mM Tris-HCl, pH 8, 150 mM KCl, 1 mM $MgCl_2$) and frozen. Next day, pellets were resuspended in Lysis Buffer (100–500 µL, according to the size of pellet) and disrupted by sonication 2 × 1 min, with 1 min pause on ice between the pulses. After centrifugation (5 min, 4 °C) to remove cell debris, the amounts of proteins were measured with the Bradford protein assay and 5 µg was resolved by SDS-PAGE and analyzed by Western blotting, using mouse monoclonal antibodies against the β subunit of RNAP (clone name 8RB13, dilution 1:1000, Genetex, Irvine, CA, USA) and anti-mouse secondary antibody conjugated with HRP (dilution 1:800,000, Sigma, Munich, Germany). Subsequently, the blot was incubated for 5 min with SuperSignal™ West Femto PLUS Chemiluminiscent substrate (Thermo scientific, Waltham, MA, USA), exposed on film and developed.

### 2.8. Proteins and DNA for Transcription In Vitro

### 2.8.1. Strain Construction

Genes encoding $\sigma^B$, $\sigma^E$, $\sigma^F$ and $\sigma^H$ were amplified from genomic wt DNA by PCR with Expand High Fidelity PCR System (Roche) with respective primers (Table 1, Material and Methods section) and cloned into pET-22b(+) via *NdeI/XhoI* restriction sites and verified by sequencing. Primers for cloning of $\sigma^E$ were designed for the active form of protein, as its first 27 AA are in the cell posttranslationally removed [31,32]. The resulting plasmids were transformed into expression strain BL21(DE3), yielding strains LK1207 ($\sigma^B$), LK2580 ($\sigma^E$), LK1425 ($\sigma^F$), and LK1208 ($\sigma^H$). His-SUMO-$\sigma^N$ fusion protein in an expression plasmid pBM05 [25] was transformed to BL21(DE3), resulting in strain LK2531.

### 2.8.2. Protein Purification

Wild type RNAP, containing a His10x-tagged β' subunit was purified from LK1723 as described [26].

The SigA subunit of RNAP (LK22) was overproduced a purified as described [27].

$\sigma^B$, $\sigma^E$, $\sigma^F$, $\sigma^H$ expression strains were grown to $OD_{600}$ ~0.5 when IPTG was added to a final concentration of 0.8 mM. Cells were allowed to grow for 3 h at room temperature, cells were harvested, washed and resuspended in P buffer (300 mM NaCl, 50 mM $Na_2HPO_4$, 3 mM β-mercaptoethanol, 5% glycerol). Cells were then disrupted by sonication and the supernatant was mixed with 1 mL Ni-NTA agarose (QIAGEN, Hilden, Germany) and incubated for 1 h at 4 °C with gentle shaking. Ni-NTA agarose with the bound protein was loaded on a Poly-Prep® Chromatography Column (Bio-Rad, Berkeley, CA, USA), washed with P buffer and subsequently with the P buffer with the 30 mM imidazole. The protein was eluted with P buffer containing 400 mM imidazole and fractions containing σ factor were pooled together and dialyzed against storage buffer (50 mM Tris-HCl, pH 8.0, 100 mM NaCl, 50% glycerol and 3 mM β-ME). The proteins were stored at −20 °C.

$\sigma^D$ was purified from inclusion bodies as described in [28].

Cells containing the plasmid for overproduction of $\sigma^N$ were grown to $OD_{600}$ ~0.5 and IPTG was added to final concentration 0.3 mM. Cells were then allowed to grow for 3 h at room temperature; afterwards the cells were harvested, washed, and resuspended in P buffer. All purification steps were done in P2 buffer (the same composition as P buffer, but pH 9.5). Cells were then disrupted by sonication and the supernatant was mixed with 1 mL Ni-NTA agarose (QIAGEN) and incubated for 1 h at 4 °C with gentle shaking. Ni-NTA agarose with the bound His-SUMO-$\sigma^N$ was loaded on a Poly-Prep® Chromatography Column (Bio-Rad), washed with P2 buffer and subsequently with the P2 buffer with the 30 mM imidazole. The protein was eluted with P2 buffer containing 400 mM imidazole and fractions containing His-SUMO-$\sigma^N$ were pooled together and dialyzed against P2 buffer.

The SUMO tag was subsequently removed by using SUMO protease (Invitrogen). The cleavage reaction mixture was again mixed with the 1 mL Ni-NTA agarose and allowed to bind for 1 h at 4 °C and centrifuged to pellet the resin. Supernatant was removed, the resin was washed once more with P2 buffer with 3 mM β-ME. The supernatants (containing $\sigma^N$) were pooled together and dialysed against storage P2 buffer (P2 buffer and 50% glycerol). The protein was stored at −20 °C.

The purity of all proteins was checked by SDS-PAGE.

### 2.8.3. Promoter DNA Construction

Promoter regions of alternative σ-dependent genes were amplified from genomic wt DNA of *B. subtilis* with primers listed in Table 2 (Material and Methods section) by PCR. All fragments were then cloned into p770 (pRLG770 [33]) using *Eco*RI/*Hind*III restriction sites and transformed into DH5α. All constructs were verified by sequencing.

**Table 2.** List of primers.

| Primer No (#) | Sequence $5' \rightarrow 3'$ | |
| --- | --- | --- |
| #1001 | GGAATTCCATATGAATCTACAGAACAACAAGG | Primers for *sigH* cloning into pET-22b(+) |
| #1002 | CCGCTCGAGCTATTACAAACTGATTTCGCG | |
| #1004 | GGAATTCCATATGACACAACCATCAAAAAC | Primers for *sigB* cloning into pET-22b(+) |
| #1006 | CCGCTCGAGCATTAACTCCATCGAGGGATC | |
| #1069 | CCGGAATTCATTCCGGAGTCATTCTTACGG | Primers for P*trxA* cloning into pRLG770 |
| #1070 | CCCAAGCTTCACTGTCATGTACTTTACCATG | |
| #1075 | CCGGAATTCCTTTACACTTTTTTAAGGAGG | Primers for P*motA* cloning into pRLG770 |
| #1076 | CCCAAGCTTCTAGCTTGTCTATGGTTAATATC | |
| #1079 | CCGGAATTCTTTATGACCTAATTGTGTAAC | Primers for P*spoVG* cloning into pRLG770 |
| #1080 | CCCAAGCTTATAAAAGCATTAGTGTATC | |
| #1309 | GGAATTCCATATGGATGTGGAGGTTAAGAAAAAC | Primers for *sigF* cloning into pET-22b(+) |
| #1311 | CCGCTCGAGGCCATCCGTATGATCCATTTG | |
| #1425 | CCGGAATTCCATTCCATCCGGTCTTCAGG | Primers for P*spoIIQ* cloning into pRLG770 |
| #1426 | CCCAAGCTTCATCACCTCAGCAACATTCTG | |
| #2973 | CAGTAACTTCCACAGTAGTTCACCAC | universal reverse primer for PE and qPCR |
| #2974 | TCTAAGCTTCTAGGATCCCC | test RNA-specific forward primer for PE and qPCR |
| #2975 | GTCGCTTTGAGAGAAGCACA | RM RNA-specific forward primer for PE and qPCR |
| #3109 | GCGAATTCCGTGTCGGTCAACATAATAAAGG | Primers for *sigN* P2+P3 cloning into pRLG770 |
| #3110 | GCAAGCTTCGGCAAAAATCTTTCTCTCACC | |
| #3111 | GCGAATTCGCGATGAATGAAGAGACACGG | Primers for P*zpaB* cloning into pRLG770 |
| #3112 | GCAAGCTTAGTCCATCTCGAAGATCTGGT | |
| #3113 | GCGAATTCGACTCCAACATTTCTATTCC | Primers for P*zpbY* cloning into pRLG770 |
| #3114 | GCAAGCTTGGTCTTCTTCACTTAATTCA | |
| #3117 | GCGAATTCTCAAAGATCTTCTAACTTGT | Primers for P*zpdG* cloning into pRLG770 |
| #3118 | GCAAGCTTGGCAGTAATCAATCAATTCT | |
| #3166 | CGGCATATGTACATAGGCGGGAGTGAAGCC | Primers for *sigE* active form cloning into pET-22b(+) |
| #3167 | CCGCTCGAGCACCATTTTGTTGAACTCTTTTC | |
| #3170 | GGCGAATTCGCTTATTTCATTTTACAGGAG | Primers for P*spoIIID* cloning into pRLG770 |
| #3171 | CCGAAGCTTTGTTAGGTTTGTAACAGTGT | |
| PRIMER A | GGGAATTCATGGACATCAATGATATCTC | Primers for *rrn*O P1+P2 cloning into pRLG770 |
| PRIMER B | GGAAGCTTTCAAAGCGACTACTTAATAG | |

Supercoiled plasmids (SC) were obtained using the Wizard® Plus Midipreps DNA Purification System, for higher yields Wizard® Plus Maxipreps DNA Purification System (both Promega, Madison, WI, USA) were used and subsequently phenol-chloroform extracted, precipitated with ethanol, and dissolved in water. Aliquots of plasmids were linearized with the *Pst*I restriction enzyme (TaKaRa, Saint-Germain-en-Laye, Francie), resulting in linear form (LIN), and again precipitated with ethanol to remove salts.

The state of DNA topology (linear, supercoiled) was checked on agarose gels.

### 2.8.4. List of Primers

*2.9. Transcription In Vitro*

Transcription experiments were performed with the *B. subtilis* RNAP core reconstituted with a saturating concentration of $\sigma^A$ (ratio 1:5) in storage buffer (50 mM Tris-HCl, pH 8.0, 0.1 M NaCl, 50% glycerol) for 15 min at 30 °C. The 1:5 ratio was used also for $\sigma^B$, $\sigma^D$, $\sigma^E$, $\sigma^F$, and $\sigma^H$. For $\sigma^N$, the ratio was 1:8. Multiple round transcription reactions were carried out in 10 µL reaction volumes with 30 nM RNAP holoenzyme. The transcription buffer contained 40 mM Tris-HCl pH 8.0, 10 mM $MgCl_2$, 1 mM dithiothreitol (DTT), 0.1 mg/mL BSA and 150 mM KCl, and all four NTPs and 2 µM radiolabeled [$\alpha$-$^{32}$P] UTP.

In $K_{GTP}$ determination experiments, the amount of DNA (SC or LIN form) was 100 ng, ATP, CTP were 200 µM; UTP was 10 µM and GTP was titrated from 0 to 2000 µM. To determine the affinity of RNAP to DNA, ATP, CTP were at 200 µM; UTP was 10 µM, GTP was 1000 µM and DNA (SC/LIN) was titrated from 0 to 900 ng per reaction. In reactions with alternative $\sigma$, DNA (SC or LIN form) was 100 ng, CTP were at 200 µM; UTP was 10 µM and GTP/ATP was 1000 µM, depending on the identity of the base in the +1 position of the transcript.

All transcription reactions were allowed to proceed for 15 min at 30 °C and then stopped with equal volumes of formamide stop solution (95% formamide, 20 mM EDTA, pH 8.0). Samples were loaded onto 7 M urea-7% polyacrylamide gels and electrophoresed. The dried gels were scanned with Molecular Imager FX (Bio-Rad) and were visualized and analysed using the Quantity One software (Bio-Rad).

## 3. Results

*3.1. The Activity of rrnB P1 Decreases during Entry into Stationary Phase*

As the main model rRNA promoter, we selected the *rrn*B P1 promoter as it is one of the best-characterized rRNA promoters in *B. subtilis* that is regulated by [iNTP], [11,34–36]. Furthermore, the dynamic range of the activity of *rrn*B P1 is wide, which facilitated the design and interpretation of the experiments. As the main control promoter, we selected the strong P*veg* promoter that forms stable open complexes and is saturated with a relatively low level of its iNTP. This promoter drives transcription of the *veg* gene that is involved in biofilm formation [37,38]. Promoter sequences are shown in Figure 1A.

To monitor promoter activities, we used core promoter-*lacZ* fusions. The endogenous copy of P*veg* initiates transcription with ATP (+1A). Here, we used a +1G variant of P*veg* so that both transcripts (from *rrn*B P1-*lacZ* and P*veg*-*lacZ*) were identical, excluding any effects due to, e.g., potentially differential decay of the transcripts. The +1G P*veg* promoter variant behaves identically with the +1A variant [11]. Throughout the study, promoter activity was determined by quantitative primer extension (qPE) or reverse transcription followed by quantitative PCR (RT-qPCR).

We used defined rich MOPS medium to grow the cells and measured (i) relative GTP level ([GTP]) and (ii) relative promoter activity (*rrn*B P1 and P*veg*) from early exponential phase till approximately two hours into stationary phase by qPE (Figure 1).

We detected a moderate decrease in [GTP] already during exponential phase (Figure 1B). This moderate decrease was followed by a precipitous decline during the transition between the two phases. This correlated with a sharp spike in the (p)ppGpp level (Supplementary Figure S1). However, early on in the stationary phase, [GTP] even slightly increased and then remained at the same level till the end of the experiment. The activities of both *rrn*B P1 and P*veg* decreased during the time course of the experiment—the activity of the former more than of the latter, consistent with the behavior of these promoters as reported in previous studies [10,11].

Surprisingly and interestingly, the activity *rrn*B P1 decreased even after the relative GTP concentration had been stabilized at a constant level. This strongly suggested that another mechanism, besides rRNA promoter regulation by [GTP], exists in the cell. DNA supercoiling is known to change between growth phases, typically the negative supercoiling from exponential phase becomes more relaxed in stationary phase, as demonstrated for *Escherichia coli* [39] and also *B. subtilis* [40]. Also, we noticed that the activity of P*veg* significantly decreased, although the decrease was not as pronounced as that of the ribosomal promoter. As DNA topology is an important factor for gene expression regulation, we decided to address the potential of *B. subtilis* rRNA promoters to be regulated by the level of supercoiling.

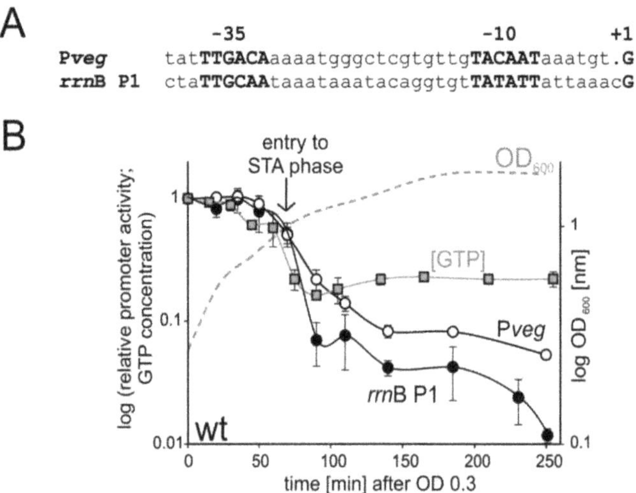


**Figure 1.** Correlation between GTP concentration and promoter activity after entry into the stationary phase. (**A**) Sequences of P*veg* and *rrn*B P1. (**B**) Relative promoter activities of *rrn*B P1 (black circles) and P*veg* promoters (open circles) after entry into stationary phase, relative GTP concentration (green squares), and optical density (dashed grey line). Promoter activities and GTP concentrations were normalized to 1 at time 0. Promoter activities were measured by qPE from wt *B. subtilis* strains: *rrn*B P1 (LK134), P*veg* (LK135). Promoter activities were calculated from three independent experiments, the error bars show ±SD. The GTP concentrations are from two independent experiments, showing the mean, the bars show the range. A representative bacterial growth curve is shown. The vertical arrow indicates the entry into stationary phase.

### 3.2. Chromosome Relaxation Inhibits Total RNA Synthesis In Vivo

To test whether DNA topology could affect rRNA expression in vivo, we used novobiocin. Novobiocin is an antimicrobial compound that binds to the β subunit of gyrase and blocks its function by inhibiting ATP hydrolysis [41–43]. Gyrase relieves tension in DNA caused by transcribing RNAPs or helicases by creating more negatively supercoiled DNA. Hence, the inhibition of gyrase causes DNA in the cell to be more relaxed [44].

In this experiment, we first used total RNA as a proxy for rRNA synthesis as in exponential phase most of RNA synthesis comes for rRNA operons (~80% of RNA in cell is rRNA and tRNA [29,45]). We treated early-exponentially growing cells ($OD_{600}$ ~0.3) with novobiocin or mock-treated them, and measured the rates of total RNA synthesis by following incorporation of radiolabeled $^{3}$H-uridine into RNA. As a positive control, where we expected cessation of RNA synthesis, we treated cells with rifampicin, a well-characterized inhibitor of bacterial RNAP.

Figure 2 shows that in the presence of novobiocin the synthesis of total RNA decreased/stopped, similarly as in the presence of rifampicin, suggesting that relaxation of the chromosome affects total RNA synthesis in the cell (Figure 2A).

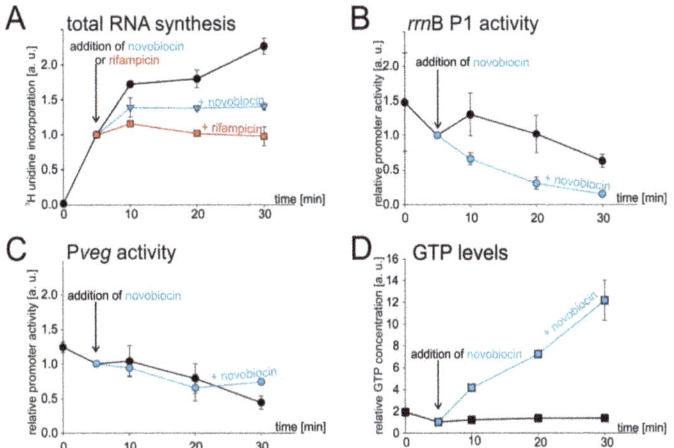

**Figure 2.** Effect of novobiocin-induced relaxation of chromosome on total RNA synthesis, selected promoter activities, and GTP level. (**A–D**) Cells were grown to early exponential phase (OD$_{600}$ ~0.3), and at time 5 min they were treated with novobiocin (5 µg/mL). (**A**) Total RNA synthesis after novobiocin treatment. After $^3$H-uridine had been added (time 0), the culture was divided into three flasks. At time 5 min the cells were treated with novobiocin (blue line) or with rifampicin (red line) as a control, or left untreated (black line). The amount of radiolabeled RNA at 5 min was set as 1. Black circles, mock-treated; blue triangles, treated with novobiocin; red squares, treated with rifampicin. The values are averages of three independent experiments ±SD. (**B,C**) The relative activities of *rrn*B P1 and P*veg* promoters after novobiocin treatment. Cells were grown and at 5 min treated with novobiocin or not. RNA was extracted and determination of promoter activity was done by RT-qPCR. Promoter activities were set as 1 at time 5 min. Blue lines are novobiocin-treated samples, black lines are untreated samples. The experiment was performed three times. The error bars show ±SD. (**D**) GTP concentration after novobiocin treatment. Cells were grown in the presence of [$^{32}$P] H$_3$PO$_4$ and treated with novobiocin. Levels of GTP were determined by TLC. The GTP level at 5 min was set as 1. Results are averages from two measurements. The error bars show the range.

### 3.3. Novobiocin-Induced Relaxation of DNA Affects the Activity of rrnB P1 In Vivo

Next, by RT-qPCR we monitored the response of *rrn*B P1 and P*veg* to novobiocin treatment, using the same conditions as in the previous experiment. We grew cells carrying the appropriate fusions (*rrn*B P1-*lacZ* (LK134) and P*veg-lacZ* (LK135)) to early-exponential phase (OD$_{600}$ ~ 0.3) and either treated them with novobiocin or mock-treated them. In the case of *rrn*B P1, the promoter activity decreased after novobiocin treatment (as opposed to mock treatment), but in the case of P*veg*, the promoter activity displayed the same moderate decline regardless of the novobiocin treatment, suggesting that *rrn*B P1 is more sensitive to changes in DNA topology (Figure 2B,C).

We also measured the GTP levels in novobiocin treated cells. We observed that novobiocin-induced relaxation resulted in a massive increase in the GTP level in cell (Figure 2D). The levels of ATP increased only slightly (Supplementary Figure S2). Thus, the activity of *rrn*B P1 and the level of GTP became uncoupled. These experiments suggested that DNA topology might affect the activity rRNA promoters, but it was also possible that unknown, secondary effects of the novobiocin treatment could be the cause.

### 3.4. Changes in DNA Topology Affect the Affinity of RNAP for iNTP In Vitro

To test directly whether DNA topology affects the activity of rRNA promoters, we performed in vitro transcription experiments. We had speculated that the in vivo decrease in the activity of *rrn*B P1 during stationary phase and in response to novobiocin treatment could be due to altered affinity of RNAP for iGTP at this promoter (induced by changes in supercoiling levels): the GTP level does not change but the open promoter becomes less stable, requiring more iGTP for maximal transcription. To address this hypothesis experimentally, we performed in vitro transcriptions with defined components. We used promoter core variants of *rrn*B P1 and *Pveg* cloned in the pRLG770 plasmid [11] (for details see Table 1 in Material and Methods section). The DNA templates were used in two different topological forms—in the negatively supercoiled plasmid form (SC), and in the relaxed form (LIN), using the same DNA construct but linearized with the *Pst*I restriction enzyme (Supplementary Figure S3).

We performed multiple round transcriptions in vitro with increasing [GTP] (Figure 3). The GTP concentration required for half-maximal transcription ($K_{GTP}$) was used as a measure of the affinity of RNAP for iGTP at the promoter. A characteristic of rRNA promoters is their requirement for relatively high levels of iGTP for maximal transcription (due to unstable open complexes), reflected in high values of $K_{GTP}$ in vitro. *Pveg*, to the contrary, has a low value of $K_{GTP}$.

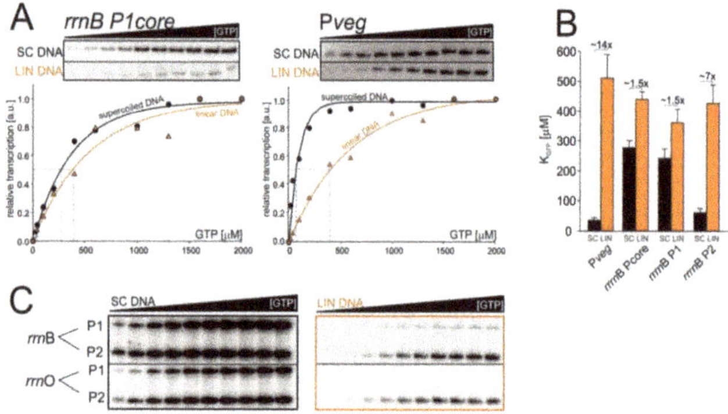


**Figure 3.** The affinity of RNAP for iNTP in vitro changes on different DNA templates. (**A**) Multiple-round transcriptions as a function of GTP concentration: representative primary data and their graphical comparison for *rrn*B P1core and *Pveg*. The maximum signal was set as 1. (**B**) Graphical comparison of $K_{GTP}$ values for SC and LIN DNA templates. The values are calculated from at least four experiments, the error bars show ±SD. (**C**) Low affinity for LIN templates of full-length *rrn* promoter variants. Representative primary data are shown.

Experiments with SC templates confirmed previously published results [46], the $K_{GTP}$ for *rrn*B P1 was 277 ± 24 µM, and for *Pveg* 36 ± 9 µM. Experiments with the LIN templates then revealed that $K_{GTP}$ values for both promoters increased (*rrn*B P1 = 440 ± 25 µM, *Pveg* = 511 ± 78 µM). In the case of *rrn*B P1 the $K_{GTP}$ increased from SC to LIN ~1.5x, and in the case of *Pveg* $K_{GTP}$ ~14x. Surprisingly, the $K_{GTP}$ value of LIN *Pveg* was even higher than the value for *rrn*B P1 (Figure 3B).

Importantly, the experiments showed that the strength (the maximal level of transcription) of the *rrn*B P1 promoter dramatically decreased on the LIN template whereas in the case of *Pveg* the maximal level of transcription was comparable for both types of the template (Figure 3A, primary data), confirming the hypothesis that DNA relaxation decreases the activity of *rrn*B P1 more than the activity of *Pveg*.

As the preceding experiments were done with the core version of the *rrn*B P1 promoter, we also decided to use an extended version of the promoter region to assess whether the surrounding sequence has significant effects. Therefore, we used a DNA fragment containing both *rrn*B P1 and *rrn*B P2 promoters in their native tandem arrangement. Each of them contained their respective native -60 to -40 regions encompassing the UP elements. UP elements are A/T-rich sequences that enhance promoter activity by binding the C-terminal domains of α-subunits of RNAP [47–49]. Although their stimulatory effect on rRNA promoters in *B. subtilis* is less pronounced than, e.g., in *E. coli* (~30x), it is still significant [11]. Experiments with these promoter versions yielded virtually the same results as with the core version (Figure 3C). The $K_{GTP}$ for *rrn*B P1 (from the tandem promoter fragment) was 242 ± 31 μM for SC and 361 ± 46 μM for LIN. $K_{GTP}$ for *rrn*B P2 was 62 ± 13 μM for SC and 427 ± 61 μM for LIN (see Supplementary Table S1 and Supplementary Figure S4A,B). Similar results were obtained also with *rrn*O P1 and *rrn*O P2 promoters (Supplementary Figure S4C,D).

Hence, we concluded that for transcription from LIN templates higher concentrations of GTP are needed, regardless of the promoter. The increased $K_{GTP}$ of *Pveg* suggested that this change in RNAP affinity for the substrate iNTP might be responsible, at least in part, for the decrease in its activity during the transition from exponential to stationary phase. However, the moderate increase in $K_{GTP}$ of *rrn*B P1 suggested that other factor(s) must be involved in the decrease of this promoter's activity in vivo. A likely candidate factor was the affinity of RNAP for promoter DNA, i.e., formation of the closed complex or/and the intracellular level of RNAP.

### 3.5. Pveg and rRNA Promoter Affinities for RNAP Change with DNA Relaxation In Vitro

We tested the relative affinity of RNAP for promoter DNA by performing in vitro transcriptions as a function of increasing promoter DNA concentration. We used the tandem *rrn*B P1+P2 DNA fragment and *Pveg*. The GTP concentration was set to 1 mM to ensure high efficiency of open complex formation for the tested promoters. Affinity for RNAP of both rRNA promoters was unchanged or slightly decreased on relaxed templates, but this effect was not statistically significant (Figure 4). Therefore, it is possible that the observed decrease in bulk transcription from *rrn*B P1 (SC vs LIN) in vitro could be due to yet another factor (e.g., promoter escape).

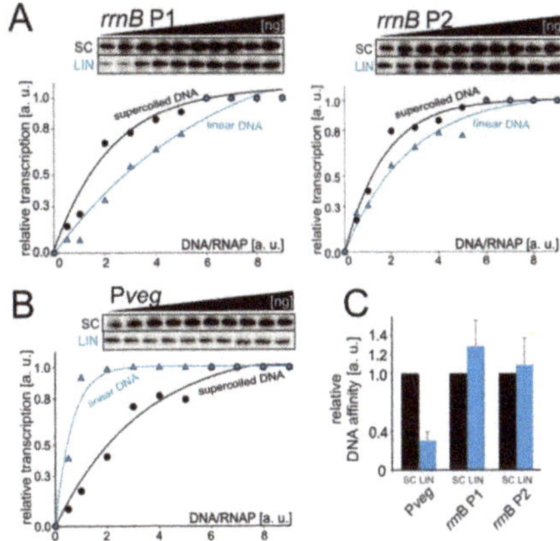

**Figure 4.** The affinity of RNAP for promoter DNA. Multiple-round transcriptions were carried as a function of the increasing DNA/RNAP ratio. The tested promoters were *rrn*B P1+P2 (**A**) and *Pveg* (**B**). Primary data are shown above the graphs. The maximum signal in the plateau phase was set as 1. SC—supercoiled and LIN—linear DNA templates. The experiments were conducted at least four times with similar results. Representative primary data are shown. (**C**) Graphical comparison of relative affinities of RNAP for *Pveg* and *rrn*B P1+P2 promoters. The bars show relative concentrations of promoter DNA at which the activity of RNAP was 50%. The affinity of RNAP for SC promoter DNA was set as 1 for each promoter.

The opposite trend was observed with P*veg*: a relatively low level of the relaxed promoter DNA was able to saturate RNAP compared to the supercoiled template. This behavior could then explain why the activity of P*veg* decreased less than the activity of *rrn*B P1 both in vitro and in vivo. Importantly, it was previously reported that the levels of RNAP subunits decrease from exponential to stationary phase [50,51] and we also observed this trend (Figure 5).

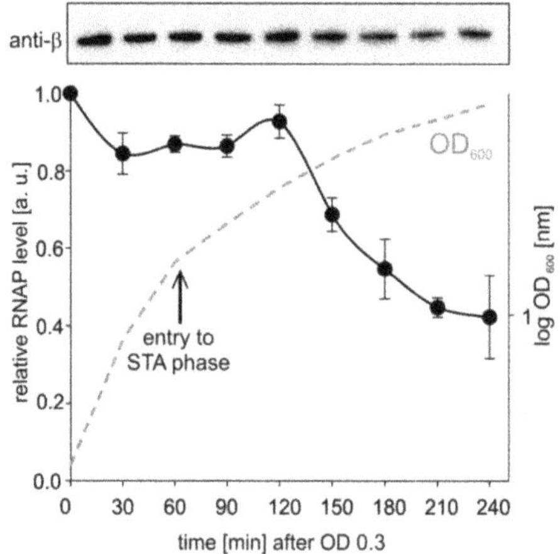


**Figure 5.** RNAP levels during bacterial growth. Amounts of RNAP were detected by Western blotting from 5 µg of total protein per lane. Representative primary data are shown above the graph. The RNAP level from time point 1 was set as a 1. STA—stationary phase (indicated with the arrow). The experiment was conducted in two independents replicas. The points are averages, the error bars show the range. The dashed line shows a representative bacterial growth.

### 3.6. The Effect of Supercoiling on Transcription In Vitro with Alternative Sigma Factors

To extend the study, we tested the effect of supercoiling on transcription from promoters dependent on alternative sigma factors: $\sigma^B$, $\sigma^D$, $\sigma^E$, $\sigma^F$ and $\sigma^H$. $\sigma^B$ is a general stress response sigma factor [52,53], $\sigma^D$ transcribes genes linked with the cell motility and flagella formation [54]. $\sigma^E$ and $\sigma^F$ are sigma factors of early sporulation [55,56]. $\sigma^H$ is responsible for transcription of early stationary genes [57].

We tested also $\sigma^N$ (ZpdN) that is present only in the *B. subtilis* NCIB 3610 strain. This strain possesses a large, low-copy-number plasmid pBS32, which was lost during domestication of the commonly used laboratory strains [58]. pBS32 carries genes responsible for cell death after mitomycin C (MMC) treatment, and this effect is dependent on $\sigma^N$ [24,25]. MMC is an antitumor antibiotic that induces DNA strand scission by DNA alkylation leading to crosslinking [59–61]. This DNA damage could lead to the formation of linear DNA fragments.

Sequences of respective promoters are listed in Supplementary Table S2. We performed transcriptions in vitro on SC and LIN DNA templates with saturating concentration of iNTP. In all but one cases it was the SC DNA that was the better template for transcription, similarly to what we observed with $\sigma^A$ (Figure 6).

The exception was $\sigma^N$, which displayed about the same or higher activity on LIN DNA than on SC DNA, depending on the promoter (Figure 6B). To show that this effect was not due to some unknown properties of the plasmid DNA bearing these promoters, we also tested a longer *sigN* promoter construct (*sigN* P2+P3). This construct contains $\sigma^A$-dependent *sigN* P2 and $\sigma^N$-dependent *sigN* P3 promoters [25] and allowed us to test the effect of SC vs LIN topology for two sigmas with the same template. The results are shown in Figure 6C: $\sigma^A$-dependent P2 is more active on SC DNA whereas $\sigma^N$-dependent P3 prefers LIN DNA for efficient transcription.

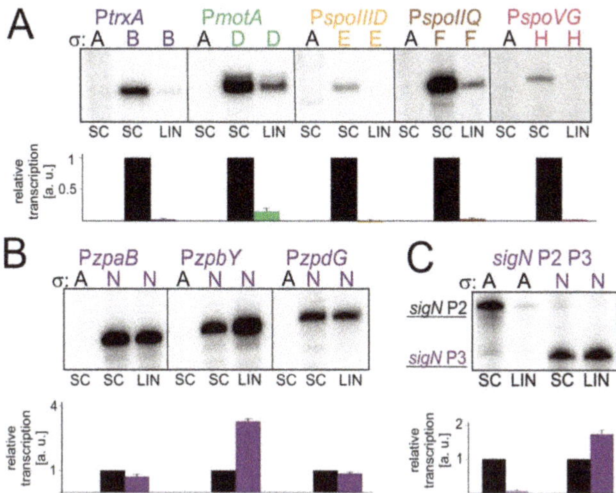


**Figure 6.** Transcription in vitro with alternative σ factors on different DNA templates. Representative primary data are shown (radioactively labelled transcripts resolved by polyacrylamide electrophoresis). SC stands for supercoiled promoter DNA, LIN for linear DNA. Letters above the gels indicate the sigma factor used—A for σ$^A$, B for σ$^B$ etc. Each sigma factor is depicted with different color (σ$^A$, black; σ$^B$, dark blue; σ$^D$, green; σ$^E$, yellow; σ$^F$, brown; σ$^H$, red and σ$^N$, purple). For each promoter three independent experiments were performed. Transcription from SC was set as 1 for each promoter. Quantitation of results is shown in graphs below the respective primary data. The graphs show averages ±SD. The reactions with σ$^A$ on all promoter fragments were used to show that the observed transcription was due to the addition of the specific σ factors and not due to (theoretical) contamination of the core with σ$^A$. (**A**) Transcription in vitro from selected σ$^B$, σ$^D$, σ$^E$, σ$^F$ and σ$^H$ -dependent promoters. (**B**) Transcription in vitro from σ$^N$-dependent promoters. (**C**) Transcription in vitro using a longer construct, *sigN* P2+P3. P2 is σ$^A$-dependent, P3 is σ$^N$-dependent.

## 4. Discussion

In this study we have identified the supercoiling level of DNA as a factor affecting the ability of *Bacillus subtilis* RNAP to transcribe from σ$^A$-dependent rRNA promoters as well as from selected promoters depending on alternative σ factors.

### 4.1. rRNA Promoters and Pveg

In our experiments, the drop in rRNA promoter activity during transition to stationary phase was pronounced and concurrent with the onset of stationary phase. A decrease in the activity of *B. subtilis* rRNA promoters in stationary phase was also observed in [62]. However, they used promoter constructs fused with GFP and monitored promoter activity by measuring the intensity of fluorescent signal. GFP is relatively stable, so the decreases they reported were less pronounced than those observed in our experiments.

Here, we propose an updated model of regulation of *B. subtilis* rRNA promoters, revealing supercoiling as a factor involved in their control. The more negatively supercoiled DNA in exponential phase contributes to the high activity of *B. subtilis* rRNA promoters. As this negative supercoiling becomes more relaxed when the cell transitions into stationary phase, this likely contributes to the decrease in the activity of RNAP at rRNA promoters. This is in part due to the decreased affinity of RNAP at rRNA promoters for the initiating GTP but also to a so far unknown step during transcription initiation (e.g., isomerization, promoter escape). The activity of rRNA promoters in stationary phase is also likely affected by the decreased RNAP concentration. The decrease in the available RNAP pool is further exacerbated by the association of the RNAP:σ$^A$ holoenzyme with 6S-1 RNA that sequesters it in an inactive form in stationary phase [63]. The combined effect results in the shut-off of rRNA synthesis. Previously, for *E. coli* rRNA promoters, the decreased stability of the open complex was identified as the main kinetic intermediate affected by supercoiling [64]. We also note that in *S. aureus* in post-exponential growth phase the downregulation of rRNA is independent of ppGpp or NTP pools [17], and it is possible that DNA topology might be a factor contributing to this downregulation.

In *B. subtilis*, correlations between the supercoiling level and rRNA activity could be found also in the forespore. Within the developing spore, DNA becomes more negatively supercoiled then in stationary phase [40] and this correlates with an increase in rRNA activity in the forespore [62]. Interestingly, during novobiocin treatment the GTP level increases in *B. subtilis* and the changes in DNA topology override its stimulatory effect so that the net result is a decrease in the activity of *rrn*B P1. This is the first observation of a situation where the GTP level and rRNA promoter activity do not correlate in *B. subtilis*. We note that supercoiling was also reported to be involved in rRNA expression in yeast although the mechanistic aspects of this regulation are less understood [65].

The activity of the control P*veg* promoter also decreases from exponential to stationary phase but the decrease is not as pronounced as in the case of *rrn*B P1. The decrease in the activity of P*veg* can be attributed, at least in part, to its increased requirement for the concentration of the iNTP when DNA supercoiling relaxes. Nevertheless, the affinity of P*veg* for RNAP seems to increase with DNA relaxation and this likely partially counteracts the negative effect on open complex formation.

### 4.2. Transcription with Selected Alternative σ Factors

Transcription experiments with promoters dependent on alternative σ factors revealed that linear templates are poorer substrates for the majority of them ($\sigma^B$, $\sigma^D$, $\sigma^E$, $\sigma^F$, and $\sigma^H$). This trend was previously reported also for RNAP:$\sigma^H$ transcribing from the *spoIIA* promoter [66]. For forespore-specific $\sigma^F$, this is consistent with the DNA supercoiling increase in the forespore [40]. For $\sigma^H$ and $\sigma^E$ that are active in stationary phase, although activities of respective promoters strongly decreased with reduced supercoiling in vitro, this likely reflects the physiologically relevant requirements for their activities in the cell in stationary phase. Also, the decrease in the level of supercoiling in stationary phase is likely not as extreme as in our in vitro experiments where it was used to better visualize the effects.

The exception was $\sigma^N$, where transcription (SC vs. LIN) is either relatively unaffected or even increased on linear templates. This is likely physiologically important as mitomycin, which induces $\sigma^N$ expression [24], causes also DNA relaxation and $\sigma^N$ may have evolved to be most active under such conditions. The proficiency of RNAP:$\sigma^N$ on linear templates then may stem from the relatively short spacers of $\sigma^N$ dependent promoters (15 bp compared to 17 bp for $\sigma^A$, [67]), analogously to $\sigma^{70}$ and $\sigma^S$ of *E. coli* where the different σ activities were proposed to be due to preferences for differently DNA supercoiled templates [68–70].

### 5. Conclusions

To conclude, our findings extend the current model of rRNA promoter regulation in *B. subtilis* and reveal the effect of supercoiling on transcription with main and alternative σ factors.

**Supplementary Materials:** The following are available online at https://www.mdpi.com/2076-2607/9/1/87/s1, Figure S1: Relative GTP and (p)ppGpp levels after entry into stationary phase. Figure S2: Effect of novobiocin-induced relaxation of chromosome on ATP levels. Figure S3: SC and LIN promoter DNA on agarose gel. Figure S4: The affinity of RNAP for iNTP in vitro changes on different DNA templates. Table S1: The $K_{GTP}$ values for the promoters tested in the transcriptions in vitro. Table S2: Alternative σ factor-dependent promoters used in the study.

**Author Contributions:** L.K. supervised the project; L.K. and P.S. conceptualized the experiments; L.K. performed qPE and TLC experiments; P.S. performed novobiocin studies (RT-PCR, TLC), purified proteins, and performed transcriptions in vitro; O.R., M.B., and H.Š. cloned and purified alternative σ factors and their respective promoters. M.K. cloned, purified, and performed experiments with $\sigma^N$ and its respective promoters; P.S. and L.K. wrote the manuscript. All authors have read and agreed to the published version of the manuscript.

**Funding:** This work was supported by 20-12109S to LK from the Czech Science Foundation.

**Acknowledgments:** We would like to acknowledge the Czech Research Infrastructure for Systems Biology C4SYS (project LM2015055). We thank J.D. Helmann for providing the $\sigma^D$ overproducing strain, and D. Kearns for the pBM05 plasmid used for construction of $\sigma^N$ overproducing strain..

**Conflicts of Interest:** The authors declare no conflict of interest. The funders had no role in the design of the study; in the collection, analyses, or interpretation of data; in the writing of the manuscript, or in the decision to publish the results.

## References

1. Gourse, R.L.; Gaal, T.; Bartlett, M.S.; Appleman, J.A.; Ross, W. rRNA Transcription and growth rate–dependent regulation of ribosome synthesis in *Escherichia Coli. Annu. Rev. Microbiol.* **1996**, *50*, 645–677. [CrossRef] [PubMed]
2. Lee, J.; Borukhov, S. Bacterial RNA Polymerase-DNA interaction-the driving force of gene expression and the target for drug action. *Front. Mol. Biosci.* **2016**, *3*, 73. [CrossRef] [PubMed]
3. Helmann, J.D. Where to begin? Sigma factors and the selectivity of transcription initiation in bacteria. *Mol. Microbiol.* **2019**, *112*, 335–347. [CrossRef] [PubMed]
4. Gross, C.A.; Chan, C.; Dombroski, A.; Gruber, T.; Sharp, M.; Tupy, J.; Young, B. The functional and regulatory roles of sigma factors in transcription. *Cold Spring Harb. Symp. Quant. Biol.* **1998**, *63*, 141–155. [CrossRef] [PubMed]
5. Paget, M.S. Bacterial sigma factors and anti-sigma factors: Structure, function and distribution. *Biomolecules* **2015**, *5*, 1245–1265. [CrossRef] [PubMed]
6. Feklistov, A.; Darst, S.A. Structural basis for promoter -10 element recognition by the bacterial RNA polymerase σ subunit. *Cell* **2011**, *147*, 1257–1269. [CrossRef]
7. Mustaev, A.; Roberts, J.; Gottesman, M. Transcription elongation. *Transcription* **2017**, *8*, 150–161. [CrossRef]
8. Schneider, D.A.; Murray, H.D.; Gourse, R.L. Measuring control of transcription initiation by changing concentrations of nucleotides and their derivatives. *Methods Enzymol.* **2003**, *370*, 606–617. [CrossRef]
9. Turnbough, C.L. Regulation of bacterial gene expression by the NTP substrates of transcription initiation. *Mol. Microbiol.* **2008**, *69*, 10–14. [CrossRef]
10. Murray, H.D.; Schneider, D.A.; Gourse, R.L. Control of rRNA expression by small molecules is dynamic and nonredundant. *Mol. Cell* **2003**, *12*, 125–134. [CrossRef]
11. Krásný, L.; Gourse, R.L. An alternative strategy for bacterial ribosome synthesis: *Bacillus subtilis* rRNA transcription regulation. *EMBO J.* **2004**, *23*, 4473–4483. [CrossRef] [PubMed]
12. Bittner, A.N.; Kriel, A.; Wang, J.D. Lowering GTP Level increases survival of amino acid starvation but slows growth rate for *Bacillus subtilis* cells lacking (p)ppGpp. *J. Bacteriol.* **2014**, *196*, 2067–2076. [CrossRef] [PubMed]
13. Kriel, A.; Bittner, A.N.; Kim, S.H.; Liu, K.; Tehranchi, A.K.; Zou, W.Y.; Rendon, S.; Chen, R.; Tu, B.P.; Wang, J.D. Direct regulation of GTP homeostasis by (p)ppGpp: A critical component of viability and stress resistance. *Mol. Cell* **2012**, *48*, 231–241. [CrossRef] [PubMed]
14. Henkin, T.M.; Yanofsky, C. Regulation by transcription attenuation in bacteria: How RNA provides instructions for transcription termination/antitermination decisions. *BioEssays* **2002**, *24*, 700–707. [CrossRef]
15. Krásný, L.; Tišerová, H.; Jonák, J.; Rejman, D.; Šanderová, H. The identity of the transcription +1 position is crucial for changes in gene expression in response to amino acid starvation in *Bacillus subtilis. Mol. Microbiol.* **2008**, *69*, 42–54. [CrossRef]
16. Natori, Y.; Tagami, K.; Murakami, K.; Yoshida, S.; Tanigawa, O.; Moh, Y.; Masuda, K.; Wada, T.; Suzuki, S.; Nanamiya, H.; et al. Transcription activity of individual rrn operons in *Bacillus subtilis* mutants deficient in (p)ppGpp synthetase genes, relA, yjbM, and ywaC. *J. Bacteriol.* **2009**, *191*, 4555–4561. [CrossRef]
17. Kästle, B.; Geiger, T.; Gratani, F.L.; Reisinger, R.; Goerke, C.; Borisova, M.; Mayer, C.; Wolz, C. rRNA regulation during growth and under stringent conditions in S taphylococcus aureus. *Environ. Microbiol.* **2015**, *17*, 4394–4405. [CrossRef]
18. Zechiedrich, E.L.; Khodursky, A.B.; Bachellier, S.; Schneider, R.; Chen, D.; Lilley, D.M.J.; Cozzarelli, N.R. Roles of topoisomerases in maintaining steady-state DNA supercoiling in *Escherichia coli. J. Biol. Chem.* **2000**, *275*, 8103–8113. [CrossRef]
19. Higgins, C.F.; Dorman, C.J.; Stirling, D.A.; Waddell, L.; Booth, I.R.; May, G.; Bremer, E. A physiological role for DNA supercoiling in the osmotic regulation of gene expression in *S. typhimurium* and *E. coli. Cell* **1988**, *52*, 569–584. [CrossRef]
20. Richardson, S.M.; Higgins, C.F.; Lilley, D.M. The genetic control of DNA supercoiling in Salmonella typhimurium. *EMBO J.* **1984**, *3*, 1745–1752. [CrossRef]
21. McClure, W.R. Mechanism and control of transcription initiation in prokaryotes. *Annu. Rev. Biochem.* **1985**, *54*, 171–204. [CrossRef] [PubMed]

22. Schnetz, K.; Wang, J.C. Silencing of the *Escherichia Coli* bgl promoter: Effects of template supercoiling and cell extracts on promoter activity in vitro. *Nucleic Acids Res.* **1996**, *24*, 2422–2428. [CrossRef] [PubMed]
23. Sioud, M.; Boudabous, A.; Cekaite, L. Transcriptional responses of Bacillus subtillis and thuringiensis to antibiotics and anti-tumour drugs. *Int. J. Mol. Med.* **2009**, *23*, 33–39. [CrossRef] [PubMed]
24. Myagmarjav, B.-E.; Konkol, M.A.; Ramsey, J.; Mukhopadhyay, S.; Kearns, D.B. ZpdN, a Plasmid-encoded sigma factor homolog, induces pBS32-dependent cell death in *Bacillus subtilis*. *J. Bacteriol.* **2016**, *198*, 2975–2984. [CrossRef]
25. Burton, A.T.; DeLoughery, A.; Li, G.W.; Kearns, D.B. Transcriptional regulation and mechanism of sigN (ZpdN), a pBS32-encoded sigma factor in bacillus subtilis. *MBio* **2019**, *10*, e01899-19. [CrossRef]
26. Qi, Y.; Hulett, F.M. PhoP~P and RNA polymerase sigma(A) holoenzyme are sufficient for transcription of Pho regulon promoters in *Bacillus subtilis*: PhoP~P activator sites within the coding region stimulate transcription in vitro. *Mol. Microbiol.* **1998**, *28*, 1187–1197. [CrossRef]
27. Chang, B.-Y.; Doi, R.H. Overproduction, Purification, and Characterization of *Bacillus subtilis* RNA Polymerase SigA Factor. *J. Bacteriol.* **1990**, *172*, 3257–3263. [CrossRef]
28. Chen, Y.-F.; Helmann, J.D. The Bacillus subtilis Flagellar Regulatory Protein SigmaD: Overproduction, Domain Analysis and DNA-binding Properties. *J. Mol. Biol.* **1995**, *249*, 743–753. [CrossRef]
29. Paul, B.J.; Ross, W.; Gaal, T.; Gourse, R.L. rRNA Transcription in Escherichia coli. *Annu. Rev. Genet.* **2004**, *38*, 749–770. [CrossRef]
30. Panova, N.; Zborníková, E.; Šimák, O.; Pohl, R.; Kolář, M.; Bogdanová, K.; Večeřová, R.; Seydlová, G.; Fišer, R.; Hadravová, R.; et al. Insights into the mechanism of action of bactericidal Lipophosphonoxins. *PLoS ONE* **2015**, *10*, e0145918. [CrossRef]
31. Imamura, D.; Zhou, R.; Feig, M.; Kroos, L. Evidence that the *Bacillus subtilis* SpoIIGA protein is a novel type of signal-transducing aspartic protease. *J. Biol. Chem.* **2008**, *283*, 15287–15299. [CrossRef] [PubMed]
32. LaBell, T.L.; Trempy, J.E.; Haldenwang, W.G. Sporulation-specific σ factor σ29 of *Bacillus subtilis* is synthesized from a precursors protein, P31. *Proc. Natl. Acad. Sci. USA* **1987**, *84*, 1784–1788. [CrossRef] [PubMed]
33. Ross, W.; Thompson, J.F.; Newlands, J.T.; Gourse, R.L.E. *E. coli* Fis protein activates ribosomal RNA transcription in vitro and in vivo. *EMBO J.* **1990**, *9*, 3733–3742. [CrossRef] [PubMed]
34. Deneer, H.G.; Spiegelman, G.B. *Bacillus subtilis* rRNA promoters are growth rate regulated in Escherichia coli. *J. Bacteriol.* **1987**, *169*, 995–1002. [CrossRef]
35. Samarrai, W.; Liu, D.X.; White, A.M.; Studamire, B.; Edelstein, J.; Srivastava, A.; Widom, R.L.; Rudner, R. Differential responses of *Bacillus subtilis* rRNA promoters to nutritional stress. *J. Bacteriol.* **2011**, *193*, 723–733. [CrossRef]
36. Wellington, S.R.; Spiegelman, G.B. The kinetics of formation of complexes between Escherichia coli RNA polymerase and the rrnB P1 and P2 promoters of *Bacillus subtilis*. Effects of guanosine tetraphosphate on select steps of transcription initiation. *J. Biol. Chem.* **1993**, *268*, 7205–7214.
37. Fukushima, T.; Ishikawa, S.; Yamamoto, H.; Ogasawara, N.; Sekiguchi, J. Transcriptional, functional and cytochemical analyses of the veg gene in *Bacillus subtilis*. *J. Biochem.* **2003**, *133*, 475–483. [CrossRef]
38. Lei, Y.; Oshima, T.; Ogasawara, N.; Ishikawa, S. Functional analysis of the protein veg, which stimulates biofilm formation in *Bacillus subtilis*. *J. Bacteriol* **2013**, *195*, 1697–1705. [CrossRef]
39. Conter, A.; Menchon, C.; Gutierrez, C. Role of DNA supercoiling and RpoS sigma factor in the osmotic and growth phase-dependent induction of the gene osmE of Escherichia coli K12. *J. Mol. Biol.* **1997**, *273*, 75–83. [CrossRef]
40. Nicholson, W.L.; Setlow, P. Dramatic increase in negative superhelicity of plasmid DNA in the forespore compartment of sporulating cells of *Bacillus subtilis*. *J. Bacteriol.* **1990**, *172*, 7–14. [CrossRef]
41. Alice, A.F.; Sanchez-Rivas, C. DNA supercoiling and osmoresistance in *Bacillus subtilis* 168. *Curr. Microbiol.* **1997**, *35*, 309–315. [CrossRef] [PubMed]
42. Lewis, R.J.; Singh, O.M.P.; Smith, C.V.; Skarzynski, T.; Maxwell, A.; Wonacott, A.J.; Wigley, D.B. The nature of inhibition of DNA Gyrase by the Coumarins and the Cyclothialidines revealed by X-ray crystallography. *Embo J.* **1996**, *15*, 1412–1420. [CrossRef] [PubMed]
43. Sugino, A.; Higginst, N.P.; Brownt, P.O.; Peeblesf, C.L.; Cozzarellitf, N.R. Energy coupling in DNA gyrase and the mechanism of action of novobiocin. *Biochemistry* **1978**, *75*, 4838–4842. [CrossRef] [PubMed]
44. Gellert, M.; O'Dea, M.H.; Itoh, T.; Tomizawa, J.I. Novobiocin and coumermycin inhibit DNA supercoiling catalyzed by DNA gyrase. *Proc. Natl. Acad. Sci. USA* **1976**, *73*, 4474–4478. [CrossRef]
45. Dennis, P.P.; Bremer, H. Modulation of chemical composition and other parameters of the cell at different exponential growth rates. *EcoSal Plus* **2008**, *3*, 1553–1569. [CrossRef]
46. Sojka, L.; Kouba, T.; Barvík, I.; Šanderová, H.; Maderová, Z.; Jonák, J.; Krásný, L. Rapid changes in gene expression: DNA determinants of promoter regulation by the concentration of the transcription initiating NTP in *Bacillus subtilis*. *Nucleic Acids Res.* **2011**, *39*, 4598–4611. [CrossRef]
47. Estrem, S.T.; Gaal, T.; Ross, W.; Gourse, R.L. Identification of an UP element consensus sequence for bacterial promoters. *Proc. Natl. Acad. Sci. USA* **1998**, *95*, 9761–9766. [CrossRef]
48. Meng, W.; Belyaeva, T.; Savery, N.J.; Busby, S.J.W.; Ross, W.E.; Gaal, T.; Gourse, R.L.; Thomas, M.S. UP element-dependent transcription at the Escherichia coli rrnB P1 promoter: Positional requirements and role of the RNA polymerase α subunit linker. *Nucleic Acids Res.* **2001**, *29*, 4166–4178. [CrossRef]

49. Rao, L.; Ross, W.; Appleman, J.A.; Gaal, T.; Leirmo, S.; Schlax, J.P.; Record, M.; Thomas, J.; Gourse, R.L. Factor independent activation of rrnB P1: An "extended" promoter with an upstream element that dramatically increases promoter strength. *J. Mol. Biol.* **1994**, *235*, 1421–1435. [CrossRef]

50. Klumpp, S.; Hwa, T. Growth-rate-dependent partitioning of RNA polymerases in bacteria. *Proc. Natl. Acad. Sci. USA* **2008**, *105*, 20245–20250. [CrossRef]

51. Liang, S.T.; Bipatnath, M.; Xu, Y.C.; Chen, S.L.; Dennis, P.; Ehrenberg, M.; Bremer, H. Activities of constitutive promoters in Escherichia coli. *J. Mol. Biol.* **1999**, *292*, 19–37. [CrossRef] [PubMed]

52. Haldenwang, W.G.; Losick, R. Novel RNA polymerase σ factor from *Bacillus subtilis*. *Proc. Natl. Acad. Sci. USA* **1980**, *77*, 7000–7004. [CrossRef] [PubMed]

53. Hecker, M.; Völker, U. General stress response of *Bacillus subtilis* and other bacteria. *Adv. Microb. Physiol.* **2001**, *44*, 35–91. [CrossRef] [PubMed]

54. Jaehning, J.A.; Wiggs, J.L.; Chamberlin, M.J. Altered promoter selection by a novel form of *Bacillus subtilis* RNA polymerase. *Proc. Natl. Acad. Sci. USA* **1979**, *76*, 5470–5474. [CrossRef] [PubMed]

55. Haldenwang, W.G.; Lang, N.; Losick, R. A sporulation-induced sigma-like regulatory protein from b. subtilis. *Cell* **1981**, *23*, 615–624. [CrossRef]

56. Partridge, S.R.; Foulger, D.; Errington, J. The role of σF in prespore-specific transcription in *Bacillus subtilis*. *Mol. Microbiol.* **1991**, *5*, 757–767. [CrossRef] [PubMed]

57. Johnson, W.C.; Moran, C.P.; Losick, R. Two RNA polymerase sigma factors from *Bacillus subtilis* discriminate between overlapping promoters for a developmentally regulated gene. *Nature* **1983**, *302*, 800–804. [CrossRef]

58. Earl, A.M.; Losick, R.; Kolter, R. *Bacillus subtilis* genome diversity. *J. Bacteriol.* **2007**, *189*, 1163–1170. [CrossRef]

59. Burby, P.E.; Simmons, L.A. A bacterial DNA repair pathway specific to a natural antibiotic. *Mol. Microbiol.* **2019**, *111*, 338–353. [CrossRef]

60. Lee, Y.J.; Park, S.J.; Ciccone, S.L.M.; Kim, C.R.; Lee, S.H. An in vivo analysis of MMC-induced DNA damage and its repair. *Carcinogenesis* **2006**, *27*, 446–453. [CrossRef]

61. Ueda, K.; Morita, J.; Komano, T. Phage inactivation and DNA strand scission activities of 7-N-(p-hydroxyphenyl)mitomycin C. *J. Antibiot.* **1982**, *35*, 1380–1386. [CrossRef] [PubMed]

62. Rosenberg, A.; Sinai, L.; Smith, Y.; Ben-Yehuda, S. Dynamic expression of the translational machinery during *Bacillus subtilis* life cycle at a single cell level. *PLoS ONE* **2012**, *7*, 41921. [CrossRef] [PubMed]

63. Burenina, O.Y.; Hoch, P.G.; Damm, K.; Salas, M.; Zatsepin, T.S.; Lechner, M.; Oretskaya, T.S.; Kubareva, E.A.; Hartmann, R.K. Mechanistic comparison of *Bacillus subtilis* 6S-1 and 6S-2 RNAs-commonalities and differences. *RNA* **2014**, *20*, 348–359. [CrossRef]

64. Glaser, G.; Sarmientos, P.; Cashel, M. Functional interrelationship between two tandem *E. coli* ribosomal RNA promoters. *Nature* **1983**, *302*, 74–76. [CrossRef] [PubMed]

65. Schultz, M.C.; Brill, S.J.; Ju, Q.; Sternglanz, R.; Reeder, R.H. Topoisomerases and yeast rRNA transcription: Negative supercoiling stimulates initiation and topoisomerase activity is required for elongation. *Genes Dev.* **1992**, *6*, 1332–1341. [CrossRef] [PubMed]

66. Bird, T.; Burbulys, D.; Wu, J.; Strauch, M.; Hoch, J.; Spiegelman, G. The effect of supercoiling on the in vitro transcription of the spoIIA operon from *Bacillus subtilis*. *Biochimie* **1992**, *74*, 627–634. [CrossRef]

67. Helmann, J.D. Compilation and analysus of Bacillus Subtilis σA-dependent promoter sequences: Evidence for extended contact between RNA polymerse and upstream promoter DNA. *Nucleic Acids Res.* **1995**, *23*, 2351–2360. [CrossRef]

68. Typas, A.; Hengge, R. Role of the spacer between the -35 and -10 regions in sigmaS promoter selectivity in Escherichia coli. *Mol. Microbiol.* **2006**, *59*, 1037–1051. [CrossRef]

69. Bordes, P.; Conter, A.; Morales, V.; Bouvier, J.; Kolb, A.; Gutierrez, C. DNA supercoiling contributes to disconnect σS accumulation from σS-dependent transcription in Escherichia coli. *Mol. Microbiol.* **2003**, *48*, 561–571. [CrossRef]

70. Kusano, S.; Ding, Q.; Fujita, N.; Ishihama, A. Promoter selectivity of Escherichia coli RNA Polymerase E and E Holoenzymes. *J. Biol. Chem.* **1996**, *271*, 1998–2004. [CrossRef]

 *microorganisms*

Article

# Kinetic Modeling and Meta-Analysis of the *Bacillus subtilis* SigB Regulon during Spore Germination and Outgrowth

Jiri Vohradsky [1,*], Marek Schwarz [1], Olga Ramaniuk [2], Olatz Ruiz-Larrabeiti [2,3], Viola Vaňková Hausnerová [2], Hana Šanderová [2] and Libor Krásný [2]

1   Laboratory of Bioinformatics, Institute of Microbiology of the Czech Academy of Sciences, Vídeňská 1083, 14220 Prague, Czech Republic; marek.schwarz@biomed.cas.cz
2   Laboratory of Microbial Genetics and Gene Expression, Institute of Microbiology of the Czech Academy of Sciences, Vídeňská 1083, 14220 Prague, Czech Republic; ramaniuk@biomed.cas.cz (O.R.); olatz.ruiz@biomed.cas.cz (O.R.-L.); viola.hausnerova@gmail.com (V.V.H.); sanderova@biomed.cas.cz (H.Š.); krasny@biomed.cas.cz (L.K.)
3   Bacterial Stress Response Research Group, Department of Immunology, Microbiology and Parasitology, University of the Basque Country UPV/EHU, 48940 Leioa, Spain
*   Correspondence: vohr@biomed.cas.cz

**Abstract:** The exponential increase in the number of conducted studies combined with the development of sequencing methods have led to an enormous accumulation of partially processed experimental data in the past two decades. Here, we present an approach using literature-mined data complemented with gene expression kinetic modeling and promoter sequence analysis. This approach allowed us to identify the regulon of *Bacillus subtilis* sigma factor SigB of RNA polymerase (RNAP) specifically expressed during germination and outgrowth. SigB is critical for the cell's response to general stress but is also expressed during spore germination and outgrowth, and this specific regulon is not known. This approach allowed us to (i) define a subset of the known SigB regulon controlled by SigB specifically during spore germination and outgrowth, (ii) identify the influence of the promoter sequence binding motif organization on the expression of the SigB-regulated genes, and (iii) suggest additional sigma factors co-controlling other SigB-dependent genes. Experiments then validated promoter sequence characteristics necessary for direct RNAP–SigB binding. In summary, this work documents the potential of computational approaches to unravel new information even for a well-studied system; moreover, the study specifically identifies the subset of the SigB regulon, which is activated during germination and outgrowth.

**Keywords:** *Bacillus subtilis*; SigB; gene regulatory networks; computational modeling; promoter sequence analysis

**Citation:** Vohradsky, J.; Schwarz, M.; Ramaniuk, O.; Ruiz-Larrabeiti, O.; Vaňková Hausnerová, V.; Šanderová, H.; Krásný, L. Kinetic Modeling and Meta-Analysis of the *Bacillus subtilis* SigB Regulon during Spore Germination and Outgrowth. *Microorganisms* **2021**, *9*, 112. https://doi.org/10.3390/microorganisms9010112

Received: 10 December 2020
Accepted: 29 December 2020
Published: 5 January 2021

**Publisher's Note:** MDPI stays neutral with regard to jurisdictional claims in published maps and institutional affiliations.



## 1. Introduction

Transcription and expression of the physiologically relevant genes is essential for adaptation of organisms to changing environmental conditions. Uncovering the nature of gene regulatory networks is one of the core tasks of systems biology. Identifying direct regulons (group of regulated genes) of sigma factors can be considered as a basic element of this task for prokaryotic organisms where sigma factors are subunits of RNA polymerase (RNAP) that are critical for recognition of promoters (DNA sequences where gene transcription starts [1]). A number of tools for gene regulatory network inference were developed in the last 20 years (a comprehensive review of the methods for gene networks inference can be found in Wang et al. [2] and Loskot et al. [3]). Several such tools (ARACNE [4], and network BMA [5]), including our tool CyGenexpi [6], were integrated into the systems biology platform Cytoscape (http://www.cytoscape.org/). Advances and limitations of network inference methods were reviewed by [7], and substantial work on reconstruction of sigma factor-controlled networks was also performed by Tiwari and

Chauhan [8,9], as well as in *Bacillus*, particularly by Nannapaneni [10]. However, it has been shown that using only one source of data for network inference (e.g., only chromatin immunoprecipitation sequencing [ChIP-seq], RNA sequencing [RNA-seq], or literature mining) can be misleading. Therefore, combining multiple sources is necessary [11].

*Bacillus subtilis* is a Gram-positive model organism that survives unfavorable conditions as an endospore. Subsequent spore germination and outgrowth are complex processes [12] that require extensive changes in gene expression that involve a number of sigma factors. *B. subtilis* contains one main (primary, housekeeping) sigma factor—SigA, which regulates gene expression mostly in exponential phase [13,14], as well as 18 alternative sigma factors [15–18] and one sigma-like factor—Xpf [19].

SigB is the general stress response factor, helping the cell resist oxidative stress; moreover, it also protects cells against heat, acid, alkaline, or osmotic stress [20]. RNAP holoenzyme containing SigB recognizes GTTTaa and GGG(A/T)A(A/T) sequences as the $-35$ and $-10$ regions (with respect to +1 transcription start). These two sequences are separated typically by 13 to 15 nucleotides [18]. The *sigB* gene itself is transcribed from two promoters, one SigA-dependent and the other SigB-dependent [21]. The activity of the SigB protein is regulated by a partner-switching signaling network that involves anti-sigma, anti-anti-sigma factors, as well as phosphatases that act upstream of the "anti" factors. Under non-stressing conditions, SigB is typically in complex with its anti-sigma factor, RsbW, and inactive. RsbV, the anti-anti-sigma factor, is under these conditions phosphorylated and unable to interact with RsbW. When stress is detected by the cell, phosphatases RsbQP and/or RsbTU dephosphorylate RsbV, which subsequently interacts with RsbW, and SigB is released and activates its target genes, including its own operon, which also includes genes for RsbW and RsbV [22]. However, as experimental data show [23], SigB is expressed also during germination and outgrowth where the stress conditions are not expected.

Currently, according to SubtiWiki (http://www.subtiwiki.uni-goettingen.de/[24]), there are 217 genes known to be in the SigB regulon. Other works identified additional SigB-regulated genes by using various methods mostly in stress-induced systems [18,25–28]. Combining all the available data sources, the SigB regulon currently consists of 411 genes. It is most likely that not all of these genes are activated during particular conditions.

To advance our understanding of the SigB regulon and the germination-outgrowth process in *B. subtilis*, we applied a combination of computational and experimental approaches. First, we extracted and combined the data from the previous experiments obtained both from literature and the SubtiWiki database. Subsequently, in order to identify genes regulated by SigB, we computationally modelled the gene expression profiles recording the activity of the cell regulatory networks under given experimental conditions (here the germination and outgrowth) using SigB as a regulator, and those satisfying selection criteria were identified as members of the active regulon. Furthermore, for the genes of the SigB regulon reported as alternatively controlled by other sigma factors, we also computed the models of their regulation, and, where the SigB profile alone could not explain the regulation, other regulators (i.e., sigma factors) were suggested. Analysis of the promoter site binding motifs with respect to the above-mentioned analyses allowed us to define more precisely the structure of the sequence, which is necessary for SigB to bind and activate transcription. The results were then validated experimentally. Using the results of the kinetic modeling and promoter sequence binding motifs analysis, we were able to identify a core set of genes controlled by SigB, which is specifically expressed during germination and outgrowth.

## 2. Materials and Methods

### 2.1. Data Acquisition

#### 2.1.1. Time Series of Gene Expression

We downloaded the *B. subtilis* transcriptomic microarray data from GEO (http://www.ncbi.nlm.nih.gov/geo/query/acc.cgi?acc=GSE6865), consisting of 14 time points (0, 5, 10, 15, 20, 25, 30, 40, 50, 60, 70, 80, 90, and 100 min) during germination and outgrowth [23]. The dataset contains time series of expression of 4008 genes. Briefly, spores of *Bacillus subtilis* 168 were generated by growing cells in a defined MOPS medium at 37 °C and shaking for 4 days. Spores were then thermally activated at 70 °C for 30 min in germination medium. The release of dipicolinic acid in the medium during spore germination was monitored using the terbium fluorescence assay. Samples for RNA isolation were drawn at regular intervals during germination and outgrowth. RNA was isolated from spores and outgrowing spores and Cy-labeled cDNA was produced by reverse transcription using Cy-labeled dUTP. Samples were hybridized to microarray slides; microarrays were scanned using an Agilent G2505 scanner. Data from replicates were averaged, and the original $\log_2$-based data were exponentiated.

### 2.1.2. SigB Regulon

The SigB regulon was compiled from the published data [18,25,28–30] and from the SubtiWiki database (http://www.subtiwiki.uni-goettingen.de/ [31]) The data in SubtiWiki contain a collection of experimentally validated regulatory relations of *B. subtilis* genes constructed by surveying literature references, including interactions found regardless of the experimental conditions under which they were obtained. The literature data are mostly results of binding experiments obtained under stress conditions. All this results in overlaps of different sigma factors and cofactor regulons within the regulon of SigB. Such interactions were also considered in the analysis below.

### 2.2. Kinetic Model of Gene Expression

We used the model originally developed by Vohradsky [32] and further extended in the works of to and Vu [33–35]. The model was implemented as a Cytoscape plugin (www.cytoscape.org) as an R package and a command line tool [6]. In this paper, we used the command line version executed from a Matlab script. All further computations were performed in the Matlab environment. The script executables and the Cytoscape plugin can be downloaded from https://github.com/cas-bioinf/genexpi/wiki. The model as implemented in this paper was described in detail in our previous work [36]. Here, we only briefly mention its principle and a model of constant rate of expression used in the data preprocessing.

The relation between the rate of accumulation of the transcribed mRNA and the sigma factor amount can be described mathematically by a sigmoid with parameters reflecting the strength of binding, reaction delay, and mRNA degradation rate. The model used in this study has the following form:

$$\frac{dy_i}{dt} = k_{1i}\frac{1}{1 + \exp[-(\sum_j w_{ij}R_j + b_i)]} - k_{2i}y_i \tag{1}$$

where $y_i$ represents the amount of the target genes mRNA, and $R_j$ is the amount of the *j*-th sigma factor modulated by parameter $w_{ij}$, corresponding to the binding strength to the promoter. The $b_i$ corresponds to the reaction delay. Accumulation of the gene's mRNA is diminished by degradation described by the term $k_{2i}y_i$. In the data preprocessing step, we also considered a constant rate of expression model where

$$\frac{dy_i}{dt} = k_{3i} - k_{4i}y_i \tag{2}$$

Here, $y$ is the expression of the target gene as a function of time, and $k_3$ and $k_4$ are mRNA synthesis and degradation rate constants, respectively. When a gene expression profile is fitted by the constant synthesis model, it means that its synthesis is not affected by amount changes in the sigma factor. Such genes have to be excluded from the analysis, as when using the Equation (1) they can be fitted with any profile and introduce false positive results.

Since the expression data is noisy, we interpolated and smoothed them prior to computation with a Sawitzky–Golay filter. The smoothing achieved more robust results with respect to high-frequency phenomena expected in gene expression, while preserving the characteristic low-frequency phenomena. A further advantage to smoothing is that it let us subsample the fitted curve at arbitrary resolution. We subsampled the profiles at 1 min time steps, which allowed us to integrate Equation (1) accurately using the computationally cheap Euler method. Optimization of the parameters of the model for individual sigma factor-transcribed gene combination was performed using a simulated annealing scheme by minimization of an objective function

$$E = \sqrt{\sum (y - \widetilde{y})^2} \tag{3}$$

where $y$ represents the experimental mRNA's amount time series and $\widetilde{y}$ represents the time series computed using the model Equation (1). Furthermore, a regularization term was added to the objective function to penalize biologically implausible values of the parameters. The regularization is non-zero if either (a) $k_{1i}$ allows maximal measured transcript level to be achieved in less than 1 min starting from zero, or (b) regulatory response is very steep with $|w_i(t)| > 10$ for some t, or (c) the regulatory interaction is never saturated with $|-w_i(t) + b_i| > 0.5$ for all t. For each profile, the optimization was repeated 256 times with random values of initial parameters estimates. From the 256 runs, the parameters giving the smallest E were selected. The goal was to identify parameters that would fit the measured expression profiles of the given regulated gene with the SigB profile as the regulator within the confidence interval. Where such parameters were found, regulatory interaction between a sigma factor and a gene was considered proven.

### 2.3. Data Preprocessing

Prior to any computations, as mentioned in the preceding paragraph, raw gene expression time series were interpolated to 1 min intervals and smoothed using Golay filters (Matlab function smoothdata, method sgolay; Supplementary File S1). Then, several constraints were introduced: (1) time series of gene expression of the genes whose maximum in expression profile was smaller than an arbitrarily chosen value of 300 were excluded, as the low values of expression profiles bear large variance that can lead to misinterpretation of the modeling results. The threshold level was based on the observation that the variance of the microarray quantified expression values rapidly increases with decreasing magnitude of the signal and for low levels of expression can reach values higher than 50% of the mean (defined as coefficient of variation (CV)). (2) The genes that could be modeled with a constant rate of synthesis model (Equation (2)) were excluded, as they could be modeled by any profile and could bring false positive predictions. (3) A random expression profile created by randomization of the SigB expression profile was used as the regulator profile to model the expression profiles that were not excluded in the previous steps. The modeling with the randomized profile was performed 10 times for all genes that were not excluded in the previous step, and those gene expression profiles that were modelled at least once by the random profile were excluded. The purpose was the same as in the previous case-exclusion of the genes that could lead to false positive predictions.

### 2.4. Promoter Binding Motif Analysis

For identification of the SigB binding motifs, we used the information from Table S4 of the Supplementary Materials of the paper of Nicolas et al. [18]. Table S4 contains sequences of the promoter regions containing two motifs ($-35$ and $-10$) found to be present for different sigma factors including SigB. The table contains also "upshift" locations (proxy for transcription start site) for all transcription units (operons). Using the upshift locations, we extracted the sequences $<-80; +40>$ nucleotide range from the upshift position. In these sequences, we used two methods for identification of the binding site motifs: exact match and consensus motif search. For the latter, we first identified the consensus motif from the motifs in Table S4 of Nicolas et al. The computed logo is shown in Figure 1. Each motif was converted to meme format, and the fimo utility (meme-suite.org/doc/fimo.html) was used to search the motifs within the extracted sequences (fimo—norc—thresh 0.001).

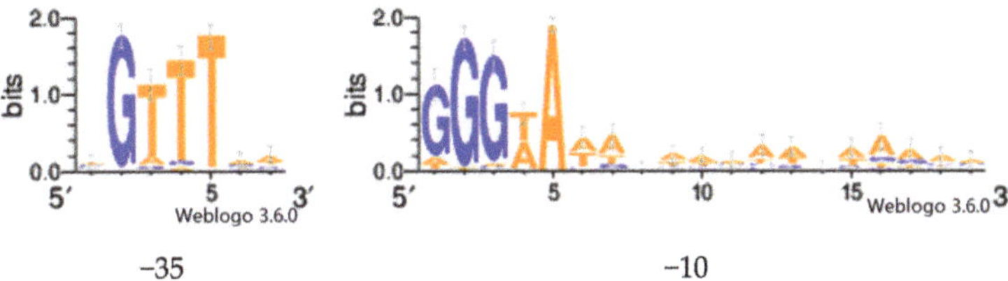



**Figure 1.** Binding motifs found in genes of SigB regulon. The motifs are ordered from −35 to −10 with spacers of 5–20 nucleotides.

For the exact match search, we directly used the sequences of the motif as indicated in the promoter sequences of Table S4 of Nicolas et al. The two results were merged and combined with the information about corresponding transcription units.

### 2.5. Transcription In Vitro

#### 2.5.1. SigB Cloning

The *sigB* gene was amplified using *B. subtilis* BaSysBio genomic DNA [18] with the primers 1004/sigB_F 5′-ggaattcCATATGacacaaccatcaaaaac-3′ and 1006/sigB_R_His 5′-ccgCTCGAGcattaactccatcgagggatc-3′. The resulting DNA fragment was cloned into expression vector pET-22b (comprising inducible promoter and 6xHis-tag) using NdeI and XhoI restriction enzymes. The resulting construct was validated by sequencing transformed into *E. coli* BL21 (DE3) competent cells, and the strain was named LK#1207.

#### 2.5.2. Media, Growth Conditions, Protein Purification

For protein purification, the strains were cultured in Luria–Bertani (LB) medium at 37 °C with continuous shaking. *Bacillus subtilis* RNAP *rpoE* (LK 637, [37]) with a His10-tagged β′ subunit was purified as described previously [38]. SigB was overexpressed from strain LK#1207. The strain was grown in 2L of liquid LB medium at 37 °C, and when the bacterial culture reached $OD_{600}$ 0.6–0.7 (mid-logarithmic phase), it was transferred to room temperature, and SigB overexpression was induced with 0.8 mM isopropyl β-D-1-thiogalactopyranoside (IPTG) for 3 hours under constant shaking (120 rpm). Purification of SigB via 6x His-tag using affinity chromatography was performed as described previously [38]. Delta subunit (RLG7023 [39]) of *B. subtilis* RNAP was purified as described [39].

#### 2.5.3. PCR of DNA Templates

Linear PCR products of putative promoter regions were used as templates for in vitro transcription assays. The primers are listed in Supplementary File S2. Putative promoter sequences were PCR-amplified using wild type (wt) *B. subtilis* gDNA as the template. All PCR reactions were performed using the Expand High Fidelity System (Roche). The purification of PCR constructs was performed using QIAquick Gel Extraction Kit from Qiagen according to the manufacturer's protocol.

#### 2.5.4. In Vitro Transcription Assays

The *B. subtilis* RNAP core was reconstituted with saturating concentrations of SigB and delta. Reconstitutions were performed in a glycerol storage buffer (50 mM Tris-HCl (pH 8.0), 0.1 M NaCl, 50% glycerol) for 10 min at 37 °C. *E. coli* RNAP holoenzyme was purchased from NEB (cat# M0551S).

Multiple round transcription reactions were carried out in 10 μL reaction volumes with 60 nM *B. subtilis* RNAP holoenzyme and 50 ng of linear DNA template. The transcription buffer contained 40 mM Tris-HCl (pH 8.0), 10 mM MgCl2, 1 mM dithiothreitol (DTT),

0.1 mg/mL bovine serum albumin (BSA), and 150 mM KCl. ATP and GTP were 400 $\mu$M, CTP was 200 $\mu$M, and UTP was 10 $\mu$M plus 2 $\mu$M radiolabeled [$\alpha$-$^{32}$P]UTP.

All transcription experiments were performed at 37 °C. Transcription was induced by adding reconstituted RNAP holoenzyme and allowed to proceed for 15 min. Transcriptions were stopped with equal volumes (10 $\mu$L) of formamide stop solution (95% formamide, 20 mM EDTA (pH 8.0)). To generate molecular size RNA marker, we used the same conditions. The only differences were the usage of *E. coli* RNAP and the templates were DNA fragments of the *hcr* gene and plasmid pLK1 [40]. These templates were combined in one reaction and yielded a ladder of 108, 145, 201, 253, 300, 357, and 407 nt. Samples were loaded onto 7 M urea–7% polyacrylamide gels and electrophoresed. The dried gels were scanned with Molecular Imager FX (Bio-Rad) and were visualized and analyzed using the Quantity One software (Bio-Rad).

## 3. Results and Discussion

### 3.1. Modeling of the SigB Regulon

In this paragraph, we discuss the use of the expression profile of SigB to identify genes that could be, from the kinetic point of view, controlled by SigB. This was achieved by modelling of their expression profiles using Equations (1) and (3). Those genes for which the model well fitted their expression profiles were identified as potentially controlled by SigB.

Out of the 411 genes (217 genes from SubtiWiki, an additional 194 compiled from the literature), 260 were reported to be controlled exclusively by SigB (the following results are summarized in Supplementary File S3). For the remaining 151 genes, numerous other regulators and sigma factors were found to participate in the control of their expression, depending on the conditions. According to SubtiWiki and the bibliography, 2 regulators, including SigB, were in the reported SigB regulon found for 88 genes, 3 regulators for 39 genes, 4 regulators for 13 genes, 5 regulators for 7 genes, and 6 regulators for 3 genes. Altogether, 37 regulators including sigma factors were reported to participate in the control of the SigB regulon; the regulon overlap was rather high. Most of the genes reported in the literature as co-controlled by other factors also belonged to the SigA regulon (87 genes) (the competition between SigB and SigA was reported [18]). Other sigma factors are (listed in descending order of number of genes reported to be co-controlled by them) SigM (20), SigG and SigF (16), SigW (14), SigX (10), SigH (6), SigE (5), and SigD and SigI (1). Of the other transcription factors, CcpA, involved in carbon catabolite repression [41] was found 13 times, while the others appeared between 1 to 9 times. We emphasize that these observations were compiled from numerous articles, where the results were obtained under diverse experimental conditions. Our goal then was to identify those regulators (i.e., sigma factors including SigB) that could participate in the control of the SigB regulon under the conditions reflected in the time series analyzed here. We applied the model defined in Equation (1) to the germination and outgrowth of gene expression data from a GEO database (http://www.ncbi.nlm.nih.gov/geo/query/acc.cgi?acc=GSE6865) that contained expression profiles of 4008 genes measured at 14 time points (the experimental conditions are briefly summarized in the Materials and Methods section), the pre-processing excluded 1349 genes with low expression, and 135 genes with constant rate of expression. Of the remaining 2524 genes, another 327 genes with profiles that had been at least once modelled with the random expression profiles were removed (see Materials and Methods, data preprocessing).

For each target gene, all combinations of its reported regulators (single and/or multiple) were modeled and the ability of the model to fit the experimental data was assessed (see Materials and Methods, Section 3.2., Equation (1)). Although the experimental data list some genes with four, five, and even six regulators, we modeled only the combinations with maximally three regulators, as more than three regulators concomitantly acting on one gene at a time is improbable. Furthermore, using more regulators could lead to overfitting, with weights of some regulators close to zero. Selection of the regulators controlling each gene was then performed with respect to the best goodness of fit of the model to the experimental expression profile. Where an equivalent fit quality was achieved for different regulators, or a combination of regulators, we used the following rules: (i) the minimal number of regulators satisfying goodness of fit was selected (e.g., if an equal fit was achieved for SigB only and a combination of SigB plus SigA, then the single SigB regulator was selected); (ii) if SigB was found as one of the regulators, this regulation was preferred; and (iii) if the error of fit was the same for some regulators or a combination of regulators, they were listed as alternatives.

Modeling results are summarized in Supplementary File S3. Of the 411 genes of the documented SigB regulon, 51 were found to be expressed at low levels (see constraints in Materials and Methods Section 3.3), and 123 did not pass the other constraints defined in the Materials and Methods section. For 63, we did not find parameters that could successfully model their expression profiles; these genes were considered to not be controlled by SigB under our conditions. For 25 genes, SigA, instead of SigB, was found as their most probable regulator (*apt, atpC, csbA, hemA, hemX, nhaC, opuD, pnpA, queA, rbfA, recO, rpsO, tgt, tmk, uvrA, uvrB, uvrC, yhbJ, yhcA, yhcC, yhdH, yhgE, yocJ, yozB, yybT*). For 17 genes (*ctc, gtaB, menC, menE, nadE, yebG, yitT, ykuT, ywsA, infB, smpB, ydaJ, ydaK, ydaL, yebE, yoaA, ypuB*), the best fit was obtained when SigB and SigA acted together. In summary, the modeling confirmed 148 (36%) genes that were during germination and outgrowth controlled by SigB alone (94) or with a possible participation of other regulators (54).

Finally, we found that of the 46 genes reported as controlled by three or more regulators, 11 were equally well modeled by SigB alone (*gabD, katX, rsbU, V, W, X, yfhE, F, yxjI, yugU*) or by SigB together with one or more reported regulators. No model could be found for six genes (*ylxP, Q, R, S, spo0E, yqjL*). Generally, in all remaining cases, two regulators were always sufficient to model the target gene expression profile equally well as if more regulators were employed.

### 3.2. Binding Motif Analysis of Predicted SigB-Dependent Transcription of the Genes Expressed during Outgrowth

In order to identify the genes that are actually controlled by SigB during germination and outgrowth, we combined promoter binding site composition (the presence of −35, −10, and their spacing) of the known SigB-dependent genes with their expression kinetics analysis.

We took all genes of the documented SigB regulon (Supplementary File S3) and we searched for the −35 and −10 motifs in the <−80; +40> region (numbering according to a putative transcription start sites, [18]). We evaluated their mutual position and their distance from each other (Supplementary File S4). The same analysis was performed on subsets of genes that we found by the kinetic modelling to be under the control of SigB only (type 1), genes where other regulators besides SigB were found (type 2), and genes where no regulation was identified or that were excluded during preprocessing (type 3). Figure 2 shows the relative representation of the binding motifs and spacer length distributions in these three groups. The distance between the promoter motifs (spacer length) ranged from 1 to 50 bp, where the highest deviation was observed for the excluded (type 3) genes.

In total, 95% of type 1 genes possessed the −35 motif, 78% the −10 motif, and 73% of these genes contained both motifs. Similarly, 97% of type 2 genes had the −35 motif and 66% the −10 motif; 63% contained both motifs. To the contrary, only 53% of type 3 genes had the −10 motif and only 40% of them had both. Importantly, the distance between the binding motifs was more consistent for type 1 and 2 genes where the mean distance was around 15 bp, while for the excluded genes, the mean distance was 18 bp. The largest deviations were observed for the genes for which the kinetic modeling did not find any regulator, or those that were excluded for low expression or flat expression profile (type 3). Therefore, this analysis clearly specified genes whose binding site composition determines that they could be regulated by SigB only and those that require during germination and outgrowth additional/different regulators.

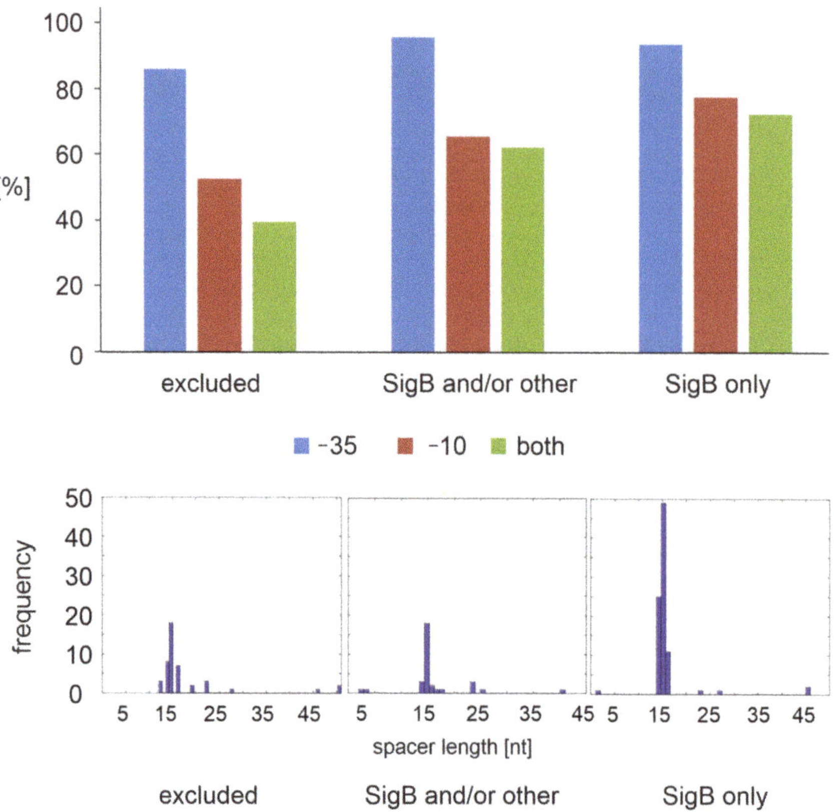



**Figure 2.** Relative occurrence of −35 and −10 binding motifs and the distance between them in the different SigB regulatory groups. Excluded (type 3)—regulator not found or the genes excluded during preprocessing; SigB and/or other (type 2)—genes for which besides SigB, also another regulator was found; SigB only (type 1)—genes for which SigB was found as the only regulator. The figure shows that the best characteristics were found for type 1 genes where the modeling results were consistent with the binding motifs analysis.

### 3.3. Experimental Verification of Selected Promoters

Finally, we performed in vitro transcriptions in a defined cell-free system for selected genes with different organization of SigB-binding motifs in the promoter region. The genes were selected according to the motif organization and the shape of its gene expression profile (see Figure 3). We used *B. subtilis* RNAP and DNA fragments containing the promoter regions. Altogether, 16 promoter regions were selected. For the full list, see Table 1. These promoters were divided into three classes (class I, II, and III). Class I promoters (10 promoters) contained canonically spaced (by 13–15 bp) −35 and −10 elements. Class II promoters (two promoters) contained only the −10 element, highly resembling the consensus sequence. Class III promoters (four promoters) contained both −35 and −10 elements, but their spacing was more than 14 bp. The genes were also chosen for the shape of their gene expression profile. Most of them were correlated with the shape of the SigB expression profile, with the exception of *pgcA* and *yqhY*, for which the model was fitted with the negative value of the parameter $w$, suggesting inverse correlation with the profile of SigB.

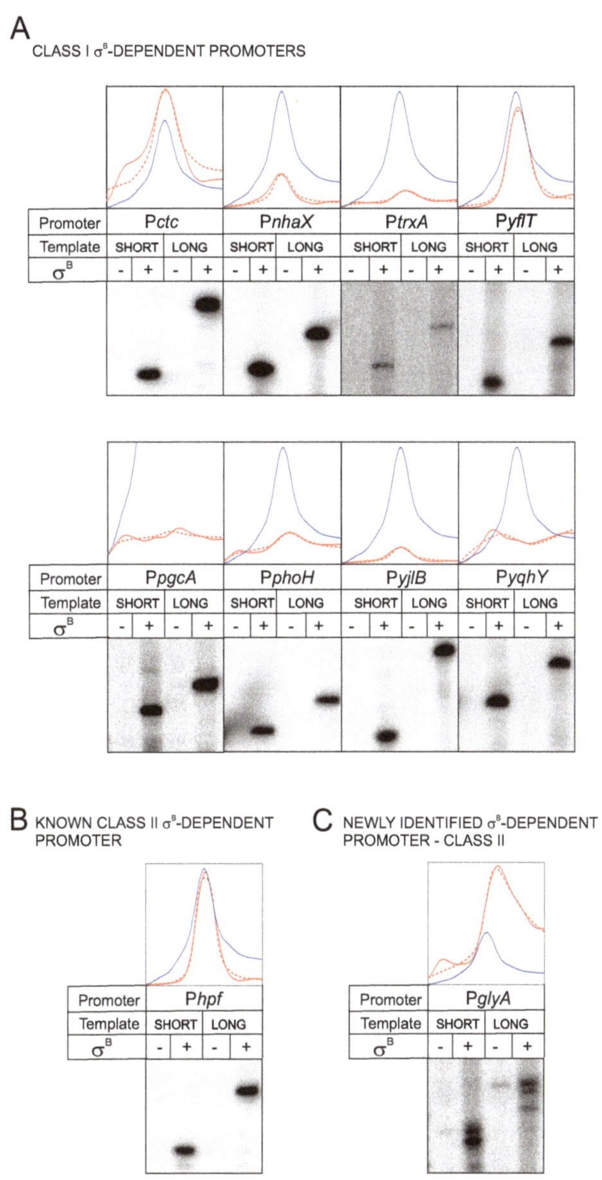



**Figure 3.** Experimental verification of SigB dependence of selected promoters. The validation was done by in vitro transcription with *B. subtilis* RNAP complexed with SigB (σB). (**A**)–Class I promoters [contain canonically spaced (by 13–15 bp) −35 and −10 elements]. (**B**)–A known Class II promoter contains only the −10 element, highly resembling the consensus sequence. (**C**)–genes of the Class II promoter. Curves represent modeled (red dashed line) and experimental (red solid line) expression profiles of the given gene; blue line is the expression profile of SigB. SHORT and LONG refer to template length (two sizes for each promoter region) to distinguish the orientation of the promoter within the template. −/+ indicate the absence/presence of SigB. In the absence of SigB, only the RNAP core was used. *trxA* was used as a standard.

**Table 1.** Experimentally verified genes. Motifs −35 and −10, refer to the binding motifs as defined in Section 2.2.

| Gene | | CLASS | Transcription | Verified TNX | Spacer Length | Motifs | −35 | Spacer | −10 |
|---|---|---|---|---|---|---|---|---|---|
| *pgcA* (*yhxB,gtaC,gtaE*) | BSU09310 | I | 1 | 1 | 16 | −35, −10 | CGTTTA | TTTTTTGA<br>TATCAATT | **GGGTAA**GAA<br>CATATAAAGA |
| *yjlB* | BSU12270 | I | 1 | 1 | 15 | −35, −10 | TGTTTG | GCGAACC<br>GCTATATG | **TGGAAG**ACAA<br>AAAAGGGAG |
| *nhaX* (*yheK*) | BSU09690 | I | 1 | 1 | 15 | −35, −10 | AGGTTA | ATTGTGC<br>TCAAATTC | **GGGTAG**TAG<br>TGTTGTAAGA |
| *cadA* (*yvgW*) | BSU33490 | | 0 | | 15 | −35, −10 | TGTTTT | TCATTGA<br>CACTTTCT | **TGGAAA**ACA<br>ACATATAATA |
| *yqhY* | BSU24330 | I | 1 | 1 | 16 | −35, −10 | GGTTTC | GCTTGCTA<br>ATGAAATT | **GGGTAT**CCT<br>GTAATTATAA |
| *yflT* | BSU07550 | I | 1 | 1 | 14 | −35, −10 | TGTTTC | AGGTACA<br>GACGATC | **GGGTAT**GAAA<br>GAAATATAG |
| *rsgA* (*cpgA, yloQ*) | BSU15780 | | 0 | | 14 | −35, −10 | AGATTG | AACCAGG<br>CCAAAAA | **GGGTAC**TATC<br>AAGTAATGG |
| *ctc* | BSU00520 | I | 1 | 1 | 15 | −35, −10 | GGTTTA | AATCCTT<br>ATCGTTAT | **GGGTAT**TGTT<br>TGTAATAGG |
| *phoH* (*yqfE*) | BSU25340 | I | 1 | 1 | 15 | −35, −10 | AGTTCA | AGAAGGC<br>ATTAAATT | **GGGTAA**ACAG<br>GATGTAGAG |
| *hpf* (*yvil,yvyD*) | BSU35310 | I | 1 | | 15 | −35, −10 | TGTTTC | AGCAGGAA<br>TTGTAAA | **GGGTAA**AAGA<br>GAAATAGAT |
| *glyA* (*glyC,ipc-34d*) | BSU36900 | II | 1 | | - | −10 | - | - | **TGGTAA**AAA<br>CAAAGAACAG |
| *yhfP* | BSU10320 | II | 0 | | - | −10 | - | - | **AGGAAG**AAAT<br>AAGATGAAC |
| *xynA* | BSU18840 | III | 0 | | 28 | −35, −10 | TGTTTT | AAATGTATAC<br>GAGTGCTAC<br>CTCAAAGTC | **GGAAAAAA**<br>TATTATAGGAG |
| *yhfl* | BSU10240 | III | 0 | | 24 | −35, −10 | TGTTTA | AAACATGCTTTT<br>TTCAAGAAAAAT | **GGGTAT**ATTG<br>AAGGAGGAC |
| *pfkA* (*pfk*) | BSU29190 | III | 0 | | 35 | −35, −10 | GGTTTC | ATAGGGAG<br>GATGGA<br>GATCCCTTTTCAT<br>TGTTTTTA | **GGGCAA**TGA<br>TCATGTTATG |
| *uppS* (*yluA*) | BSU16530 | III | 0 | | 46 | −35, −10 | TGTTTA | CAGGGGGTTTT<br>TTTGTTAATACTG<br>TTGATTACATTG<br>ATTATCAGCA | **GGGAAT**GT<br>AACCTTTTTGG |

Figure 3 shows transcriptions from eight class I and two class II promoters, in all cases with a clear positive result. However, two class I (cadA, rsgA), one class II (yhfP), and all class III promoters (xynA, yhfl, pfkA, uppS) were inactive. The transcriptional inactivity of class III promoters with RNAP–SigB was anticipated—the −10 elements were not strong enough to act as promoters on their own and the spacer distances to their respective −35 elements were too long. Hence, for genes whose promoters were inactive in vitro, it is apparent that additional transcription factors not included in the in vitro assays are required.

Of the 10 class I genes, the *yqhY* gene was selected for its inversely correlated profile with the expression profile of SigB. The SigB binding motif was present and the promoter was active in vitro (Figure 3). In the *yqhY* promoter region, we also identified a SigA-dependent promoter-like sequence overlapping that one of SigB. However, this promoter was not active in vitro (data not shown). The phenomenon of inversely correlated profiles was also observed for some other genes (see Supplementary File S3) as well reported in a previous study [36]. The exact mechanistic interplay of RNAP holoenzymes/transcription factors at such promoter regions is currently unknown.

Finally, Figure 3D shows SigB-dependent transcription from *PglyA*, a newly found class II promoter not previously known to be SigB-dependent. The *glyA* gene encodes serine hydroxymethyltransferase involved in purine nucleotide metabolism. Its expression is known to be controlled by SigA and the PurR repressor [42].

Taken together, both computational and experimental analyses show that for a gene to be controlled by SigB, it is necessary that the promoter sequence contains both $-10$ and $-35$ binding motifs with the spacer length in the range of $15+/-2$ nucleotides (with the exception of *glyA*, where the $-35$ motif was not identified as being significant, but transcription was confirmed). Figure 2 shows substantial differences between the genes that were previously proposed to be under the control of SigB, and those genes that were excluded on the basis of our kinetic modeling. The kinetics of the genes that were excluded were not coherent with the gene expression model, and, importantly, their promoter sequences did not contain full binding motifs with the correct spacing. If we combine all the above mentioned criteria and select the genes that satisfy them, we obtain a core set of 146 genes representing 115 operons that are controlled by SigB during spore germination and outgrowth representing 35% of all genes reported as being the SigB regulon (*aag, aldY, atpC, bmrU, clpC, copB, cpgA, csbB, csbC, csbD, csbX, csoR, ctc, ctsR, cypC, dps, galK, galT, gspA, gtaB, katE, katX, malS, mcsA, mcsB, menC, mgsR, nagBA, nhaX, ohrB, opuD, opuE, pgcA, phoH, pth, radA, rbfA, rnr, rpe, rpmEB, rsbRD, rsbV, rsbW, rsbX, rsoA, sigB, tmk, truB, trxA, ung, yaaI, ybyB, ycbP, ycdF, ycdG, yczO, ydaC, ydaD, ydaE, ydaG, ydaJ, ydaK, ydaL, ydaM, ydaN, ydaP, ydaS, ydaT, ydbD, ydfO, ydjJ, yerD, yetO, yfhD, yfhF, yfhK, yfhO, yfkM, yflA, yflD, yflT, yhcM, yhdF, yhdN, yhxD, yjgB, yjgC, yjgD, yjlB, yjzE, ykgA, ykzN, ylxP, ymaE, yocB, yocK, yojJ, yorA, yoxB, yoxC, ypuD, yqbM, yqgC, yqhB, yqhY, yqjF, yqxL, yrhK, yrvD, ysnF, ytaB, ytxG, ytxH, ytxJ, yugU, yuzH, yvaA, yvaG, yvaK, yvbG, yvgN, yvgO, yvrE, yvyD, yvyI, ywiE, ywjC, ywkB, ywlB, ywmE, ywmF, ywsB, ywtG, ywzA, yxaB, yxbG, yxkA, yxkO, yxnA, yxxB, yxzF, yybO, yycD, yyzG, yyzH*; the full list of the genes and their operons is given in tabular form in the Supplementary File S5). Their functional analysis unsurprisingly showed that they mostly code stress proteins (72%, $n = 106$), and the second largest group belonged to a category "membrane proteins" with 27 genes (18%). The remaining genes of the reported SigB regulon may be expressed under different conditions, require additional factors, or are controlled by different sigma factors or cofactors (the alternatives to SigB control for specific genes are shown in Supplementary File S3). Some of these genes may not be true targets of SigB regulation.

## 4. Conclusions

In this study, using a combined approach of static and dynamic information analyses, we defined the SigB regulon in *B. subtilis* that is active during spore germination and outgrowth. The used approach combined a meta-analysis of literature data with the information about the kinetics of gene expression and promoter sequence analysis. The analysis showed that out of the 411 genes of the theoretical SigB regulon, 146 (35%) were expressed and controlled by SigB during normal growth conditions; most of them coded for stress and/or membrane proteins. The remaining genes of the reported SigB regulon may be expressed under different conditions, are controlled by different sigma factors or in combination with them (mainly SigA), or they require additional factor or cofactors for their expression. The analysis also showed the importance of the organization of the promoter binding sequence, especially the spacing of $-35$ and $-10$ elements for promoter recognition by the RNAP–SigB holoenzyme. Consistently, experimentally confirmed class III promoters, which have not been assigned to SigB-dependent regulation by the other analyses presented here (although they were identified as SigB-dependent in the literature), failed to be recognized by the RNAP–SigB holoenzyme in our study. The presented approach shows that in order to identify the regulon active during a specific biological process or specific conditions, the static information (e.g., binding experiments) is not sufficient and other additional source of information is necessary to employ. It also shows that the vast amount of data accumulated in literature and databases can be effectively used to discover new relations even in already well studied systems. The presented approach is general enough to be applied to other systems for which sufficient amounts of data are available.

**Supplementary Materials:** The following are available online at https://www.mdpi.com/2076-260 7/9/1/112/s1.

**Author Contributions:** Conceived the study, designed the models, made computational analyses, and wrote the manuscript, J.V.; made computations associated with the promoter binding motif analysis, M.S.; made the experiments for in vitro verifications, O.R., O.R.-L., V.V.H., and H.Š.; designed the experiments and wrote the corresponding part of the manuscript, L.K. All authors have read and agreed to the published version of the manuscript.

**Funding:** This work was funded by ELIXIR CZ research infrastructure projects (MEYS grant no. LM2018131) including access to computing and storage facilities. European Regional Development Fund–Project "ELIXIR-CZ: Budování kapacit" (no. CZ.02.1.01/0.0/0.0/16_013/0001777) to J.V., Czech Science Foundation—grant no. 20-12109S to L.K., and post-doctoral grant POS_2019_1_0033 (Basque Government) to O.R.-L.

**Data Availability Statement:** Data is contained within the article or supplementary material.

**Conflicts of Interest:** The authors declare no conflict of interest.

# References

1. Paget, M.S. Bacterial Sigma Factors and Anti-Sigma Factors: Structure, Function and Distribution. *Biomolecules* **2015**, *5*, 1245. [CrossRef] [PubMed]
2. Loskot, P.; Atitey, K.; Mihaylova, L. Comprehensive Review of Models and Methods for Inferences in Bio-Chemical Reaction Networks. *Front. Genet.* **2019**, *10*, 549. [CrossRef]
3. Wang, Y.R.; Huang, H. Review on statistical methods for gene network reconstruction using expression data. *J. Theor. Biol.* **2014**, *362*, 53–61. [CrossRef]
4. Margolin, A.A.; Nemenman, I.; Basso, K.; Wiggins, C.; Stolovitzky, G.; Dalla-Favera, R.; Califano, A. ARACNE: An Algorithm for the Reconstruction of Gene Regulatory Networks in a Mammalian Cellular Context. *BMC Bioinform.* **2006**, *7*, 1. [CrossRef] [PubMed]
5. Yeung, K.Y.; Dombek, K.M.; Lo, K.; Mittler, J.E.; Zhu, J.; Schadt, E.E.; Bumgarner, R.; Raftery, A.E. Construction of regulatory networks using expression time-series data of a genotyped population. *Proc. Natl. Acad. Sci. USA* **2011**, *108*, 19436–19441. [CrossRef] [PubMed]
6. Modrak, M.; Vohradsky, J. Genexpi: A toolset for identifying regulons and validating gene regulatory networks using time-course expression data. *BMC Bioinform.* **2018**, *19*, 137. [CrossRef]
7. De Smet, R.; Marchal, K. Advantages and limitations of current network inference methods. *Nat. Rev. Genet.* **2010**, *8*, 717–729. [CrossRef]
8. Tiwari, A.; Balázsi, G.; Gennaro, M.L.; Igoshin, O.A. The interplay of multiple feedback loops with post-translational kinetics results in bistability of mycobacterial stress response. *Phys. Biol.* **2010**, *7*, 036005. [CrossRef]
9. Chauhan, R.; Ravi, J.; Datta, P.; Chen, T.; Schnappinger, D.; Bassler, T.C.K.E.; Balázsi, G.; Gennaro, M.L. Reconstruction and topological characterization of the sigma factor regulatory network of Mycobacterium tuberculosis. *Nat. Commun.* **2016**, *7*, 11062. [CrossRef]
10. Nannapaneni, P.; Hertwig, F.; Depke, M.; Hecker, M.; Mäder, U.; Völker, U.; Steil, L.; Van Hijum, S.A.F.T. Defining the structure of the general stress regulon of Bacillus subtilis using targeted microarray analysis and random forest classification. *Microbiology* **2012**, *158*, 696–707. [CrossRef]
11. MacQuarrie, K.L.; Fong, A.P.; Morse, R.H.; Tapscott, S.J. Genome-wide transcription factor binding: Beyond direct target regulation. *Trends Genet.* **2011**, *27*, 141–148. [CrossRef] [PubMed]
12. Xing, Y.; Harper, W.F. Bacillus spore awakening: Recent discoveries and technological developments. *Curr. Opin. Biotechnol.* **2020**, *64*, 110–115. [CrossRef] [PubMed]
13. Shorenstein, R.G.; Losick, R. Purification and properties of the sigma subunit of ribonucleic acid polymerase from vegetative Bacillus subtilis. *J. Biol. Chem.* **1973**, *248*, 6163–6169. [PubMed]
14. Price, C.W.; Doi, R.H. Genetic mapping of rpoD implicates the major sigma factor of Bacillus subtilis RNA polymerase in sporulation initiation. *Mol. Genet. Genom.* **1985**, *201*, 88–95. [CrossRef] [PubMed]
15. Haldenwang, W.G. The sigma factors of Bacillus subtilis. *Microbiol. Rev.* **1995**, *59*, 1–30. [CrossRef]
16. Matsumoto, T.; Nakanishi, K.; Asai, K.; Sadaie, Y. Transcriptional analysis of the ylaABCD operon of Bacillus subtilis encoding a sigma factor of extracytoplasmic function family. *Genes Genet. Syst.* **2005**, *80*, 385–393. [CrossRef]
17. Helmann, J.D. Bacillus subtilis extracytoplasmic function (ECF) sigma factors and defense of the cell envelope. *Curr. Opin. Microbiol.* **2016**, *30*, 122–132. [CrossRef]
18. Nicolas, P.; Mäder, U.; Dervyn, E.; Rochat, T.; LeDuc, A.; Pigeonneau, N.; Bidnenko, E.; Marchadier, E.; Hoebeke, M.; Aymerich, S.; et al. Condition-Dependent Transcriptome Reveals High-Level Regulatory Architecture in Bacillus subtilis. *Science* **2012**, *335*, 1103–1106. [CrossRef]

19. McDonnell, E.G.; Wood, H.; Devine, K.M.; McConnell, D.J. Genetic control of bacterial suicide: Regulation of the induction of PBSX in Bacillus subtilis. *J. Bacteriol.* **1994**, *176*, 5820–5830. [CrossRef]
20. Hecker, M.; Pané-Farré, J.; Uwe, V. SigB-Dependent General Stress Response inBacillus subtilisand Related Gram-Positive Bacteria. *Annu. Rev. Microbiol.* **2007**, *61*, 215–236. [CrossRef]
21. Wise, A.A.; Price, C.W. Four additional genes in the sigB operon of Bacillus subtilis that control activity of the general stress factor sigma B in response to environmental signals. *J. Bacteriol.* **1995**, *177*, 123–133. [CrossRef] [PubMed]
22. Locke, J.C.W.; Young, J.W.; Fontes, M.; Jiménez, M.J.H.; Elowitz, M.B. Stochastic Pulse Regulation in Bacterial Stress Response. *Science* **2011**, *334*, 366–369. [CrossRef] [PubMed]
23. Keijser, B.J.F.; Ter Beek, A.; Rauwerda, H.; Schuren, F.; Montijn, R.; Van Der Spek, H.; Brul, S. Analysis of Temporal Gene Expression during Bacillus subtilis Spore Germination and Outgrowth. *J. Bacteriol.* **2007**, *189*, 3624–3634. [CrossRef] [PubMed]
24. Michna, R.H.; Commichau, F.M.; Tödter, D.; Zschiedrich, C.P.; Stülke, J. SubtiWiki—A database for the model organism Bacillus subtilis that links pathway, interaction and expression information. *Nucleic Acids Res.* **2014**, *42*, D692–D698. [CrossRef]
25. Price, C.W.; Fawcett, P.; Cérémonie, H.; Su, N.; Murphy, C.K.; Youngman, P. Genome-wide analysis of the general stress response in Bacillus subtilis. *Mol. Microbiol.* **2002**, *41*, 757–774. [CrossRef]
26. Helmann, J.D.; Wu, M.F.W.; Kobel, P.A.; Gamo, F.-J.; Wilson, M.; Morshedi, M.M.; Navre, M.; Paddon, C. Global Transcriptional Response of Bacillus subtilis to Heat Shock. *J. Bacteriol.* **2001**, *183*, 7318–7328. [CrossRef]
27. Petersohn, A.; Brigulla, M.; Haas, S.; Jörg, D.; Völker, U.; Hecker, M. Global Analysis of the General Stress Response of Bacillus subtilis Global Analysis of the General Stress Response of Bacillus subtilis. *J. Bacteriol.* **2001**, *183*, 5617–5631. [CrossRef]
28. Arrieta-Ortiz, M.L.; Hafemeister, C.; Bate, A.R.; Chu, T.; Greenfield, A.; Shuster, B.; Barry, S.N.; Gallitto, M.; Liu, B.; Kacmarczyk, T.; et al. An experimentally supported model of the Bacillus subtilis global transcriptional regulatory network. *Mol. Syst. Biol.* **2015**, *11*, 839. [CrossRef]
29. Helmann, J.D.; Márquez, L.M.; Chamberlin, M.J. Cloning, sequencing, and disruption of the Bacillus subtilis sigma 28 gene. *J. Bacteriol.* **1988**, *170*, 1568–1574. [CrossRef]
30. Petersohn, A.; Bernhardt, J.; Gerth, U.; Höper, D.; Koburger, T.; Völker, U.; Hecker, M.; Bernhardt, R.G.; Gerth, U.L.F. Identification of ς B -Dependent Genes in Bacillus subtilis Using a Promoter Consen-sus-Directed Search and Oligonucleotide Hybridization Identification of B -Dependent Genes in Bacillus subtilis Using a Promoter Consensus-Directed Search and Oligonucleo. *J. Bacteriol.* **1999**, *181*, 5718–5724. [CrossRef]
31. Zhu, B.; Stülke, J. SubtiWiki in 2018: From genes and proteins to functional network annotation of the model organism Bacillus subtilis. *Nucleic Acids Res.* **2018**, *46*, D743–D748. [CrossRef] [PubMed]
32. Vohradsky, J. Neural network model of gene expression. *FASEB J.* **2001**, *15*, 846–854. [CrossRef]
33. To, C.C.; Vohradsky, J. Supervised inference of gene-regulatory networks. *BMC Bioinform.* **2008**, *9*, 2. [CrossRef] [PubMed]
34. Vu, T.T.; Vohradsky, J. Nonlinear differential equation model for quantification of transcriptional regulation applied to microarray data of Saccharomyces cerevisiae. *Nucleic Acids Res.* **2006**, *35*, 279–287. [CrossRef] [PubMed]
35. Vu, T.T.; Vohradsky, J. Inference of active transcriptional networks by integration of gene expression kinetics modeling and multisource data. *Genomics* **2009**, *93*, 426–433. [CrossRef] [PubMed]
36. Ramaniuk, O.; Černý, M.; Krásný, L.; Vohradsky, J. Kinetic modelling and meta-analysis of the B. subtilis SigA regulatory network during spore germination and outgrowth. *Biochim. Biophys. Acta Bioenerg.* **2017**, *1860*, 894–904. [CrossRef]
37. Rabatinová, A.; Šanderová, H.; Matějčková, J.J.; Korelusová, J.; Sojka, L.; Barvík, I.; Papoušková, V.; Sklenář, V.; Žídek, L.; Krásný, L. The Subunit of RNA Polymerase Is Required for Rapid Changes in Gene Expression and Competitive Fitness of the Cell. *J. Bacteriol.* **2013**, *195*, 2603–2611. [CrossRef]
38. Qi, Y.; Hulett, F.M. PhoP-P and RNA polymerase sigmaA holoenzyme are sufficient for transcription of Pho regulon promoters in Bacillus subtilis: PhoP-P activator sites within the coding region stimulate transcription in vitro. *Mol. Microbiol.* **1998**, *28*, 1187–1197. [CrossRef]
39. De Saro, F.J.L.; Woody, A.-Y.M.; Helmann, J.D. Structural Analysis of theBacillus subtilisδ Factor: A protein Polyanion which Displaces RNA from RNA Polymerase. *J. Mol. Biol.* **1995**, *252*, 189–202. [CrossRef]
40. Sojka, L.; Kouba, T.; Barvík, I.; Šanderová, H.; Maderová, Z.; Jonák, J.; Krásný, L. Rapid changes in gene expression: DNA determinants of promoter regulation by the concentration of the transcription initiating NTP in Bacillus subtilis. *Nucleic Acids Res.* **2011**, *39*, 4598–4611. [CrossRef]
41. Meyer, F.M.; Jules, M.; Mehne, F.M.P.; Le Coq, M.; Landmann, J.J.; Görke, B.; Aymerich, S.; Stülke, J. Malate-Mediated Carbon Catabolite Repression in Bacillus subtilis Involves the HPrK/CcpA Pathway. *J. Bacteriol.* **2011**, *193*, 6939–6949. [CrossRef] [PubMed]
42. Saxild, H.H.; Brunstedt, K.; Nielsen, K.I.; Jarmer, H.; Nygaard, P. Definition of the Bacillus subtilisPurR Operator Using Genetic and Bioinformatic Tools and Expansion of the PurR Regulon with glyA, guaC,pbuG, xpt-pbuX, yqhZ-folD, and pbuO. *J. Bacteriol.* **2001**, *183*, 6175–6183. [CrossRef] [PubMed]

*microorganisms*

*Article*

# Non-B DNA-Forming Motifs Promote Mfd-Dependent Stationary-Phase Mutagenesis in *Bacillus subtilis*

Tatiana Ermi [1,†], Carmen Vallin [1,†], Ana Gabriela Regalado García [2], Moises Bravo [1], Ismaray Fernandez Cordero [1], Holly Anne Martin [1], Mario Pedraza-Reyes [2] and Eduardo Robleto [1,*]

[1] School of Life Sciences, University of Nevada, Las Vegas, 4505 S Maryland Pkwy, Las Vegas, NV 89154, USA; tatianaermi@gmail.com (T.E.); vallincarmen@gmail.com (C.V.); moises.bravo.jr@gmail.com (M.B.); fernai1@unlv.nevada.edu (I.F.C.); holly.martin2@ucsf.edu (H.A.M.)

[2] Department of Biology, Division of Natural and Exact Sciences, University of Guanajuato, P.O. Box 187, Guanajuato Gto. 36050, Mexico; ag.regaladogarcia@gmail.com (A.G.R.G.); pedrama@ugto.mx (M.P.-R.)

* Correspondence: eduardo.robleto@unlv.edu; Tel.: +1-702-895-2496

† These authors contributed equally to this work.

**Abstract:** Transcription-induced mutagenic mechanisms limit genetic changes to times when expression happens and to coding DNA. It has been hypothesized that intrinsic sequences that have the potential to form alternate DNA structures, such as non-B DNA structures, influence these mechanisms. Non-B DNA structures are promoted by transcription and induce genome instability in eukaryotic cells, but their impact in bacterial genomes is less known. Here, we investigated if G4 DNA- and hairpin-forming motifs influence stationary-phase mutagenesis in *Bacillus subtilis*. We developed a system to measure the influence of non-B DNA on *B. subtilis* stationary-phase mutagenesis by deleting the wild-type *argF* at its chromosomal position and introducing IPTG-inducible *argF* alleles differing in their ability to form hairpin and G4 DNA structures into an ectopic locus. Using this system, we found that sequences predicted to form non-B DNA structures promoted mutagenesis in *B. subtilis* stationary-phase cells; such a response did not occur in growing conditions. We also found that the transcription-coupled repair factor Mfd promoted mutagenesis at these predicted structures. In summary, we showed that non-B DNA-forming motifs promote genetic instability, particularly in coding regions in stressed cells; therefore, non-B DNA structures may have a spatial and temporal mutagenic effect in bacteria. This study provides insights into mechanisms that prevent or promote mutagenesis and advances our understanding of processes underlying bacterial evolution.

**Keywords:** mutagenesis; non-B DNA; hairpins; G4 DNA; *B. subtilis*; stationary phase

**Citation:** Ermi, T.; Vallin, C.; García, A.G.R.; Bravo, M.; Cordero, I.F.; Martin, H.A.; Pedraza-Reyes, M.; Robleto, E. Non-B DNA-Forming Motifs Promote Mfd-Dependent Stationary-Phase Mutagenesis in *Bacillus subtilis*. *Microorganisms* **2021**, *9*, 1284. https://doi.org/10.3390/microorganisms9061284

Academic Editor: Imrich Barák

Received: 12 May 2021
Accepted: 9 June 2021
Published: 12 June 2021

**Publisher's Note:** MDPI stays neutral with regard to jurisdictional claims in published maps and institutional affiliations.

## 1. Introduction

Experiments in the 1950s established that mutations are replication-dependent events that occur randomly during growth [1]. However, studies in the past few decades have uncovered transcription-dependent mutagenic mechanisms and their underlying factors. These types of mechanisms affect mutagenesis at specific genomic regions as a function of transcriptional activity [2–6]. One enzyme important in transcription-dependent mutagenesis is Mfd, the transcription-repair coupling factor [4,7,8]. Mfd gets recruited to stalled RNA Polymerase (RNAP) at damaged DNA sites and initiates high-fidelity transcription-coupled repair (TCR). Recent structural studies indicate that Mfd is a more dynamic protein than previously described and can function outside of TCR and in the absence of exogenous damage. Mfd was shown to bind [9] and translocate on double-stranded DNA [10] as well as to discriminate between different transcription elongation-stalling events [11]. Moreover, Mfd is proposed to be a pro-mutagenic/evolvability factor [12–14] that can associate with RNAP [9] and interact with different repair proteins to promote mutations [15,16]. More recently, it has been described as a factor affecting global transcription [17,18]. An exciting aspect raised by the transcription-induced mutagenesis mechanism is whether intrinsic sequences with the potential to form non-B DNA structures influence such a process in an Mfd-dependent manner.

Non-B DNA sequences can form structures that deviate from the canonical right-handed Watson and Crick structure. In vitro studies found that non-B DNA structures are promoted and stabilized by transcription [19,20] and can block the Mammalian and T7 phage RNA polymerases [21,22]. In eukaryotic cells, non-B DNA-forming sequences associate with genomic instability [23,24] as well as transcription and replication disruptions [25–27]. Many studies have shown that these types of events lead to recognition by and recruitment of several DNA repair factors such as CSB, the functional human homolog of Mfd, helicases, and error-prone polymerases [26,28,29]. However, we know less about the role of non-B DNA in bacteria, mainly whether it influences genome instability and the factors that promote or prevent mutagenesis at these sites.

Research on the influence of non-B DNA structures in mutagenesis in bacteria has focused on two types of sequence motifs predicted to form G-quadruplex (G4) DNA and hairpin structures [30,31]. A G4 DNA sequence has four equal guanine tracts separated by three loop regions: $G_{3+}N_{Y1}G_{3+}N_{Y2}G_{3+}N_{Y3}G_{3+}$. The four guanine tracts come together to form a tetrad plane via Hoogsteen base-pairing and stack vertically, creating a four-stranded structure [32]. Hairpin sequences are formed by inverted repeats, making a paired stem and a single-stranded DNA loop region. A recent study on *E. coli* growing cells found that the presence of a G4 DNA sequence in an actively transcribed reporter gene strongly increased mutation rates. Interestingly, they found that Hfq promoted this process by binding and stabilizing the G4 structure [31]. Hfq is a regulatory nucleic acid-binding protein whose function has broadened recently [33], and now extends to interactions with alternate DNA structures and a role in *B. subtilis* stationary-phase survival [34]. Similarly, hairpin sequences can promote genetic instability through slippage events that result in insertions or deletions (indels) during replication [35]. Notably, when reporter genes containing hairpin-forming sequences were transcriptionally active, mutation frequencies increased and correlated with promoter strength [36]. These studies in *E. coli* suggest that non-B DNA structures may have a role in mutagenesis, particularly in transcribed regions, but does this occur in other bacterial species? If so, what factors, besides Hfq, promote such a mechanism?

This study investigated if G4 DNA and hairpin-forming motifs influence stationary-phase mutagenesis (SPM) in *Bacillus subtilis*. *B. subtilis* is a Gram-positive sporulating bacterium adapted to feast and famine cycles in the soil's upper layers [37]. Upon the onset of stress, such as starvation, *B. subtilis* cells activate gene expression programs, enter a non-growing state, and differentiate into distinct subpopulations [38]. Our previous observations have demonstrated that increased mutagenesis occurs in highly transcribed regions [39] and in distinct cell subpopulations [40,41]. Mfd also acts in coordination with nucleotide excision repair (NER), base excision repair (BER), and the transcription factor GreA to promote mutagenesis during stationary phase [15,42]. These observations make *B. subtilis* an attractive model to test our hypothesis that non-B DNA-forming motifs influence Mfd-dependent stationary-phase mutagenesis at transcribed regions.

We developed a system to measure non-B DNA's influence on *B. subtilis* stationary-phase mutagenesis by deleting the wild-type *argF* at its chromosomal position and introducing IPTG-inducible *argF* alleles differing in their ability to form hairpin and G4 DNA structures into the *amyE* locus. To design these *argF* alleles, we used in silico tools to find endogenous non-B DNA sequences in the *argF gene* and introduced point mutations to the open reading frame to either increase or decrease their predicted structure stability. We then introduced a nonsense codon in a region corresponding to a loop in the structure, which resulted in a non-functional truncated ArgF (confers auxotrophy) and a system to measure reversions to Arg$^+$ (Figure 1). Using this system, we found that sequences predicted to form non-B DNA structures promoted mutagenesis in stationary-phase cells; such a response did not occur in growing conditions. We also showed that Mfd promotes the formation of mutations within the predicted structures.

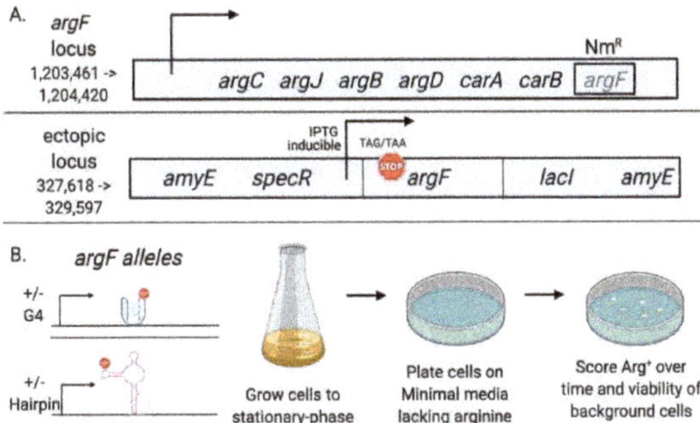

**Figure 1.** Schematic of the stationary-phase assay. (**A**) The *argF* gene was replaced with a selectable neomycin marker at its chromosomal locus. Then, IPTG-inducible *argF* alleles containing nonsense codons and differing in their potential to form non-B DNA structures were introduced into an ectopic chromosomal region. (**B**) Stationary-phase mutagenesis assay. Cells are grown to stationary phase and plated on minimal media lacking arginine. Then, we scored Arg$^+$ mutants over time.

## 2. Materials and Methods

### 2.1. Strain Construction

The bacterial strains used in this study are shown in Table 1. For detailed strain construction, see supplemental material Non-B DNA strain construction. Briefly, Mfold [43] and QGRS mapper [44] were used to find sequences with the potential to form hairpin and G4 DNA structures, respectively, in the *argF* ORF. Once the sequences were found, synonymous point mutations were introduced into positions that were predicted to strengthen or destabilize the structure. A nonsense codon was then engineered into a region of the structure corresponding to a loop. For the +/−hairpin strains, we introduced an ochre nonsense codon at position 37Q (CAA→TAA) corresponding to the loop part of the predicted hairpin before disruption. For the +/−G4 strains, we introduced an amber nonsense codon at position 70Q (CAG→TAG) corresponding to the second loop of the predicted G4 structure. This nonsense codon resulted in a truncated ArgF protein and arginine auxotrophy. The new defective *argF* alleles were then cloned into pDR111 downstream of the IPTG-inducible *Pspac* promoter, transformed, and recombined into the *amyE* locus of the *argF* deletion strain CV4000 (see Figure 1). The Mfd$^-$ strains were constructed by transforming genomic DNA from YB9801 [8] into the +/−hairpin and +/−G4 strains and selecting on TBAB plates containing tetracycline.

**Table 1.** Strains and plasmids used in this study.

| Strain or Plasmid | Relevant Genotype | Source or Reference |
|---|---|---|
| CV1000 (+Hairpin) | *metB5, hisC952, leuC427, argF::neo, amyE::pHS− argF+SLS* | This study |
| CV2000 (−Hairpin) | *metB5, hisC952, leuC427, argF::neo, amyE::pHS− argF−SLS* | This study |
| CV1001 (+Hairpin Mfd$^-$) | *metB5, hisC952, leuC427, argF::neo, mfd::tc, amyE::pHS− argF+SLS* | This study |
| CV2009 (−Hairpin Mfd$^-$) | *metB5, hisC952, leuC427, argF::neo, mfd::tc, amyE::pHS− argF−SLS* | This study |
| TE300 (+G4) | *metB5, hisC952, leuC427, argF::neo, amyE::pHS− argF+G4* | This study |
| TE302 (−G4) | *metB5, hisC952, leuC427, argF::neo, amyE::pHS− argF−G4* | This study |
| TE400 (+G4 Mfd$^-$) | *metB5, hisC952, leuC427, argF::neo, mfd::tc amyE::pHS− argF+G4* | This study |
| YB9801 (Mfd$^-$) | *metB5, hisC952, leuC427, mfd::tet* | [8] |

**Table 1.** *Cont.*

| Strain or Plasmid | Relevant Genotype | Source or Reference |
|---|---|---|
| CV4000 (*argF-*) | *metB5, hisC952, leuC427, argF::neo* | [45] |
| TE402 (−G4 Mfd⁻) | *metB5, hisC952, leuC427, argF::neo, mfd::tc amyE::pHS−argF−G4* | This study |
| pBEST502 | neomycin resistance (Nm$^r$) gene | [46] |
| pDR111 | *pdr111 amyE-hyper-SPANK (spec)* | Rudner lab |

## 2.2. Stationary-Phase Mutagenesis Assay

We performed a stationary-phase mutagenesis assay on each strain [47]. Strains were streaked out on TBAB plates with the appropriate antibiotic and incubated overnight at 37 °C. The next day, a colony from the TBAB plate was used to inoculate 2 mL of Penassay Broth (PAB). The cells were grown overnight at 37 °C with aeration. An aliquot of the overnight was then added to a 250 mL Erlenmeyer culture flask containing 10 mL of PAB medium and 10 µL of 1000× Ho-Le trace elements. Cells were grown at 37 °C with aeration (250 rpm) to stationary phase. Growth was monitored with a spectrophotometer measuring O.D. at 600 nm (OD$_{600}$). Cells were harvested 90 min after cessation of exponential growth and centrifuged at 4000 rpm for 5 min, the supernatant was decanted, and the cells were resuspended in Spizizen minimal salts (SMS); this process was repeated twice. To determine initial cell titers, we serial diluted and spread plated 100 µL of cells on minimal media supplemented with 100 µg/mL arginine. Colonies were counted after 48 h of incubation at 37 °C. Then, 100 µL aliquots were plated on minimal media containing 0.1 mM IPTG and no arginine (*n* = 5). The plates were incubated for nine days at 37 °C. Each day plates were scored for the appearance of Arg⁺ colonies. The initial titers were used to normalize the cumulative number of revertants per day to the number of the CFU plated. To assay the background survival of cells, the viability was tracked by taking agar plugs every other day, serial diluting, and plating on minima media supplemented with 100 µg/mL arginine.

## 2.3. Arg⁺ Mutant Analysis

To characterize Arg⁺ mutants, we performed a tRNA suppressor analysis and DNA sequencing [47]. Suppressor analysis was conducted by first patching between 20 and 25 Arg⁺ colonies collected from days 5–9 of each replicate trial onto the same media used in the SPM experiment (minimal media lacking arginine supplemented with 0.1 mM IPTG) to eliminate transient Arg⁺ prototrophs. Previous studies determined that analyzing mutants that arose after day 5 would avoid selecting slow-growing mutants from growth [47]. Arg⁺ mutants that grew were stocked and then patched on minimal media lacking either methionine or histidine. Our background *B. subtilis* YB955 strain is also auxotrophic for methionine and histidine due to ochre and amber nonsense codon in the *metB* and *hisC* genes, respectively [47]. Growth of the Arg⁺ mutants on either trace met and trace his plates was scored after 24 h and 48 h. If growth occurred, those revertants were considered tRNA suppressors. For those Arg⁺ mutants determined not to be tRNA suppressors, we performed a colony PCR amplifying and sequencing the complete *argF* gene.

## 2.4. Fluctuation Test

We conducted a fluctuation test to determine the mutation rates for arginine prototrophy during growth for each strain. Briefly, an overnight aerated culture grown overnight in 2 mL of PAB at 37 °C were diluted 1:10 in 1× Spizizen minimal salts (SMS), vortexed, and diluted 1:10 again in this solution. Aliquots of 0.5 mL of the second 1× SMS solution were added to 49.5 mL of PAB supplemented with 0.1 mM IPTG. The cells were vortexed, and 1 mL of the $10^{-4}$ mixture was dispensed into 40 different 18 mm test tubes. Cells were grown for 12 h at 37 °C with aeration (250 rpm). Each test tube culture was pelleted, washed, resuspended in 100 µL of 1× SMS, and individually plated on minimal media lacking arginine with 0.1 mM IPTG. In addition, three individual test tubes were serial diluted and plated on minimal media supplemented with 100 µg/mL arginine to determine cell count. Plates were incubated at 37 °C for 48 h, and then colonies were counted. The mutation rates were determined using the MSS maximum likelihood method [48,49].

### 2.5. Statistical Analysis

To determined what type of significance test to use, we first checked the data for normality and variance using a Shapiro–Wilk test and Brown–Forsythe test, respectively. Significance was determined by performing a Student's *t*-test to compare two strains or a one-way ANOVA with a post-hoc Tukey's test to compare more than two strains.

### 2.6. Protein Alignment Analysis

The *B. subtilis* 319 amino acid ArgF protein sequence was inputted into pBLAST as the query sequence and blasted into the UniProtKB/Swiss-Prot (SwissProt) database. The output identified 100 ArgF amino acid sequences. The sequences were exported and aligned. Then, we analyzed position 37 and 70 using Weblogo. Weblogo reported the conservation at that site. In addition, Weblogo reported the frequency of each amino acid found at that site within the 100 ArgF sequences.

## 3. Results

### 3.1. Non-B DNA-Forming Motifs Promote Mutagenesis

3.1.1. Hairpins and G4 DNA-Forming Motifs Accumulate Mutations in Stationary Phase *B. subtilis* Cells

Our stationary-phase mutagenesis assay revealed that the strains containing *argF* alleles with sequences predicted to form non-B DNA structures accumulated significantly more mutants over nine days than those *argF* alleles less likely to form such structures (Figure 2). Specifically, by day 6, the +G4 strain containing the sequence $G_3N_{(1-13)}G_3N_{(1-13)}G_3N_{(1-13)}G_3$ accumulated ~3 times more mutants than the −G4, which was not predicted to form any G4 structure. Arg$^+$ accumulation for the +G4 strain continued after day 6, but the number of revertants became too numerous to count compared to what was observed in the −G4. The +hairpin (120 b) strain showed a significant increase (~two fold) in the accumulation of Arg+ colonies compared to the −hairpin. Since window size is an important parameter for predicting hairpin structures [36], our approach included disrupting a hairpin structure using a smaller window size in the same region of *argF*. We found mutagenesis of the −hairpin (40 b) strain indistinguishable from the W.T. in our experiments (Supplemental Figure S6). In addition, mutagenic differences observed in all the strains were not the result of viability differences during the experiment (Supplementary Figure S7). Interestingly, the strains differing in G4 stability had more mutants over time compared to the strains differing in hairpin stability. These results are consistent with our previous report [47] and we attribute this difference to the type of nonsense codon used in the different constructs (see Supplemental TAG vs. TAA section).

3.1.2. Analysis of the Arg$^+$ Population

Analysis of a sample of the Arg$^+$ colonies that arose after day 5 indicated that two categories of mutants restored arginine prototrophy in the tested strains: (1) nonsense suppressor tRNAs and (2) mutations in *argF* (Figure 3A,D). Nonsense tRNA suppressors represented <50% of the Arg$^+$ colonies sampled. Sequencing of non-tRNA suppressor mutants showed that mutations in *argF* were limited to base substitutions that changed the stop codon's first or third position (Figure 3B,E). We did not observe deletion events in our system.

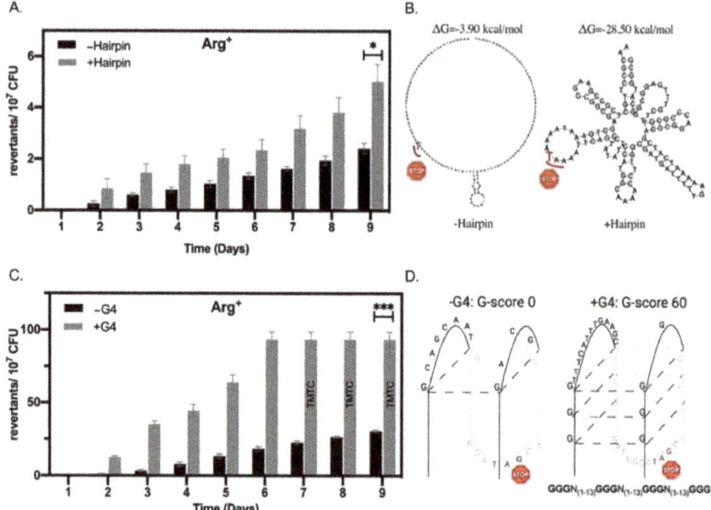

**Figure 2.** Hairpins and G4 DNA-forming motifs promote mutations in stationary phase *B. subtilis* cells. (**A**) Stationary-phase mutagenesis assay results for +/−hairpin strains. (**B**) Mfold predicted hairpin structures for 120 bp region in *argF* (positions 82–202). (**C**) Stationary-phase mutagenesis assay results for +/−G4 strains. (**D**) Schematic of predicted G4 DNA structures and G-score in *argF* (position 182–219). Please note the differences in scale between the two graphs. Bars represent the mean, and error bars are SEM ($n = 3$). Statistical significance was determined using a Student's $t$-test * $p < 0.05$, *** $p < 0.001$.

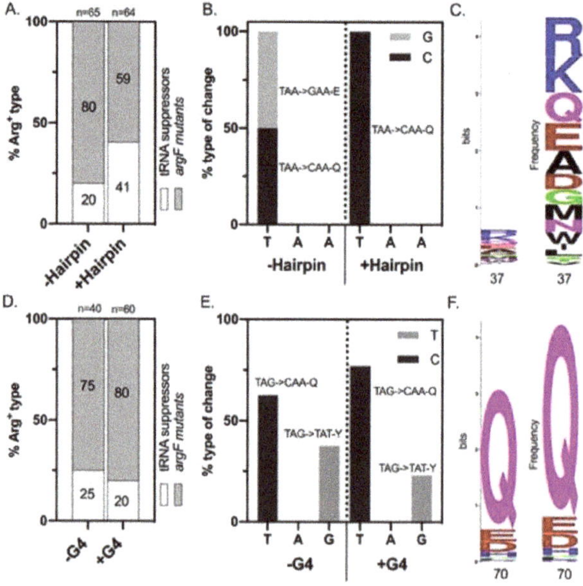

**Figure 3.** Arg$^+$ revertant analysis. (**A–C**) +/−Hairpin results; (**D–F**) +/−G4 results. (**A,D**) The proportion of tRNA suppressors from Arg$^+$ population sampled that arose from day 5–9. (**B,E**) argF sequencing results from Arg$^+$ mutants that were determined not to be tRNA suppressors. (**C,F**) Protein alignment analysis of positions containing the nonsense codons 37 and 70.

In the strains differing in hairpin stability with the ochre nonsense codon TAA at position 37, transitions in the first position from T→C restored the W.T. polar glutamine-Q codon, and T→G transversions resulted in the acidic glutamate-E codon (Figure 3B). A protein analysis that looked at 100 ArgF sequences across different phyla confirmed that this codon position had some flexibility and that glutamate-E was indeed found at this position in some bacterial species (Figure 3C). Only transitions in the first positions were observed in the +hairpin strain, and no mutations occurred in the second or third position in either of the hairpin strains. Interestingly, the proportion of tRNA suppressor mutants appeared higher in our sample of the +hairpin than in the −hairpin one (Figure 3A).

In the strains differing in G4 stability carrying the amber nonsense codon TAG at position 70, revertants displayed T→C transitions in the first codon position, which restored the W.T. polar glutamine-Q triplet, and G→T transversions in the third codon position, which resulted in the aromatic amino acid tyrosine-Y triplet (Figure 3E). Protein analysis of this codon position found that it exhibited some flexibility (Figure 3F). Additionally, glutamate-E was the second most frequent amino acid found at this position in other ArgF proteins, while tyrosine was not observed in our sample of 100 ArgF proteins. (Figure 3F). Moreover, the proportions of the revertants with these genetic changes were similar in both strains (Figure 3E). The proportion of tRNA suppressors was similar in our sample of the Arg$^+$ population between the $+/-$G4 strains (Figure 3D).

### 3.2. Hairpin- and G4 DNA-Forming Motifs Do Not Influence Mutation Rates in B. subtilis

To test whether the results observed were exclusive to stationary-phase conditions, we performed a fluctuation test on the strains employed in this study. We found that the $+/-$hairpin and $+/-$G4 DNA's predicted stability did not correlate with mutation rates in *B. subtilis* growing cells. Additionally, we did not observe differences between the sets of strains as we did in stationary phase (hairpin vs. G4) (Figure 4). This result suggests that cell physiology influences the role non-B DNA structures have in mutagenesis.

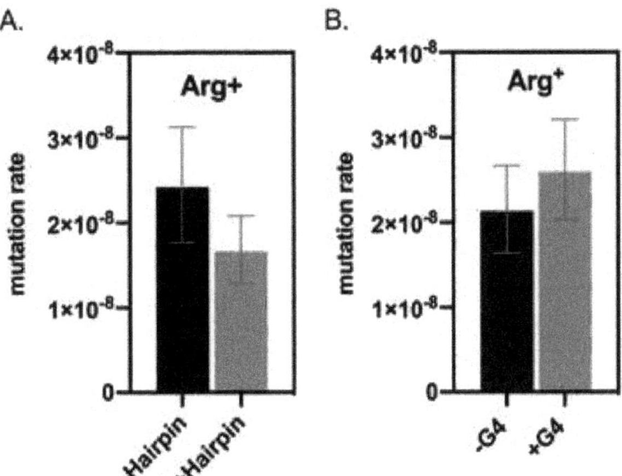


**Figure 4.** Hairpin (**A**) and G4 DNA (**B**) forming motifs did not affect mutation rates in growing *B. subtilis* cells. Bars represent mutation rates using the MSS maximum likelihood method, and error bars are confidence limits (CL$_{95}$).

### 3.3. Mfd Promotes Mutations at Non-B DNA Sequences

#### 3.3.1. Accumulation of Arg⁺ Mutations at Non-B DNA Sequences Was Decreased in the Absence of Mfd

We tested if the transcription-repair coupling factor Mfd influenced SPM at non-B DNA sequences in *B. subtilis*. We transformed the *mfd* mutation into the strains carrying *argF* alleles differing in their predicted ability to form either hairpin or G4 DNA structures and performed a stationary-phase mutagenesis assay. Disrupting *mfd* significantly decreased the accumulation of mutations by ~two-fold in the +hairpin and +G4 strains (Figure 5). This response was not observed in the −hairpin, and there was a small but significant increase in the accumulation of Arg⁺ mutants in the −G4 strain deficient in Mfd (Figure 5B). These results were not due to differences in viability (Supplementary Figure S8).

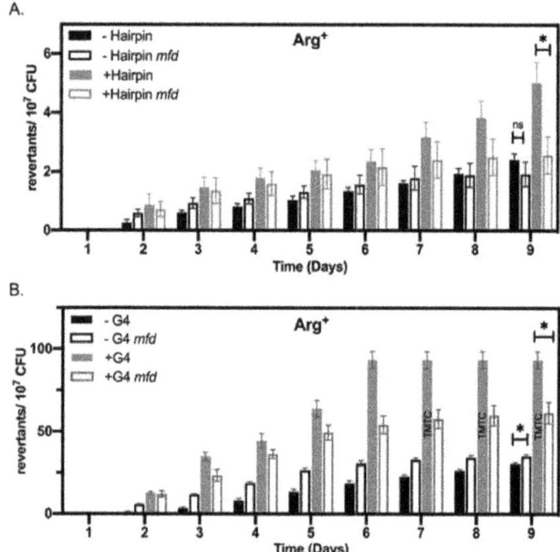


**Figure 5.** Mfd promotes mutations at hairpin (**A**) and G4 DNA (**B**) forming sequences. Bars represent the mean of at least three SPM trials, and error bars are SEM. Please note the differences in scales on the Y axis between graphs. Significance was determined using a Student's *t*-test * $p < 0.05$.

#### 3.3.2. Analysis of the Arg⁺ Population in Mfd⁻ Cells

Subsequent analysis of the Arg⁺ mutants found that Mfd deficiency abolished mutations in *argF* as almost 100% of the mutants sampled were nonsense tRNA suppressors (Figure 6A,D). Thus, the spectrum of mutations changed drastically in the Mfd deficient strains. We estimated the total number of tRNA suppressors in this Arg⁺ population based on the sampling of mutants that arose at days 5–9 during the SPM trials. This analysis suggested that nonsense tRNA suppressors significantly increase in the +/−G4 lacking but not in the +/−hairpin strains in the Mfd⁻ background (Figure 6C,F). The genetic changes that confer arginine prototrophy in *argF* (Figure 3E) suggest that mutagenesis of tyrosine (first position in the G/ATA anticodon) and glutamine (third position in the TTG anticodon) tRNA genes are increased in the absence of Mfd.

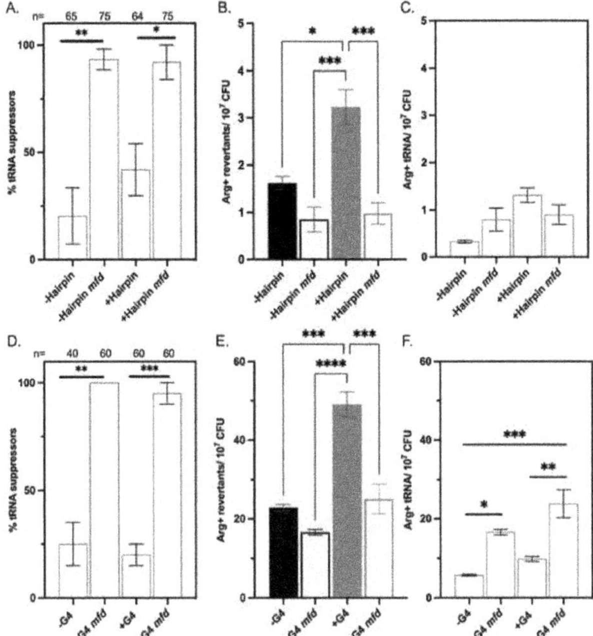

**Figure 6.** Mfd promotes mutagenesis at *argF*. (A–C) +/−Hairpin results; (D–F) +/−G4 results. (A,D) The proportion of tRNA suppressors from Arg$^+$ population sampled that arose from day 5–9. (B,E) Arg$^+$ mutants that accumulated from day 5–9. (C,F) tRNA suppressors that arose in the Arg$^+$ population after day 5. Bars represent the mean with SEM. Significance was determined using ANOVA; * denotes $p < 0.05$, ** denotes $p < 0.01$, *** denotes $p < 0.001$, and **** denotes $p < 0.0001$.

## 4. Discussion

### 4.1. An In Vivo System to Measure the Effects of Non-B DNA-Forming Motifs on Bacterial Mutagenesis

We investigated if non-B DNA-forming motifs, such as hairpins and G4 DNA sequences, contribute to stationary-phase mutagenesis in *B. subtilis*. First, we exploited the fact that this bacterium does not have a strong codon bias [50–52] to develop a system that altered endogenous non-B DNA sequences in the *argF* gene by introducing point mutations that increase or decrease their predicted stability. We then tested the effect of these changes on mutagenesis during growth and stationary phase.

Our approach to construct the non-B DNA sequence variants included the use of in silico tools that predict their stability based on extensive thermodynamic studies. We used Mfold [43] and QGRS mapper [44] to measure hairpins and G4 DNA stability, respectively. Previous in vitro [21] and in vivo studies [31] used optimized sequence motifs in tandem arrangements. These arrangements are not commonly found in coding regions, likely due to functional constraints. In contrast, we worked with an endogenous system by altering sequence motifs within a coding region. The G4 motif used in our experiments had longer asymmetrical loops ($G_3N_{(1-13)}G_3N_{(1-13)}G_3N_{(1-13)}G_3$) than those commonly used ($G_3N_{(7)}G_3N_{(7)}G_3N_{(7)}G_3$), and thermodynamic studies have shown that sequences having asymmetrical loops can still form stable G-quadruplex structures [24,53]. Additionally, G4 DNA containing a single-nucleotide loop, like the one used in this study, was shown to be thermally stable and to incite genomic instability in eukaryotic genomes [24]. In this regard, we performed a G4 motif search across the *B. subtilis* genome and found that the canonical G4 motif occurred four times in coding regions, while the G4 motif used here appeared in 39 genes (Supplemental Table S2). Further characterization of these genes showed they were distributed throughout the *B. subtilis* chromosome and many coded for membrane-bound proteins (Supplemental Figures S9 and S10).

To design the constructs that tested the effect of hairpins within coding regions on SPM, we used two window sizes because this parameter was reported to affect their stability, particularly in the context of their transcription [36]. High levels of transcription generate more supercoiling, which increases the likelihood of forming single-stranded DNA and intrastrand base pairing between distant DNA residues [54,55]. We found that disrupting a hairpin using a 40 b window size did not affect mutation levels (Supplemental Figure S6); in contrast, using a 120 b window size did. The IPTG-inducible Pspac promoter used in our study is robust, suggesting that the design using a larger window size provided a better parameter for forming single-stranded hairpins. Studies have used window sizes ranging from 30–300 bases [25,36]. In summary, the features we used to design the non-B DNA forming sequences are consistent with what is observed within gene coding regions and account for conditions in which a region is robustly transcribed.

### 4.2. Non-B DNA-Forming Motifs in Transcribed Coding Regions Promote Mutations in Nutritionally Stressed Cells

Early in silico analysis in bacteria found that non-B DNA forming sequences were enriched in noncoding regions, including promoters and untranslated regions (UTR). This finding led researchers to focus on elucidating the role of these structures in gene regulation [56,57]. Indeed, a study in a polyextremophile, *Deinococcus radiodurans*, which contains G4-motifs in promoter regions, found that adding a G4 stabilizing ligand led to repression of DNA repair genes and decreased radioresistance [58].

The effect of DNA structures found in bacterial regulatory regions continues to be well studied [59]. However, very little is known about what happens to alternate DNA structures that are within an ORF. Wright and colleagues proposed that hairpin structures formed during transcription in coding regions were hotspots for mutagenesis in *E. coli* [30]. More recently, a study examined the role of G4 structures in genome instability in *E. coli*-transcribed regions and found that these sequences increased the rate of deletions [31]. Here, we investigated if G4 DNA- and hairpin-forming motifs as intrinsic elements that influence stationary-phase mutagenesis in *B. subtilis*. We found that non-B DNA forming sequences promoted mutagenesis in stationary-phase cells. Specifically, sequences increased or decreased in the likelihood to form a hairpin, and G4 DNA were increased and decreased in the accumulation of Arg$^+$ mutants, respectively. Furthermore, unlike results in *E. coli*, this response was only observed in nutritionally stressed cells and did not occur in growing cells. This result underlines the contribution of mutagenic processes taking place in non-replicating conditions to the evolutionary process and the importance of using different model systems. Moreover, this work provides support to the concept that non-B DNA plays a role in bacterial transcription-dependent mutagenic processes.

### 4.3. Mfd Promotes Mutations at Non-B DNA Sequences

To identify factors governing the process by which non-B DNA-forming motifs promote the accumulation of mutations, we investigated if the transcription-repair coupling factor Mfd promoted mutations at non-B DNA sites. Mfd is a factor shown to promote transcription-dependent mutagenesis and is recruited to genomic regions containing non-B-DNA-forming motifs [17,28]. Disrupting *mfd* led to significant decreases in mutagenesis levels in the +hairpin and +G4 DNA strains. Subsequent analysis of the Arg$^+$ population that arose in the absence of Mfd showed that the vast majority of revertants were tRNA suppressors in all strains; *argF* mutants were abolished in the absence of Mfd. This result is consistent with our previous studies in stationary-phase mutagenesis in *B. subtilis* [8,60]. This result further supports the promutagenic role of Mfd at transcribed regions and suggests that non-B DNA sites in bacteria influence Mfd-dependent mutagenesis.

Mfd associates with RNAP when transcription elongation is blocked by a bulky DNA lesion. Upon contacting the blocked RNAP, Mfd undergoes a series of conformational changes that recruit nucleotide excision repair proteins [11]. However, recent single-molecule resolution and structural studies suggest that Mfd and RNAP interact, even in the absence of DNA damage, and that Mfd works to modulate transcription [9,10]. In this context, in vivo and in vitro reports showed that non-B DNA structures formed in the DNA or RNA could interrupt transcription, especially when they are found in the coding strand [21,61–63]. In *B. subtilis*, loss of Mfd led to increases in RNAP association at genes enriched with hairpins [17]. Similarly, deficiencies in CSB, the functional homolog of Mfd, caused RNAPII pausing at G4 DNA motifs in a neuroblastoma cell line [28]. Considering these observations, one can propose a model in which transcription promotes the formation of alternate structures, through changes in supercoiling, in the DNA or mRNA with the potential to impede transcription. We speculate that the stability of such structures defines whether the impediment is resolved by Mfd through transcription rescue (low structure stability) or disassembly of the elongation complex (high structure stability) and recruitment of error-prone repair of DNA. In *E. coli*, Mfd promotes mutagenesis in stressed cells via the formation of R-loops and recombination intermediates that generate point mutations or amplifications in a Lac$^+$ mutagenesis system [4]. This system requires recombination functions and activation of the sigma S regulon (amplification) and the DinB (point mutations) [64]. Unlike in *E. coli*, Mfd-dependent mutagenesis in *B. subtilis* operates independently of recombination functions [47] and can operate by interacting with components of the NER and BER systems [15,16,65]. It is also possible that the loss of Mfd causes depression in mutagenesis indirectly. A recent RNAseq study showed profound changes in the transcriptome of Mfd$^-$ cells compared to wild-type stationary-phase cells. Almost half the genes in the *B. subtilis* genome were dysregulated, including factors previously identified to promote SPM, such as GreA [42], PolY [66], PolX [67], and others [15]. Therefore, it is possible that the Mfd effects observed in mutagenesis are the indirect product of dysregulation of a factor(s), such as a helicase or error-prone polymerase involved in mutagenesis at non-B DNA forming sequences. Interestingly, a previous study identified Hfq as a factor that promotes mutation at G4-DNA motifs in *E. coli* growing cells [31]. Like Mfd, the role of Hfq has been broadening [33], and like Mfd, cells deficient in Hfq have transcriptome changes and decreased survival during stress in *B. subtilis* [34]. Future work will determine if Hfq, along with Mfd, contribute to stationary-phase mutagenesis at non-B DNA motifs.

## 5. Conclusions

Non-B DNA sequences are present in all domains of life. These structures have a role in genome instability and disease in humans, but we know less about their function in bacteria. Here, we showed that non-B DNA-forming motifs promote genetic variation in *B. subtilis*, particularly in transcribed coding regions in stressed cells in an Mfd-dependent manner. Future work studying SPM in *B. subtilis* will discern between the competing but not necessarily exclusive hypotheses that Mfd works directly or indirectly to promote the formation of mutations at non-B DNA sequences in actively transcribed genes. This type of study provides insight into transcription-induced mutagenic mechanisms limiting mutations to times of stress and defined genomic regions, and it improves our view of evolution.

**Supplementary Materials:** The following are available online at https://www.mdpi.com/article/10.3390/microorganisms9061284/s1, Figure S1: Prediction of hairpin structures across the *argF* gene using a sliding window analysis and in silico tools. Figure S2: Hairpin-forming motifs and predicted structures of the WT and +/−Hairpin (120) strains. Figure S3: Hairpin-forming motif and predicted structure of WT and −Hairpin (40) strains. Figure S4: G4-forming motif and predicted structure of +G4 and −G4 strains. Figure S5: Codon analysis using the Kazusa Codon Usage Database of regions in *argF* that were changed to disrupt or stabilize an endogenous non-B DNA structure. Figure S6: The −Hairpin (40bp strain) is indistinguishable from the W.T. Figure S7: Viability of the non-revertant population for the non-B DNA strains during the SPM assay. Figure S8: Viability of the non-revertant population for the non-B DNA strains and their Mfd deficient counterpart during the SPM assay. Figure S9: A. Schematic showing the chromosomal distribution of the 39 genes that contained the G4 motif used in this study. Figure S10: Modified figure from Bsucyc.org showing the cellular function of the 39 genes that contained the G4 motif used in this study. Table S1: QGRS Pattern search in argF ORF. Table S2: QGRS Pattern search in *B. subtilis* coding regions. Table S3: QGRS Results. Table S4: Primers used in this study. Table S5: Summary of Phenotype Test.

**Author Contributions:** Conceptualization, C.V., T.E., M.P.-R. and E.R.; methodology, H.A.M., A.G.R.G., and M.B.; formal analysis, C.V., H.A.M., and T.E.; investigation, C.V., T.E., M.B., and I.F.C.; writing— original draft preparation, T.E. and C.V.; writing—review and editing, T.E., C.V., H.A.M., M.P.-R., and E.R.; visualization, C.V.; supervision, E.R.; project administration, E.R.; funding acquisition, C.V., E.R., and M.P.-R. All authors have read and agreed to the published version of the manuscript.

**Funding:** This research was funded by the National Science Foundation (NSF-GRFP), the National Institutes of Health (GM131410), and the Consejo Nacional de Ciencia y Tecnología (A-1S-27116).

**Data Availability Statement:** Data is contained within the article or supplementary material.

**Conflicts of Interest:** The authors declare no conflict of interest.

# References

1. Luria, S.E.; Delbrück, M. Mutations of Bacteria from Virus Sensitivity to Virus Resistance. *Genetics* **1943**, *28*, 491–511. [CrossRef]
2. Reimers, J.M.; Schmidt, K.H.; Longacre, A.; Reschke, D.K.; Wright, B.E. Increased transcription rates correlate with increased reversion rates in leuB and argH Escherichia coli auxotrophs. *Microbiology* **2004**, *150*, 1457–1466. [CrossRef]
3. Lang, K.S.; Hall, A.N.; Merrikh, C.N.; Ragheb, M.; Tabakh, H.; Pollock, A.J.; Woodward, J.J.; Dreifus, J.E.; Merrikh, H. Replication-Transcription Conflicts Generate R-Loops that Orchestrate Bacterial Stress Survival and Pathogenesis. *Cell* **2017**, *170*, 787–799. [CrossRef]
4. Wimberly, H.; Shee, C.; Thornton, P.C.; Sivaramakrishnan, P.; Rosenberg, S.M.; Hastings, P.J. R-loops and nicks initiate DNA breakage and genome instability in non-growing Escherichia coli. *Nat. Commun.* **2013**, *4*, 2115. [CrossRef]
5. Merrikh, H. Spatial and Temporal Control of Evolution through Replication–Transcription Conflicts. *Trends Microbiol.* **2017**, *25*, 515–521. [CrossRef] [PubMed]
6. Million-Weaver, S.; Samadpour, A.N.; Moreno-Habel, D.A.; Nugent, P.; Brittnacher, M.J.; Weiss, E.; Hayden, H.S.; Miller, S.I.; Liachko, I.; Merrikh, H. An underlying mechanism for the increased mutagenesis of lagging-strand genes in *Bacillus subtilis*. *Proc. Natl. Acad. Sci. USA* **2015**, *112*, E1096–E1105. [CrossRef] [PubMed]
7. Paul, S.; Million-Weaver, S.; Chattopadhyay, S.; Sokurenko, E.V.; Merrikh, H. Accelerated gene evolution through replication–transcription conflicts. *Nature* **2013**, *495*, 512–515. [CrossRef] [PubMed]
8. Ross, C.; Pybus, C.; Pedraza-Reyes, M.; Sung, H.-M.; Yasbin, R.E.; Robleto, E. Novel Role of mfd: Effects on Stationary-Phase Mutagenesis in *Bacillus subtilis*. *J. Bacteriol.* **2006**, *188*, 7512–7520. [CrossRef]
9. Ho, H.N.; Van Oijen, A.M.; Ghodke, H. The transcription-repair coupling factor Mfd associates with RNA polymerase in the absence of exogenous damage. *Nat. Commun.* **2018**, *9*, 1570. [CrossRef]
10. Le, T.T.; Yang, Y.; Tan, C.; Suhanovsky, M.M.; Fulbright, R.M.; Inman, J.T.; Li, M.; Lee, J.; Perelman, S.; Roberts, J.W.; et al. Mfd Dynamically Regulates Transcription via a Release and Catch-Up Mechanism. *Cell* **2018**, *172*, 344–357. [CrossRef]
11. Kang, J.Y.; Llewellyn, E.; Chen, J.; Olinares, P.D.B.; Brewer, J.; Chait, B.T.; Campbell, E.A.; Darst, S.A. Structural basis for transcription complex disruption by the Mfd translocase. *Elife* **2021**, *10*, 1–28. [CrossRef]
12. Ragheb, M.; Thomason, M.K.; Hsu, C.; Nugent, P.; Gage, J.; Samadpour, A.N.; Kariisa, A.; Merrikh, C.N.; Miller, S.I.; Sherman, D.R.; et al. Inhibiting the Evolution of Antibiotic Resistance. *Mol. Cell* **2019**, *73*, 157–165. [CrossRef] [PubMed]
13. Han, S.; Gong, Z.; Liang, T.; Chen, Y.; Xie, J. The role of Mfd in Mycobacterium tuberculosis physiology and underlying regulatory network. *Microbiol. Res.* **2021**, *246*, 126718. [CrossRef]
14. Ukkivi, K.; Kivisaar, M. Involvement of transcription-coupled repair factor Mfd and DNA helicase UvrD in mutational processes in Pseudomonas putida. *DNA Repair* **2018**, *72*, 18–27. [CrossRef]
15. Gómez-Marroquín, M.; Martin, H.A.; Pepper, A.; Girard, M.E.; Kidman, A.A.; Vallin, C.; Yasbin, R.E.; Pedraza-Reyes, M.; Robleto, E.A. Stationary-Phase Mutagenesis in Stressed Bacillus subtilis Cells Operates by Mfd-Dependent Mutagenic Pathways. *Genes* **2016**, *7*, 33. [CrossRef]
16. Suárez, V.P.; Martínez, L.E.; Leyva-Sánchez, H.C.; Valenzuela-García, L.I.; Lara-Martínez, R.; Jiménez-García, L.F.; Ramírez-Ramírez, N.; Obregon-Herrera, A.; Cuéllar-Cruz, M.; Robleto, E.A.; et al. Transcriptional coupling and repair of 8-OxoG activate a RecA—Dependent checkpoint that controls the onset of sporulation in *Bacillus subtilis*. *Sci. Rep.* **2021**, *11*, 1–12. [CrossRef] [PubMed]
17. Ragheb, M.N.; Merrikh, C.; Browning, K.; Merrikh, H. Mfd regulates RNA polymerase association with hard-to-transcribe regions in vivo, especially those with structured RNAs. *Proc. Natl. Acad. Sci. USA* **2021**, *118*, 1–10. [CrossRef] [PubMed]
18. Martin, H.A.; Sundararajan, A.; Ermi, T.S.; Heron, R.; Gonzales, J.; Lee, K.; Anguiano-Mendez, D.; Schilkey, F.; Pedraza-Reyes, M.; Robleto, E.A. Mfd Affects Global Transcription and the Physiology of Stressed Bacillus subtilis Cells. *Front. Microbiol.* **2021**, *12*, 1–14. [CrossRef]

19. Zhang, C.; Liu, H.-H.; Zheng, K.-W.; Hao, Y.-H.; Tan, Z. DNA G-quadruplex formation in response to remote downstream transcription activity: Long-range sensing and signal transducing in DNA double helix. *Nucleic Acids Res.* **2013**, *41*, 7144–7152. [CrossRef]
20. Duquette, M.L.; Handa, P.; Vincent, J.A.; Taylor, A.F.; Maizels, N. Intracellular transcription of G-rich DNAs induces formation of G-loops, novel structures containing G4 DNA. *Genes Dev.* **2004**, *18*, 1618–1629. [CrossRef]
21. Tornaletti, S.; Park-Snyder, S.; Hanawalt, P.C. G4-forming Sequences in the Non-transcribed DNA Strand Pose Blocks to T7 RNA Polymerase and Mammalian RNA Polymerase II. *J. Biol. Chem.* **2008**, *283*, 12756–12762. [CrossRef] [PubMed]
22. Pandey, S.; Ogloblina, A.M.; Belotserkovskii, B.P.; Dolinnaya, N.; Yakubovskaya, M.G.; Mirkin, S.M.; Hanawalt, P.C. Transcription blockage by stable H-DNA analogs in vitro. *Nucleic Acids Res.* **2015**, *43*, 6994–7004. [CrossRef] [PubMed]
23. Du, X.; Gertz, E.M.; Wójtowicz, D.; Zhabinskaya, D.; Levens, D.; Benham, C.J.; Schäffer, A.A.; Przytycka, T.M. Potential non-B DNA regions in the human genome are associated with higher rates of nucleotide mutation and expression variation. *Nucleic Acids Res.* **2014**, *42*, 12367–12379. [CrossRef] [PubMed]
24. Piazza, A.; Adrian, M.; Samazan, F.; Heddi, B.; Hamon, F.; Serero, A.; Lopes, J.; Teulade-Fichou, M.; Phan, A.T.; Nicolas, A. Short loop length and high thermal stability determine genomic instability induced by G-quadruplex-forming minisatellites. *EMBO J.* **2015**, *34*, 1718–1734. [CrossRef]
25. Szlachta, K.; Thys, R.G.; Atkin, N.D.; Pierce, L.C.T.; Bekiranov, S.; Wang, Y.-H. Alternative DNA secondary structure formation affects RNA polymerase II promoter-proximal pausing in human. *Genome Biol.* **2018**, *19*, 89. [CrossRef]
26. Northam, M.R.; Moore, E.A.; Mertz, T.M.; Binz, S.K.; Stith, C.M.; Stepchenkova, E.I.; Wendt, K.L.; Burgers, P.M.J.; Shcherbakova, P.V. DNA polymerases ζ and Rev1 mediate error-prone bypass of non-B DNA structures. *Nucleic Acids Res.* **2014**, *42*, 290–306. [CrossRef]
27. Teng, F.-Y.; Hou, X.-M.; Fan, S.-H.; Rety, S.; Dou, S.-X.; Xi, X.-G. Escherichia coli DNA polymerase I can disrupt G-quadruplex structures during DNA replication. *FEBS J.* **2017**, *284*, 4051–4065. [CrossRef]
28. Scheibye-Knudsen, M.; Tseng, A.; Jensen, M.B.; Scheibye-Alsing, K.; Fang, E.F.; Iyama, T.; Bharti, S.K.; Marosi, K.; Froetscher, L.; Kassahun, H.; et al. Cockayne syndrome group A and B proteins converge on transcription-linked resolution of non-B DNA. *Proc. Natl. Acad. Sci. USA* **2016**, *113*, 12502–12507. [CrossRef]
29. Huber, M.D.; Lee, D.C.; Maizels, N. G4 DNA unwinding by BLM and Sgs1p: Substrate specificity and substrate-specific inhibition. *Nucleic Acids Res.* **2002**, *30*, 3954–3961. [CrossRef]
30. Wright, B.E.; Schmidt, K.H.; Davis, N.; Hunt, A.T.; Minnick, M.F., II. Correlations between secondary structure stability and mutation frequency during somatic hypermutation. *Mol. Immunol.* **2008**, *45*, 3600–3608. [CrossRef] [PubMed]
31. Parekh, V.J.; Niccum, B.A.; Shah, R.; Rivera, M.A.; Novak, M.J.; Geinguenaud, F.; Wien, F.; Arluison, V.; Sinden, R.R. Role of Hfq in Genome Evolution: Instability of G-Quadruplex Sequences in *E. coli*. *Microorganisms* **2019**, *8*, 28. [CrossRef] [PubMed]
32. Maizels, N.; Gray, L.T. The G4 Genome. *PLoS Genet.* **2013**, *9*, e1003468. [CrossRef] [PubMed]
33. Ricardo, F.; Cecília, S.; José, M.A. New molecular interactions broaden the functions of the RNA chaperone Hfq. *Curr. Genet.* **2019**, *65*, 1313–1319.
34. Rochat, T.; Delumeau, O.; Figueroa-Bossi, N.; Noirot, P.; Bossi, L.; Dervyn, E.; Bouloc, P. Tracking the Elusive Function of Bacillus subtilis Hfq. *PLoS ONE* **2015**, *10*, e0124977. [CrossRef]
35. Sinden, R.R.; Zheng, G.X.; Brankamp, R.G.; Allen, K.N. On the deletion of inverted repeated DNA in *Escherichia coli*: Effects of length, thermal stability, and cruciform formation in vivo. *Genetics* **1991**, *129*, 991–1005. [CrossRef]
36. Schmidt, K.H.; Reimers, J.M.; Wright, B.E. The effect of promoter strength, supercoiling and secondary structure on mutation rates in Escherichia coli. *Mol. Microbiol.* **2006**, *60*, 1251–1261. [CrossRef]
37. Bernhardt, J.; Weibezahn, J.; Scharf, C.; Hecker, M. Bacillus subtilis During Feast and Famine: Visualization of the Overall Regulation of Protein Synthesis during Glucose Starvation by Proteome Analysis. *Genome Res.* **2003**, *13*, 224–237. [CrossRef]
38. López, D.; Kolter, R. Extracellular signals that define distinct and coexisting cell fates in *Bacillus subtilis*. *FEMS Microbiol. Rev.* **2010**, *34*, 134–149. [CrossRef]
39. Pybus, C.; Pedraza-Reyes, M.; Ross, C.A.; Martin, H.; Ona, K.; Yasbin, R.E.; Robleto, E. Transcription-Associated Mutation in Bacillus subtilis Cells under Stress. *J. Bacteriol.* **2010**, *192*, 3321–3328. [CrossRef]
40. Martin, H.A.; Kidman, A.A.; Socea, J.; Vallin, C.; Pedraza-Reyes, M.; Robleto, E.A. The Bacillus subtilis K-State Promotes Stationary-Phase Mutagenesis via Oxidative Damage. *Genes* **2020**, *11*, 190. [CrossRef]
41. Ambriz-Aviña, V.; Yasbin, R.E.; Robleto, E.A.; Pedraza-Reyes, M. Role of Base Excision Repair (BER) in Transcription-associated Mutagenesis of Nutritionally Stressed Nongrowing Bacillus subtilis Cell Subpopulations. *Curr. Microbiol.* **2016**, *73*, 721–726. [CrossRef]
42. Leyva-Sánchez, H.C.; Villegas-Negrete, N.; Abundiz-Yañez, K.; Yasbin, R.E.; Robleto, E.A.; Pedraza-Reyes, M. Role of Mfd and GreA in Bacillus subtilis Base Excision Repair-Dependent Stationary-Phase Mutagenesis. *J. Bacteriol.* **2020**, *202*, 1–17. [CrossRef] [PubMed]
43. Zuker, M. Mfold web server for nucleic acid folding and hybridization prediction. *Nucleic Acids Res.* **2003**, *31*, 3406–3415. [CrossRef] [PubMed]
44. Kikin, O.; D'Antonio, L.; Bagga, P.S. QGRS Mapper: A web-based server for predicting G-quadruplexes in nucleotide sequences. *Nucleic Acids Res.* **2006**, *34*, W676–W682. [CrossRef] [PubMed]

45. Martin, H.A.; Porter, K.E.; Vallin, C.; Ermi, T.; Contreras, N.; Pedraza-Reyes, M.; Robleto, E.A. Mfd protects against oxidative stress in Bacillus subtilis independently of its canonical function in DNA repair. *BMC Microbiol.* **2019**, *19*, 26. [CrossRef] [PubMed]
46. Itaya, M.; Kondo, K.; Tanaka, T. A neomycin resistance gene cassette selectable in a single copy state in the Bacillus subtilis chromosome. *Nucleic Acids Res.* **1989**, *17*, 4410. [CrossRef] [PubMed]
47. Sung, H.-M.; Yasbin, R.E. Adaptive, or Stationary-Phase, Mutagenesis, a Component of Bacterial Differentiation in *Bacillus subtilis*. *J. Bacteriol.* **2002**, *184*, 5641–5653. [CrossRef]
48. Foster, P.L. Methods for Determining Spontaneous Mutation Rates. *Methods Enzymol.* **2006**, *409*, 195–213.
49. Rosche, W.A.; Foster, P.L. Determining Mutation Rates in Bacterial Populations. *Methods* **2000**, *20*, 4–17. [CrossRef]
50. Ogasawara, N. Markedly unbiased codon usage in Bacillus subtilis. *Gene* **1985**, *40*, 145–150.
51. Li, G.-W.; Oh, E.; Weissman, J.S. The anti-Shine–Dalgarno sequence drives translational pausing and codon choice in bacteria. *Nature* **2012**, *484*, 538–541. [CrossRef]
52. Shields, D.C.; Sharp, P.M. Synonymous codon usage in Bacillus subtilis reflects both translational selection and mutational biases. *Nucleic Acids Res.* **1987**, *15*, 8023–8040. [CrossRef] [PubMed]
53. Guédin, A.; Gros, J.; Alberti, P.; Mergny, J.-L. How long is too long? Effects of loop size on G-quadruplex stability. *Nucleic Acids Res.* **2010**, *38*, 7858–7868. [CrossRef] [PubMed]
54. Kouzine, F.; Gupta, A.; Baranello, L.; Wojtowicz, D.; Ben-Aissa, K.; Liu, J.; Przytycka, M.T.; Levens, D. Transcription-dependent dynamic supercoiling is a short-range genomic force. *Nat. Struct. Mol. Biol.* **2013**, *20*, 396. [CrossRef] [PubMed]
55. Zhabinskaya, D.; Benham, C.J. Competitive superhelical transitions involving cruciform extrusion. *Nucleic Acids Res.* **2013**, *41*, 9610–9621. [CrossRef] [PubMed]
56. Du, X.; Wójtowicz, D.; Bowers, A.A.; Levens, D.; Benham, C.J.; Przytycka, T.M. The genome-wide distribution of non-B DNA motifs is shaped by operon structure and suggests the transcriptional importance of non-B DNA structures in *Escherichia coli*. *Nucleic Acids Res.* **2013**, *41*, 5965–5977. [CrossRef] [PubMed]
57. Rawal, P.; Kummarasetti, V.B.R.; Ravindran, J.; Kumar, N.; Halder, K.; Sharma, R.; Mukerji, M.; Das, S.K.; Chowdhury, S. Genome-wide prediction of G4 DNA as regulatory motifs: Role in *Escherichia coli* global regulation. *Genome Res.* **2006**, *16*, 644–655. [CrossRef]
58. Beaume, N.; Pathak, R.; Yadav, V.; Kota, S.; Misra, H.S.; Gautam, H.K.; Chowdhury, S. Genome-wide study predicts promoter-G4 DNA motifs regulate selective functions in bacteria: Radioresistance of *D. radiodurans* involves G4 DNA-mediated regulation. *Nucleic Acids Res.* **2012**, *41*, 76–89. [CrossRef] [PubMed]
59. Dey, U.; Sarkar, S.; Teronpi, V.; Yella, V.R.; Kumar, A. G-quadruplex motifs are functionally conserved in cis-regulatory regions of pathogenic bacteria: An in-silico evaluation. *Biochimie* **2021**, *184*, 40–51. [CrossRef]
60. Vidales, L.E.; Cárdenas, L.C.; Robleto, E.; Yasbin, R.E.; Pedraza-Reyes, M. Defects in the error prevention oxidized guanine system potentiate stationary-phase mutagenesis in *Bacillus subtilis*. *J. Bacteriol.* **2009**, *191*, 506–513. [CrossRef]
61. Agarwal, T.; Roy, S.; Kumar, S.; Chakraborty, T.K.; Maiti, S. In the Sense of Transcription Regulation by G-Quadruplexes: Asymmetric Effects in Sense and Antisense Strands. *Biochemistry* **2014**, *53*, 3711–3718. [CrossRef]
62. Eddy, J.; Vallur, A.C.; Varma, S.; Liu, H.; Reinhold, W.C.; Pommier, Y.; Maizels, N. G4 motifs correlate with promoter-proximal transcriptional pausing in human genes. *Nucleic Acids Res.* **2011**, *39*, 4975–4983. [CrossRef] [PubMed]
63. Endoh, T.; Kawasaki, Y.; Sugimoto, N. Suppression of Gene Expression by G-Quadruplexes in Open Reading Frames Depends on G-Quadruplex Stability. *Angew. Chem.* **2013**, *125*, 5522–5526. [CrossRef] [PubMed]
64. Rosenberg, S.M.; Shee, C.; Frisch, R.L.; Hastings, P.J. Stress-induced mutation via DNA breaks in *Escherichia coli*: A molecular mechanism with implications for evolution and medicine. *BioEssays* **2012**, *34*, 885–892. [CrossRef]
65. Gómez-Marroquín, M.; Vidales, L.E.; Debora, B.N.; Santos-Escobar, F.; Obregón-Herrera, A.; Robleto, E.A.; Pedraza-Reyes, M. Role of Bacillus subtilis DNA Glycosylase MutM in Counteracting Oxidatively Induced DNA Damage and in Stationary-Phase-Associated Mutagenesis. *J. Bacteriol.* **2015**, *197*, 1963–1971. [CrossRef] [PubMed]
66. Sung, H.M.; Yeamans, G.; Ross, C.a.; Yasbin, R.E. Roles of YqjH and YqjW, homologs of the Escherichia coli UmuC/DinB or Y superfamily of DNA polymerases, in stationary-phase mutagenesis and UV-induced mutagenesis of *Bacillus subtilis*. *J. Bacteriol.* **2003**, *185*, 2153–2160. [CrossRef]
67. Barajas-Ornelas, R.D.C.; Ramírez-Guadiana, F.H.; Juárez-Godínez, R.; Ayala-García, V.M.; Robleto, E.A.; Yasbin, R.E.; Pedraza-Reyes, M. Error-Prone Processing of Apurinic/Apyrimidinic (AP) Sites by PolX Underlies a Novel Mechanism That Promotes Adaptive Mutagenesis in *Bacillus subtilis*. *J. Bacteriol.* **2014**, *196*, 3012–3022. [CrossRef]

*Article*

# Functional Redundancy and Specialization of the Conserved Cold Shock Proteins in *Bacillus subtilis*

Patrick Faßhauer [1], Tobias Busche [2], Jörn Kalinowski [2], Ulrike Mäder [3], Anja Poehlein [4], Rolf Daniel [4] and Jörg Stülke [1,*]

1   Department of General Microbiology, GZMB, Georg-August-University Göttingen, 37077 Göttingen, Germany; patrick.fasshauer@uni-goettingen.de
2   Center for Biotechnology (CeBiTec), Bielefeld University, 33615 Bielefeld, Germany; tbusche@cebitec.uni-bielefeld.de (T.B.); joern@cebitec.uni-bielefeld.de (J.K.)
3   Interfaculty Institute for Genetics and Functional Genomics, University Medicine Greifswald, 17487 Greifswald, Germany; ulrike.maeder@uni-greifswald.de
4   Department of Genomic and Applied Microbiology, GZMB, Georg-August-University Göttingen, 37077 Göttingen, Germany; apoehle3@gwdg.de (A.P.); rdaniel@gwdg.de (R.D.)
*   Correspondence: jstuelk@gwdg.de; Tel.: +49-551-3933781

**Abstract:** Many bacteria encode so-called cold shock proteins. These proteins are characterized by a conserved protein domain. Often, the bacteria have multiple cold shock proteins that are expressed either constitutively or at low temperatures. In the Gram-positive model bacterium *Bacillus subtilis*, two of three cold shock proteins, CspB and CspD, belong to the most abundant proteins suggesting a very important function. To get insights into the role of these highly abundant proteins, we analyzed the phenotypes of single and double mutants, tested the expression of the *csp* genes and the impact of CspB and CspD on global gene expression in *B. subtilis*. We demonstrate that the simultaneous loss of both CspB and CspD results in a severe growth defect, in the loss of genetic competence, and the appearance of suppressor mutations. Overexpression of the third cold shock protein CspC could compensate for the loss of CspB and CspD. The transcriptome analysis revealed that the lack of CspB and CspD affects the expression of about 20% of all genes. In several cases, the lack of the cold shock proteins results in an increased read-through at transcription terminators suggesting that CspB and CspD might be involved in the control of transcription termination.

**Keywords:** *Bacillus subtilis*; cold shock proteins; quasi-essential

**Citation:** Faßhauer, P.; Busche, T.; Kalinowski, J.; Mäder, U.; Poehlein, A.; Daniel, R.; Stülke, J. Functional Redundancy and Specialization of the Conserved Cold Shock Proteins in *Bacillus subtilis*. *Microorganisms* **2021**, *9*, 1434. https://doi.org/10.3390/microorganisms9071434

Academic Editor: Imrich Barák

Received: 3 June 2021
Accepted: 30 June 2021
Published: 2 July 2021

**Publisher's Note:** MDPI stays neutral with regard to jurisdictional claims in published maps and institutional affiliations.

## 1. Introduction

*Bacillus subtilis* is the model organism for a large group of Gram-positive bacteria, among them serious pathogens and biotechnologically relevant bacteria. Due to this relevance, *B. subtilis* is one of the best-studied organisms. However, for about 25% of all proteins, the function is completely unknown, or the proteins are only poorly characterized [1]. Similarly, in the artificially created genome of the minimal organism *Mycoplasma mycoides* JCVI-syn3.0, about one-third of all encoded proteins are of unknown function [2]. These numbers demonstrate how far we are from a complete understanding of even seemingly simple model organisms.

We are interested in a comprehensive understanding of *B. subtilis*. Obviously, the large fraction of unknown proteins is a major challenge. Among the unknown and poorly characterized proteins, the large majority is not or only weakly expressed under standard growth conditions and probably only important under very specific conditions [3]. However, among the 100 most abundant proteins, there are six proteins for which no clear function has been identified [4]. These highly abundant unknown proteins are likely to be of major relevance for the cell, and their functional analysis should be given a high priority. Among these proteins are the cold shock proteins CspB and CspD, which are thought to act as RNA chaperones in *B. subtilis* [5]. The cold shock proteins are a large protein family that are all characterized by a conserved cold shock domain (see [6] for a phylogenetic tree). Among the bacteria, the cold shock proteins are nearly ubiquitous, and most bacteria encode multiple cold shock proteins. *B. subtilis* has three cold shock proteins, CspB, CspC, and CspD. The genes encoding the *B. subtilis* cold shock proteins are highly expressed under a wide variety of growth conditions, and with 31,200 and 21,500 protein molecules per cell, CspB and CspD, respectively, even belong to the 15 most abundant proteins in *B. subtilis* [4,7].

The presence of multiple closely related proteins in an organism always raises the question of whether they participate in the same function or if their specific activity is slightly different. There are a few examples of families of closely related proteins in *B. subtilis*. In the case of ABC transporters or classic amino acid and ion transporters, both functional overlaps, as well as specific functions, have been observed [8–10]. In some cases, as described for the three diadenylate cyclases, all proteins have the same enzymatic activity, but their expression and activities are differentially controlled [11,12]. Similarly, the four DEAD-box RNA helicases have distinct domains in addition to the active protein core, and these proteins have independent functions [13]. Finally, in some cases, regulatory systems consisting of multiple parts, such as two-component regulatory systems, ECF sigma factors, PTS-controlled RNA-binding antitermination proteins, or stress signaling proteins, specificity is achieved by co-evolving interacting partners of these systems [14–17].

For the *B. subtilis* cold shock proteins, it has been observed that the bacteria are not viable in the absence of all three proteins [5]. This, on the one hand, supports the idea that these highly abundant proteins play a major role in the *B. subtilis* cell and, on the other hand, suggests that there is at least some functional overlap between them. In *Staphylococcus aureus*, the CspA protein is unique among the three cold shock proteins since it is the only one that is strongly constitutively expressed and important for the expression of the virulence factor staphyloxanthin [18,19]. It was shown that a single amino acid in CspA, a conserved proline residue, is responsible for this regulatory effect [18].

To get more insights into the function(s) of the cold shock proteins of *B. subtilis*, we studied their expression as well as phenotypes of mutants. The deletion of *cspB* and *cspD* was found to result in the rapid acquisition of suppressor mutations that often result in increased expression of *cspC*, suggesting that CspC might at least partially take over the function of CspB and CspD if it is overexpressed. We also studied the global RNA profile of a strain lacking CspB and CspD and observed that these proteins are important for transcription elongation and termination and that they can control the read-through at transcription terminators.

## 2. Materials and Methods

### 2.1. Bacterial Strains, Growth Conditions, and Phenotypic Characterization

All *B. subtilis* strains used in this study are listed in Table 1. All strains are derived from the laboratory strain 168 (*trpC2*). *B. subtilis* was grown in Lysogeny Broth (LB medium) [20]. LB plates were prepared by addition of 17 g Bacto agar/l (Difco) [20,21]. Quantitative studies of *lacZ* expression in *B. subtilis* were performed as described previously [20]. One unit of β-galactosidase is defined as the amount of enzyme which produces 1 nmol of o-nitrophenol per min at 28 °C.

**Table 1.** *B. subtilis* strains used in this study.

| Strain | Genotype | Source or Reference |
|---|---|---|
| 168 | *trpC2* | Laboratory collection |
| GP1968 | *trpC2 ΔcspB::cat* | This study |
| GP1969 | *trpC2 ΔcspC::spec* | This study |
| GP1970 | *trpC2 ΔcspB::cat ΔcspC::spec* | This study |
| GP1971 | *trpC2 ΔcspB::cat ΔcspD::aphA3* | This study |
| GP1972 | *trpC2 ΔcspC::spec ΔcspD::aphA3* | This study |
| GP1984 | *trpC2 amyE::(P$_{cspC}$-lacZ cat)* | pGP3117→168 |
| GP1986 | *trpC2 amyE::(P$_{cspC}$[G-65A]-lacZ cat)* | pGP3119→168 |
| GP1989 [1] | *trpC2 ΔcspB::cat ΔcspD::aphA3 P$_{cspC}$-[G-65A]* | This study |
| GP1990 [1] | *trpC2 ΔcspB::cat ΔcspD::aphA3 P$_{veg}$-[G-10T]* | This study |
| GP2614 | *trpC2 ΔcspD::aphA3* | [22] |
| GP2888 | *trpC2 Δveg::ermC* | This study |
| GP2896 | *trpC2 Δveg::ermC ΔcspD::aphA3* | GP2888→GP2614 |
| GP2897 | *trpC2 Δveg::ermC ΔcspB::cat ΔcspD::aphA3* | GP1968→GP2896 |
| GP2898 | *trpC2 amyE::(P$_{veg}$-lacZ cat)* | pGP3133→168 |
| GP2899 | *trpC2 amyE::(P$_{veg}$-[G-10T]-lacZ cat)* | pGP3134→168 |
| GP2900 [1] | *trpC2 ΔcspB::cat ΔcspD::aphA3 degS[P245S]* | This study |
| GP3251 | *trpC2 ΔcspB::tet* | This study |
| GP3274 | *trpC2 cspC-A58P-spec ΔcspD::aphA3* | This study |
| GP3275 | *trpC2 cspC-A58P-spec ΔcspB::tet ΔcspD::aphA3* | GP3251→GP3274 |
| GP3283 | *trpC2 amyE::(P$_{cspB}$-lacZ cat)* | pGP3136→168 |
| GP3286 | *trpC2 amyE::(P$_{cspD}$-lacZ-cat)* | This study |

[1] The genomic DNA of these strains was analyzed by whole-genome sequencing.

## 2.2. DNA Manipulation and Genome Sequencing

*B. subtilis* was transformed with plasmids, genomic DNA, or PCR products according to the two-step protocol [20,21]. Transformants were selected on LB plates containing erythromycin (2 µg/mL) plus lincomycin (25 µg/mL), chloramphenicol (5 µg/mL), kanamycin (10 µg/mL), tetracycline (12.5 µg/mL), or spectinomycin (250 µg/mL). S7 Fusion DNA polymerase (Biozym, Hessisch Oldendorf, Germany) was used as recommended by the manufacturer. DNA fragments were purified using the QIAquick PCR purification kit (Qiagen, Hilden, Germany). DNA sequences were determined by the dideoxy chain termination method [21]. Chromosomal DNA from *B. subtilis* was isolated using the peqGOLD Bacterial DNA Kit (Peqlab, Erlangen, Germany). To identify the mutations in the suppressor mutant strains GP1989, GP1990, and GP2900, the genomic DNA was subjected to whole-genome sequencing as described previously [3]. The reads were mapped on the reference genome of *B. subtilis* 168 (GenBank accession number: NC_000964) [23]. Mapping of the reads was performed using the Geneious software package (Biomatters Ltd., Auckland, New Zealand) [24]. The resulting genome sequences were compared to that of our in-house wild-type strain. Single nucleotide polymorphisms were considered as significant when the total coverage depth exceeded 25 reads with a variant frequency of $\geq$90%. All identified mutations were verified by PCR amplification and Sanger sequencing.

## 2.3. Construction of Translational LacZ Reporter Gene Fusions

Plasmid pAC5 [25] was used to construct translational fusions of the *cspB*, *cspC*, and *veg* promoter regions with the *lacZ* gene. For this purpose, the respective promoter regions were amplified using the oligonucleotide pairs PF151/PF152 (*cspB*, corresponds to 379 bp upstream and 5 bp downstream of the ATG start codon, respectively), PF97/PF98 (*cspC*, 237 bp upstream, 125 bp downstream), and PF127/PF118 (*veg*, 280 bp upstream, 50 bp downstream), and chromosomal DNA of *B. subtilis* 168 as the template. The PCR products were digested with EcoRI and BamHI and cloned into pAC5 linearized with the same enzymes. The resulting plasmids were pGP3136, pGP3117, and pGP3133 for the *cspB*, *cspC*, and *veg* promoters, respectively. Plasmids pGP3119 and pGP3134 for the *cspC* and *veg* mutant upstream regions were constructed accordingly using chromosomal DNA of the suppressor mutants GP1986 and GP1990 as the template, respectively. The promoter region of *cspD* (194 bp upstream, 30 bp downstream) was fused to *lacZ* using a PCR product constructed with oligonucleotides PF246/PF247 and PF250/PF251 using pAC5 as a template and PF248/PF249 with chromosomal DNA from *B. subtilis* 168 as a template. The fragments were joined and amplified in a PCR as described previously for the deletion of genes [26,27]. The fusion constructs were integrated into the *amyE* site of the *B. subtilis* chromosome by the transformation of *B. subtilis* 168 with linearized plasmids or the PCR product. The resulting strains were GP3283 (*cspB* wild type promoter), GP1984 (*cspC* wild type promoter), GP1986 (*cspC* mutant promoter), GP3286 (*cspD* wild type promoter), GP2898 (*veg* wild type promoter), and GP2899 (*veg* mutant promoter).

### 2.4. Construction of Mutants

Deletion of the *csp* genes was achieved by transformation with PCR products constructed using oligonucleotides (see Table S1) to amplify DNA fragments flanking the target genes and intervening antibiotic resistance cassettes as described previously [26,27]. The strain GP3274 harboring the CspC-A58P variant was created using the multiple-mutation reaction method, as described previously [28], to generate the *cspC* fragment followed by the combination of the fragment with the downstream flanking region and a spectinomycin resistance cassette as described above.

### 2.5. RNA-Sequencing

*B. subtilis* 168 wild type and the *cspB cspD* double mutant GP1971 were grown in LB medium to an $OD_{600}$ of 0.2 and were harvested by mixing them 5:3 with frozen killing buffer (20 mM Tris-HCl pH 7.5, 6 mM $MgCl_2$, 20 mM $NaN_3$) followed by snap-freezing of the pellets in liquid nitrogen. RNA isolation and quality assessment were performed as described previously [7] with an additional DNase I treatment using TURBO DNase (Ambion). The RNA quality was checked by Trinean Xpose (Trinean, Gentbrugge, Belgium) and the Agilent RNA Nano 6000 kit using an Agilent 2100 Bioanalyzer (Agilent Technologies, Böblingen, Germany). The Ribo-Zero rRNA Removal Kit (Bacteria) from Illumina (San Diego, CA, USA) was used to remove the rRNA. The TruSeq Stranded mRNA Library Prep Kit from Illumina was applied to prepare the cDNA libraries. Final libraries were sequenced paired-end on an Illumina MiSeq system (San Diego, CA, USA) using 75 bp read length.

Trimmed reads were mapped to the *B. subtilis* 168 genome sequences (NCBI GenBank accession number AL009126.3) using Bowtie2 [29]. The annotation AL009126.3 from the NCBI RefSeq database was augmented with the RNA features previously annotated [7]. In order to perform differential gene expression analysis, DEseq2 [30] was used as a part of the software ReadXplorer v2.2 [31]. Statistically significant expression changes (adjusted *p*-value $\leq$ 0.01) with log2 fold change >1.0 or <−1.0 were used. RNA-seq data have been deposited in the ArrayExpress database at EMBL-EBI (www.ebi.ac.uk/arrayexpress (accessed on 2 July 2021)) under accession number E-MTAB-10658.

### 2.6. Qualitative PCR and Real-Time Quantitative Reverse Transcription PCR

For RNA isolation, the cells were grown in LB medium and harvested at an $OD_{600}$ of 1.2 for qualitative PCR and an $OD_{600}$ of 0.5–0.8 for quantitative RT-PCR. Preparation of total RNA was carried out as described previously [32]. cDNAs for qualitative PCR were synthesized using the RevertAid First Strand cDNA Synthesis Kit from ThermoFisher according to the manufacturer's instructions. cDNAs for qRT-PCR were synthesized using the One-Step RT-PCR kit (BioRad, Feldkirchen, Germany) as described [33]. qRT-PCR was carried out on the iCycler instrument (BioRad), following the manufacturer's recommended protocol by using the primers indicated in Table S1. The *rpsE* and *rpsJ* genes encoding constitutively expressed ribosomal proteins were used as internal controls. Data analysis and the calculation of expression ratios as fold changes were performed as described [33]. qRT-PCR experiments were performed in triplicate.

*2.7. Microscopy*

For microscopy, cells were grown at 37 °C in liquid LB medium overnight. The overnight culture was directly used for microscopy or for inoculation of LB medium, which was incubated at 37 °C to an $OD_{600}$ of 0.3–0.5. Images were acquired using an Axioskop 40 FL fluorescence microscope, equipped with digital camera AxioCam MRm and AxioVision Rel (version 4.8, Carl Zeiss, Oberkochen, Germany) software for image processing (Carl Zeiss, Göttingen, Germany) and Neofluar series objective at ×100 primary magnification.

## 3. Results

### 3.1. Relative Contribution of the Cold Shock Proteins to the Growth of B. subtilis

It has been reported that *B. subtilis* is not viable in the absence of the three cold shock proteins CspB, CspC, and CspD [5]. To get better insights into the role(s) of the individual cold shock proteins, we constructed a set of single and double mutants. In agreement with the published data, deletion of all three *csp* genes was not possible. First, we compared the growth of the strains on plates at 37 °C and 15 °C. As shown, in Figure 1A, at 37 °C, all single mutants, as well as the *cspB cspC* and *cspC cspD* double mutants, grew indistinguishable from the wild-type strain *B. subtilis* 168. In contrast, the *cspB cspD* double mutant GP1971 barely formed colonies; instead, this mutant acquired suppressor mutations that restored growth (Figure 1A, see below). At 15 °C, growth of the *cspB cspC* double mutant was also strongly impaired comparable to the *cspB cspD* double mutant (Figure 1B), whereas the *cspC cspD* double mutant showed a slight growth defect.

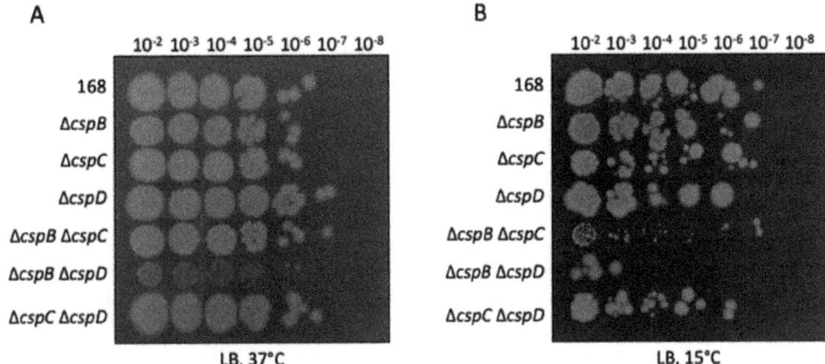


**Figure 1.** Growth of *csp* knockout mutants. The wild-type strain *B. subtilis* 168 and the *csp* mutants GP1968 (Δ*cspB*), GP1969 (Δ*cspC*), GP2614 (Δ*cspD*), GP1970 (Δ*cspB* Δ*cspC*), GP1971 (Δ*cspB* Δ*cspD*), and GP1972 (Δ*cspC* Δ*cspD*) were cultivated on LB-agar (**A**) at 37 °C for one day and (**B**) at 15 °C for 11 days. The results are representative of three biological replicates.

Microscopic analysis of the cell morphologies of the different csp mutant strains indicated that the cold shock proteins did not affect cell morphology during logarithmic growth (see Figure 2A). Similarly, loss of *cspC* or *cspD* did not affect the morphology in the stationary phase. In contrast, the deletion of *cspB* in GP1968 resulted in the formation of elongated cells in the stationary phase suggesting a defect in cell wall biosynthesis and/or cell division or in the entry to stationary phase (Figure 2B). The *cspB cspD* double mutant showed a highly aberrant cell morphology with curly cells in the stationary phase (see Figure 2B). The strong phenotype of the *cspB cspD* double mutant is in good agreement with the poor growth of this mutant.

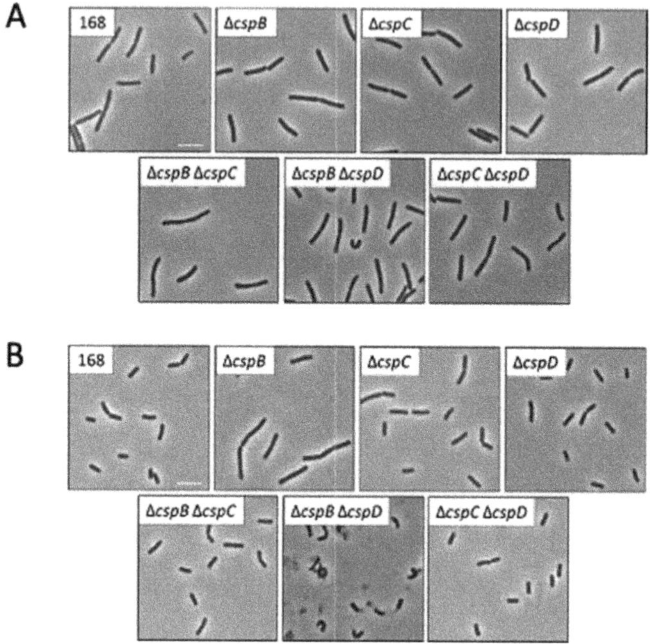

**Figure 2.** Phenotypic analysis of *B. subtilis csp* mutants. The wild-type strain *B. subtilis* 168 and the csp mutants GP1968 (Δ*cspB*), GP1969 (Δ*cspC*), GP2614 (Δ*cspD*), GP1970 (Δ*cspB* Δ*cspC*), GP1971 (Δ*cspB* Δ*cspD*), and GP1972 (Δ*cspC* Δ*cspD*) were analyzed for morphology by phase-contrast microscopy. Scale bars, 5 µM. (**A**) Cells were cultivated in liquid LB medium at 37 °C to an $OD_{600}$ of 0.3–0.5. (**B**) Cells were cultivated in liquid LB medium at 37 °C overnight to the stationary growth phase.

In our attempts to combine the different mutations by genetic transformation, we observed that the *cspB cspD* double mutant GP1971 had completely lost genetic competence, whereas all single mutants, as well as the other double mutants, were not affected.

Taken together, these data indicate that none of the individual cold shock proteins is essential for the viability of *B. subtilis*. Moreover, strains expressing either CspB or CspD exhibited few phenotypic effects. In contrast, the simultaneous loss of the latter two proteins had drastic consequences for growth, cell morphology, and genetic competence. This suggests that the two proteins have similar and overlapping activities and that their function cannot be taken over by CspC.

In several cases, such as for diadenylate cyclases and undecaprenyl pyrophosphate phosphatases of *B. subtilis*, the proteins have very similar functions but are unable to replace each other because they are expressed under different conditions [12,34,35]. In contrast, all cold shock genes are highly expressed under all conditions [1,7], and CspB and CspD do even belong to the most abundant proteins in *B. subtilis* [4]. To get more insights into the expression of the three *csp* genes in *B. subtilis*, we constructed and analyzed fusions of their promoter regions to a promoterless *lacZ* reporter gene encoding β-galactosidase. As shown in Table 2, all three genes were highly expressed at both 15 °C and 37 °C. However, the expression of the *cspB* and *cspD* genes did not respond to the temperature, whereas *cspC* expression was about five times higher at 15 °C as compared to the expression at 37 °C. Thus, based on the expression response to temperature, only *cspC* can be called a cold shock protein. The high constitutive expression of *cspB* and *cspD* is in excellent agreement with the central function of the encoded proteins in *B. subtilis*.

**Table 2.** Promoter activities of the *B. subtilis csp* genes.

| Strain | Promoter | Enzyme Activity in Units/Mg of Protein [a] | |
|---|---|---|---|
| | | 15 °C | 37 °C |
| GP3283 | *cspB* | 14,900 ± 1680 | 11,750 ± 560 |
| GP1984 | *cspC* | 30,950 ± 2900 | 6320 ± 280 |
| GP3286 | *cspD* | 11,300 ± 1600 | 19,600 ± 450 |

[a] All measurements were performed in triplicate. The standard deviations are indicated.

### 3.2. Suppressor Analysis of the CspB CspD Double Mutant

As mentioned above, the *cspB cspD* double mutant GP1971 was impaired in growth and formed suppressor mutants after two days of incubation on plates. In order to get more insights into the function of these cold shock proteins, we isolated and characterized a set of suppressor mutants that was able to grow in the absence of CspB and CspD. Three of these mutants were subjected to whole-genome sequencing. Each of the strains carried a single point mutation. In the case of strain GP1989, we identified a point mutation upstream of the *cspC* gene. Strain GP1990 had a mutation upstream of the constitutively expressed *veg* gene, and GP2900 carried a mutation affecting the sensor kinase DegS that resulted in a substitution of Pro-245 by serine. We then analyzed three additional suppressor mutants for mutations in the *cspC* and *veg* upstream regions as well as in the *degS* coding sequence. They all carried mutations upstream of *cspC*, highlighting the importance of this mutation, whereas no mutations potentially affecting *veg* and *degS* were found. Since the *csp* genes are not part of the regulon controlled by the DegS-DegU two-component system, we focused our further analyses on the mutations in the upstream regions of *cspC* and *veg*.

In four suppressor mutants, we found single point mutations in the 5′ untranslated region of the *cspC* gene (RNA feature S179 [7]). The 5′ UTRs of *cspB* and *cspC* contained two conserved regions that were dubbed cold shock boxes [5]. The identified mutations in the *cspC* upstream regions were located in one of the cold shock boxes or very close to it (see Figure 3). We assumed that these mutations might affect the expression of *cspC*. To test this idea, we fused the G-65A control region of the mutant strain GP1989 to the *lacZ* reporter gene lacking its own transcription and translation signals and compared the expression of the *lacZ* gene to that driven by the wild type control region. For this purpose, the strains GP1984 and GP1986 carrying the *lacZ* gene under the control of the wild type and mutant *cspC* upstream regions, respectively, were cultivated in LB medium at 37 °C, and the resulting β-galactosidase activities were determined. For the wild type, we observed 4360 (±210) units of β-galactosidase activity per mg of protein, whereas the activity was increased to 9450 (±710) units for the mutant *cspC* upstream region. This indicates that the mutation resulted in higher *cspC* expression, which in turn is likely to be the reason for the suppression of the *cspB cspD* double mutant.

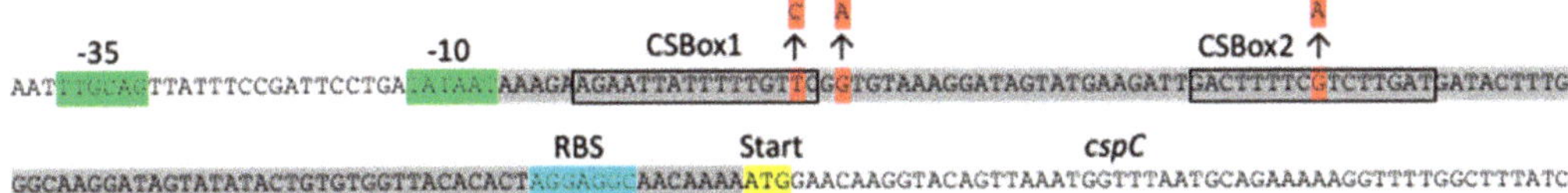


**Figure 3.** Genetic organization of the *cspC* promoter region. Sequence of the promoter region showing the −35 and −10 regions. The 5′UTR (RNA feature S179) is highlighted in grey. CSBoxes, cold shock boxes; RBS, ribosomal binding site; mutations found in Δ*cspB* Δ*cspD* suppressor mutants are highlighted in red.

The mutation in the 5′UTR of *veg* affected the ribosomal binding site (GGUGGA → UGUGGA), suggesting reduced expression of the *veg* gene in the suppressor mutant GP1990. Again, we fused the wild type and mutant control regions of *veg* to the *lacZ* reporter gene and compared the β-galactosidase activities. For *B. subtilis* GP2898 (wild type), we observed 34 (±1) units of β-galactosidase activity per mg of protein. In contrast, the activity was reduced to 2 (±1) units of β-galactosidase for the *lacZ* gene under the control of the mutant *veg* 5′ UTR in GP2899. This indicates that a reduced expression of the *veg* gene might help to overcome the deleterious effect of the simultaneous loss of the two major cold shock proteins. To test this idea, we combined the *veg* gene deletion with the *cspB cspD* deletion and observed the growth of the strains on complex medium. While the *cspB cspD* double mutant GP1971 grew only poorly, growth was restored both if the expression of the *veg* gene was reduced or if the *veg* gene was deleted (Figure 4). These observations indicate that the Veg protein may play a role in RNA metabolism in *B. subtilis*, and that it becomes toxic in the absence of the major cold shock proteins.

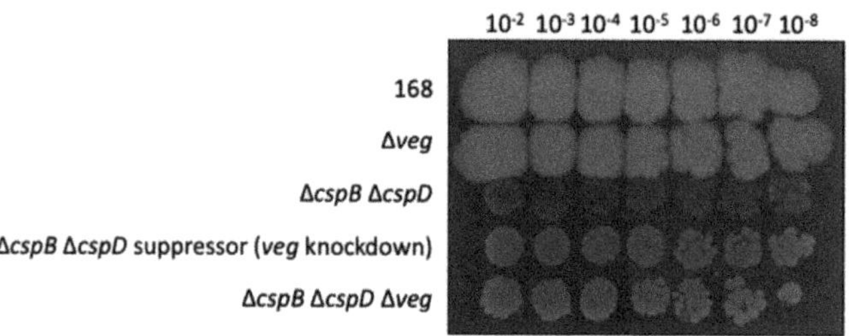


**Figure 4.** Growth of *veg* knockdown and knockout mutants in the Δ*cspB* Δ*cspD* background. The wild-type strain *B. subtilis* 168, GP2888 (Δ*veg*), GP1971 (Δ*cspB* Δ*cspD*), GP1990 (Δ*cspB* Δ*cspD* suppressor with knockdown mutation of *veg*), and GP2897 (Δ*cspB* Δ*cspD* Δ*veg*) were cultivated on LB-agar at 37 °C for 36 h. The results are representative of three biological replicates.

### 3.3. A Modified CspC Protein Can Take over the Functions of CspB and CspD

As mentioned above, the three cold shock proteins of *B. subtilis* are highly similar to each other. A recent study on the cold shock proteins of *S. aureus* identified the proline residue at position 58 in CspA as functionally important [18]. This residue is located on the protein's surface close to the RNA-binding site [18]. The corresponding residue conserved in *B. subtilis* CspB and CspD, but not in CspC. We, therefore, decided to replace the alanine present at this position in CspC with a proline residue and to assay whether the modified protein could compensate for the lack of CspB and CspD. For this purpose, we combined the *cspC* (A58P) allele with the deletions of *cspB* and *cspD*. The resulting strain was GP3275. We then compared GP3275 and the corresponding *cspB cspD* double mutant GP1971 with respect to growth and genetic competence (see Figure 5). As shown, CspC-A58P could compensate for the loss of both CspB and CspD for growth, and partial compensation was observed for genetic competence. This indicates that native expression of the CspC-A58P variant is sufficient for the suppression of the *cspB cspD* double mutant, similar to increased expression of the wild type CspC protein.

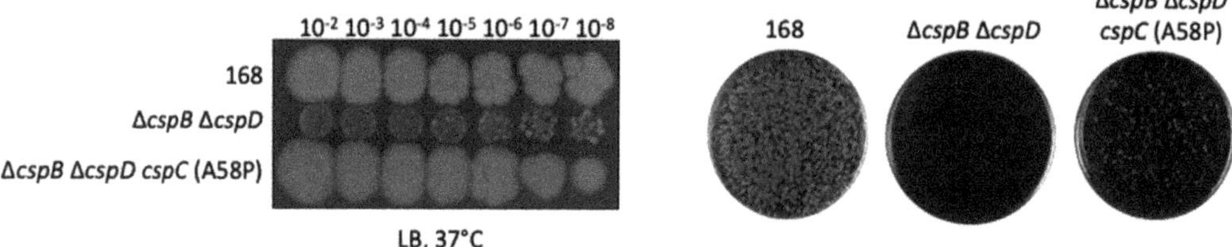


**Figure 5.** The *cspC* (A58P) variant restores growth and genetic competence in the Δ*cspB* Δ*cspD* background. The left panel shows the strains *B. subtilis* wild type 168, GP1971 (Δ*cspB* Δ*cspD*), and GP3275 (Δ*cspB* Δ*cspD* *cspC* (A58P)) that were cultivated on LB-agar at 37 °C for 36 h. The right panel shows the transformants obtained after transformation an equal number of competent cells with 200 ng of chrom. DNA. Data are representative of three biological replicates.

*3.4. RNA-Seq Analysis Suggests a Role for Cold Shock Proteins in Transcription Termination and Elongation*

While it is clear from the data presented above that the cold shock proteins play a pivotal role in the physiology of *B. subtilis*, even at 37 °C, nothing is known about the specific functions of these proteins. A study on the *E. coli* cold shock proteins suggested that they act as transcriptional antiterminators [36]. To test whether the cold shock proteins of *B. subtilis* are involved in transcription as well, we performed an RNA-seq analysis with the wild-type strain and the *cspB cspD* double mutant GP1971. This analysis allowed us to not only get insights into the regulation of the *B. subtilis* genes by the cold shock proteins but also to study their impact on the expression of intergenic regions. In total, 542 and 305 genes exhibited an increased and decreased expression in the *cspB cspD* double mutant with more than twofold changes of expression. The most strongly affected transcription units (more than 20-fold change in expression) are listed in Table 3 (see Table S2 for the complete dataset). Among the most strongly upregulated genes were many involved in the utilization of different carbon sources. The most strongly downregulated genes were dominated by genes involved in aerobic and anaerobic regulation. The genes belong to the Rex and Fnr regulons (see Table 3).

**Table 3.** Effect of the deletion of *cspB* and *cspD* on the expression of *B. subtilis* genes and operons.

| Transcription Unit | Function [1] | Remarks | Fold Regulation Upon *cspB cspD* Deletion |
|---|---|---|---|
| mRNAs with increased amounts upon deletion of *cspB* and *cspD* | | | |
| *liaIH* | Resistance against cell wall antibiotics | Activated by LiaR | 150 |
| *rbsACB* | Ribose utilization | Repressed by CcpA | 110 |
| *maeN* | Malate uptake | Activated by MalR | 62 |
| *tlpA* | Chemotaxis receptor | SigD regulon | 47 |
| *ywsB* | General stress protein | SigB regulon | 45 |
| *manPA-yjdF* | Mannose utilization | Activated by ManR | 36 |
| *yodTSR* | Spore metabolism | SigE regulon | 32 |
| *yorR* | Unknown, SPβ prophage | | 30 |
| *ybdN* | Unknown | Repressed by AbrB | 23 |
| *uxaC* | Hexuronate utilization | Repressed by ExuR and CcpA | 21 |
| mRNAs with decreased amounts upon deletion of cspB and cspD | | | |
| *cydABCD* | Respiration, cytochrome bd oxidase | Repressed by CcpA and Rex | −640 |
| *ldh-lctP* | Overflow metabolism | Repressed by Rex | −530 |
| *narGHJI* | Nitrate respiration | Activated by Fnr | −210 |
| *ywcJ* | Putative nitrate channel | Repressed by Rex | −163 |
| *arfM* | Regulation of anaerobic genes | Activated by Fnr | −90 |
| *mhqNOP* | Resistance against oxidative stress | Repressed by MhqR | −64 |
| *cotJC* | Spore coat protein | SigE regulon | −35 |

[1] Functional information is based on the *SubtiWiki* database [1].

As the *E. coli* Csp proteins act as transcriptional antiterminators, we also studied the expression of intergenic regions of the most strongly CspB/CspD responsive genes. Strikingly, in several cases, we observed an increased read-through at transcriptional terminators in the double mutant, e.g., between the *manR* and *manP* genes and the *liaH* and *liaG* genes (see Figure 6). This finding suggests that CspB and CspD support termination instead of acting as antiterminators in these intergenic regions. In contrast, the lack of CspB and CspD resulted in reduced read-through at the transcription termination structures between the *pyrR* and *pyrP* and *pyrP* and *pyrB* genes. The regulation of transcription between these genes involves the binding of the PyrR protein to the intergenic regions in the presence of UMP or UTP. PyrR binding results in the termination of transcription between *pyrR* and *pyrP* and *pyrP* and *pyrB* [37]. The reduced read-through in the absence of CspB and CspD allows two conclusions: the cold shock protein may either act as antiterminator proteins at the *pyr* operon, or they may interfere with the binding of PyrR to its target regions thus allowing a basal read-through in this operon.

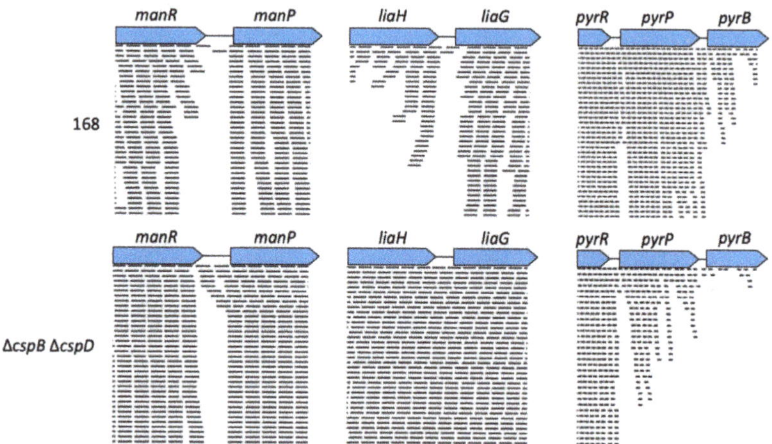

**Figure 6.** Differences in transcription in 168 and the Δ*cspB* Δ*cspD* mutant. The intergenic regions between *manR-manP*, *liaH-liaG*, *pyrR-pyrP-pyrB* in *B. subtilis* wild type 168 (upper panel) and the Δ*cspB* Δ*cspD* mutant GP1971 (lower panel) aligned with the paired reads generated by RNA-sequencing are shown. The data are representative of three biological replicates. Pictures of aligned reads were created using the Geneious Software version 2020.0.4 (Biomatters Ltd., New Zealand).

### 3.5. Analysis of Csp-Dependent Transcription Read-Through

To verify the effect of CspB and CspD on transcription read-through, we performed PCR analyses for the intergenic regions of the *man*, the *lia*, and the *pyr* operons. For this purpose, total mRNA was converted to cDNA and used as a template for PCR assays. To make sure that the analysis was specific for the intergenic regions, we amplified regions covering 400 nucleotides upstream and downstream of the transcription terminators. As positive and negative controls, we used assays with chromosomal DNA as the template and assays with mRNA that had not been subjected to reverse transcription, respectively. As shown in Figure 7A, clear products corresponding to the *manR-manR*, *liaH-liaG*, *pyrR-pyrP*, and *pyrP-pyrB* intergenic regions could be observed if the PCR analysis was performed using the cDNA as a template. In the case of the *man* and *lia* operons, the intensity of these products was strongly increased for the *cspB cspD* double mutant GP1971 as compared to the wild-type strain 168. In contrast, little effect of the inactivation of the *cspB* and *cspD* genes was detected for the intergenic regions of the *pyr* operon.

In order to get further evidence for the effect of CspB and CspD on read-through, we performed quantitative RT-PCR analyses for the intergenic regions (see Figure 7B). In agreement with the RNA-seq data and the qualitative PCR analysis, we observed increased read-through in the *cspB cspD* double mutant for the *manR-manP* and *liaH-liaG* intergenic regions. For the intergenic regions of the *pyr* operon, we found reduced read-through, again in agreement with the RNA-seq data.

Thus, our results suggest that CspB and CspD are involved in the control of transcription elongation at terminator structures in B. *subtilis*; however, the increased expression level of the gene upstream of the terminator may also play a role in reduced termination in the *cspB cspD* double mutant.

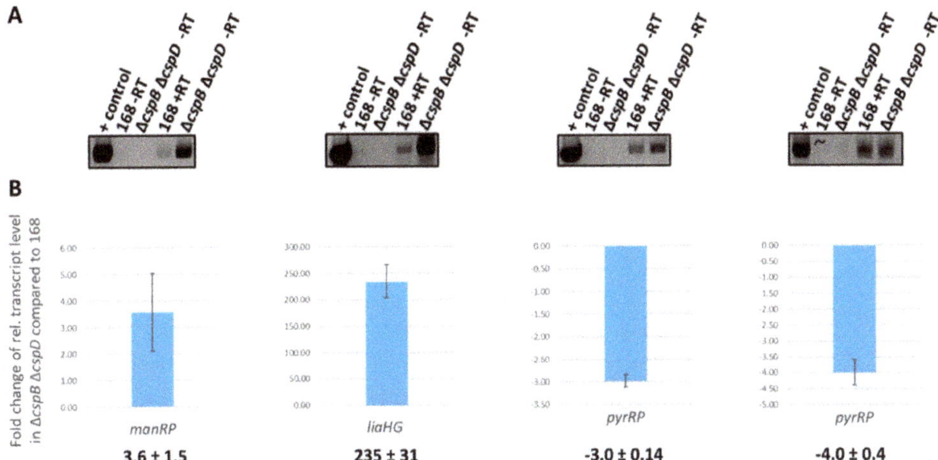

**Figure 7.** Transcriptional read-through in 168 and the Δ*cspB* Δ*cspD* mutant. (**A**) Qualitative PCR on read-through transcripts. Total RNA was extracted from 168 and GP1971. cDNA was synthesized and served as a template in a PCR with primers annealing up- and downstream of the terminators between *manR-manP*, *liaH-liaG*, *pyrR-pyrP-pyrB*. '+ control': standard PCR using chromosomal DNA of *B. subtilis* 168 as the template, '-RT': control sample without reverse transcriptase, '+RT': sample with reverse transcriptase. (**B**) Fold changes in the expression of read-through transcripts in a Δ*cspB* Δ*cspD* (GP1971) mutant relative to levels in the wild-type strain (168) are shown. RNA was purified from each strain, and quantitative RT-PCR was performed using primer sets amplifying up- and downstream of the terminators in the indicated intergenic regions. Shown values represent the mean of three biological replicates. Errors bars indicate the standard deviations.

## 4. Discussion

Cold shock proteins are widespread among bacteria, and many species encode several of these proteins. This is exemplified by the actinobacterium *Lentzea guizhouensis* DHSC013 with as many as 15 distinct *csp* genes [38]. On the other side, some bacteria do not express any cold shock proteins. In the Bacilli, the proteins are ubiquitous, indicating that they are very important for the biology of these bacteria. Our results indicate that the cold shock proteins CspB and CspD performed a quasi-essential function in the cell and suggest that at least one of these proteins should be present in a minimal genome based on *B. subtilis* [39,40].

*B. subtilis* encodes the three cold shock proteins CspB, CspC, and CspD. CspB and CspD are highly similar to each other (81% identical residues). CspC is also similar to both proteins, but the identical residues account only for 71% (with CspD) and 69% (with CspB). The evaluation of the expression as well as the analysis of mutants demonstrated that CspB and CspD are the key cold shock proteins. Actually, the term cold shock proteins is misleading since they are strongly constitutively expressed under a wide range of conditions [7]. Strains lacking one of the cold shock proteins had no or only slightly changed phenotype alterations as compared to the wild type. However, the concomitant loss of the highly similar and constitutively expressed proteins CspB and CspD had severe consequences: growth was strongly impaired, the cells had an aberrant morphology, and lost genetic competence. These observations suggest that CspB and CspD are functionally redundant: at least one of them is required, and a double mutant rapidly acquires suppressor mutations indicating that CspB and CspD together are quasi-essential for *B. subtilis*. In contrast, CspC seemed to play a major role at low temperatures: this was the only cold shock gene that was induced at low temperatures (see Table 2), and double mutants lacking CspC exhibited impaired growth at 15 °C. In the case of the *cspB cspC* double mutant, the growth defect was as severe as observed for the *cspB cspD* mutant, suggesting that CspB is more important than CspD.

Even though CspC seems to be specialized distinctly as compared to CspB and CspD, we observed suppression of the *cspB cspD* double mutant by overexpression of CspC, indicating that CspC can take over the function of the other proteins if it is present at high amounts. Similarly, we have observed suppression of a *topA* mutant lacking the quasi-essential DNA topoisomerase I by overexpression of topoisomerase IV even though the two enzymes are specialized in distinct functions [41]. Interestingly, one amino acid exchange in CspC was sufficient to allow suppression of the *cspB cspD* mutant even at a wild type expression rate: This mutation affects a conserved proline residue that was not present in CspC. Upon introduction of a proline in position 58 of CspC, the protein can functionally replace CspB and CspD, even with respect to genetic competence.

The suppressor analysis also identified the inactivation of the *veg* gene as sufficient to allow the growth of a strain lacking CspB and CspD. The precise function of the constitutively expressed Veg protein has not been identified, but a *veg* mutant was impaired in the expression of biofilm genes and biofilm formation [42]. Interestingly, the Veg protein was recently identified as an mRNA-associated protein in *Clostridium difficile* [22]. Suppression of the *cspB cspD* double mutant by inactivation of Veg suggests that the proteins might act as antagonists.

Two lines of evidence suggest a function related to transcription for CspB and CspD: First, the corresponding *E. coli* proteins play a role in transcription initiation and antitermination [36,43]. Second, a *B. subtilis* strain lacking the quasi-essential endoribonuclease RNase Y readily acquires mutations that affect the RNA polymerase and different transcription factors, such as the transcription elongation factor GreA and the RNA polymerase recycling factor RpoE. In addition to these mutations, inactivation of CspD was found to compensate for the lack of RNase Y [44]. Since all other suppressor mutants affected transcription in *B. subtilis*, it is likely that this was also the case for CspD. The analysis of the transcriptome of the *B. subtilis cspB cspD* double mutant confirmed the involvement of these proteins in the expression of about 20% of all genes. The high resolution of the RNA-seq analysis allowed us to interrogate whether the cold shock proteins might have a function at transcription terminators. Indeed, we observed altered read-through at several transcription terminators in the *cspB cspD* mutant. In contrast to the findings for *E. coli*, both our RNA-seq data as well as the PCR-based validation indicated that CspB and CspD favored transcription termination rather than antitermination.

Further studies will have to address the molecular interactions of the cold shock proteins and the specific role of CspC at optimal and low temperatures.

**Supplementary Materials:** The following are available online at https://www.mdpi.com/article/10.3390/microorganisms9071434/s1, Table S1: Oligonucleotides used in this study. Table S2. Fold changes of the RNA-seq analysis of the cspB cspD double mutant GP1971 as compared to the wild type strain *B. subtilis* 168.

**Author Contributions:** Conceptualization, P.F. and J.S.; methodology, P.F.; validation, P.F., A.P.; investigation, P.F., T.B., J.K., U.M. and R.D.; resources, J.S.; writing—original draft preparation, P.F. and J.S.; writing—review and editing, P.F., U.M. and J.S.; supervision, J.S.; funding acquisition, J.S. All authors have read and agreed to the published version of the manuscript.

**Funding:** This research was funded by the EU Horizons 2020 program (Grant Rafts4Biotech, 720776).

**Institutional Review Board Statement:** Not applicable.

**Informed Consent Statement:** Not applicable.

**Data Availability Statement:** RNA-seq data have been deposited in the ArrayExpress database at EMBL-EBI (www.ebi.ac.uk/arrayexpress, accessed on 2 July 2021) under accession number E-MTAB-10658.

**Acknowledgments:** We thank Marc Schaffer for excellent technical assistance.

**Conflicts of Interest:** The authors declare no conflict of interest.

# References


1. Michna, R.H.; Zhu, B.; Mäder, U.; Stülke, J. *Subti*Wiki 2.0—An integrated database for the model organism *Bacillus subtilis*. *Nucleic Acids Res.* **2016**, *44*, D654–D662. [CrossRef] [PubMed]
2. Hutchison, C.A.; Chuang, R.Y.; Noskov, V.N.; Assad-Garcia, N.; Deerinck, T.J.; Ellisman, M.H.; Gill, J.; Kannan, K.; Karas, B.J.; Ma, L.; et al. Design and synthesis of a minimal bacterial genome. *Science* **2016**, *351*, aad6253. [CrossRef] [PubMed]
3. Reuß, D.R.; Altenbuchner, J.; Mäder, U.; Rath, H.; Ischebeck, T.; Sappa, P.K.; Thürmer, A.; Guérin, C.; Nicolas, P.; Steil, L.; et al. Large-scale reduction of the *Bacillus subtilis* genome: Consequences for the transcriptional network, resource allocation, and metabolism. *Genome Res.* **2017**, *27*, 289–299. [CrossRef]
4. Eymann, C.; Dreisbach, A.; Albrecht, D.; Bernhardt, J.; Becher, D.; Gentner, S.; Tam, L.T.; Büttner, K.; Buurman, G.; Scharf, C.; et al. A comprehensive proteome map of growing *Bacillus subtilis* cells. *Proteomics* **2004**, *4*, 2849–2876. [CrossRef]
5. Graumann, P.; Wendrich, T.M.; Weber, M.H.W.; Schröder, K.; Marahiel, M.A. A family of cold shock proteins in *Bacillus subtilis* is essential for cellular growth and for efficient protein synthesis at optimal and low temperatures. *Mol. Microbiol.* **1997**, *25*, 741–756. [CrossRef] [PubMed]
6. Feliciano, J.R.; Seixas, A.M.M.; Pita, T.; Leitao, J.H. Comparative genomics and evolutionary analysis of RNA-binding proteins of *Burkholderia cenocepacia* J2315 and other members of the *B. cepacia* complex. *Genes* **2020**, *11*, 231. [CrossRef]
7. Nicolas, P.; Mäder, U.; Dervyn, E.; Rochat, T.; Leduc, A.; Pigeonneau, N.; Bidnenko, E.; Marchadier, E.; Hoebeke, M.; Aymerich, S.; et al. Condition-dependent transcriptome reveals high-level regulatory architecture in *Bacillus subtilis*. *Science* **2012**, *335*, 1103–1106. [CrossRef]
8. Gebhard, S. ABC transporters of antimicrobial peptides in Firmicutes bacteria—Phylogeny, function and regulation. *Mol. Microbiol.* **2012**, *86*, 1295–1317. [CrossRef] [PubMed]
9. Krüger, L.; Herzberg, C.; Rath, H.; Pedreira, T.; Ischebeck, T.; Poehlein, A.; Gundlach, J.; Daniel, R.; Völker, U.; Mäder, U.; et al. Essentiality of c-di-AMP in *Bacillus subtilis*: Bypassing mutations converge in potassium and glutamate homeostasis. *PLoS Genet.* **2021**, *17*, e1009092. [CrossRef] [PubMed]
10. Klewing, A.; Koo, B.M.; Krüger, L.; Poehlein, A.; Reuß, D.; Daniel, R.; Gross, C.A.; Stülke, J. Resistance to serine in *Bacillus subtilis*: Identification of the serine transporter YbeC and of a metabolic network that links serine and threonine metabolism. *Environ. Microbiol.* **2020**, *22*, 3937–3949. [CrossRef] [PubMed]
11. Commichau, F.M.; Heidemann, J.L.; Ficner, R.; Stülke, J. Making and breaking of an essential poison: The cyclases and phosphodiesterases that produce and degrade the essential second messenger cyclic di-AMP in bacteria. *J. Bacteriol.* **2019**, *201*, e00462-18. [CrossRef] [PubMed]
12. Mehne, F.M.P.; Schröder-Tittmann, K.; Eijlander, R.T.; Herzberg, C.; Hewitt, L.; Kaever, V.; Lewis, R.J.; Kuipers, O.P.; Tittmann, K.; Stülke, J. Control of the diadenylate cyclase CdaS in *Bacillus subtilis*: An autoinhibitory domain limits c-di-AMP production. *J. Biol. Chem.* **2014**, *289*, 21098–21107. [CrossRef] [PubMed]
13. Lehnik-Habrink, M.; Rempeters, L.; Kovács, Á.; Wrede, C.; Baierlein, C.; Krebber, H.; Kuipers, O.P.; Stülke, J. The DEAD-box RNA helicases in *Bacillus subtilis* have multiple functions and act independent from each other. *J. Bacteriol.* **2013**, *195*, 534–544. [CrossRef]
14. Procaccini, A.; Lunt, B.; Szurmant, H.; Hwa, T.; Weight, M. Dissecting the specificity of protein-protein interaction in bacterial two-component signaling: Orphans and crosstalks. *PLoS ONE* **2011**, *6*, e19729. [CrossRef] [PubMed]
15. Helmann, J.D. *Bacillus subtilis* extracytoplasmic function (ECF) sigma factors and defense of the cell envelope. *Curr. Opin. Microbiol.* **2016**, *30*, 122–132. [CrossRef] [PubMed]
16. Schilling, O.; Herzberg, C.; Hertrich, T.; Vörsmann, H.; Jessen, D.; Hübner, S.; Titgemeyer, F.; Stülke, J. Keeping signals straight in transcription regulation: Specificity determinants for the interaction of a family of conserved bacterial RNA-protein couples. *Nucleic Acids Res.* **2006**, *34*, 6102–6115. [CrossRef]
17. Pané-Farré, J.; Lewis, R.J.; Stülke, J. The RsbRST stress module in bacteria: A signalling system that may interact with different output modules. *J. Mol. Microbiol. Biotechnol.* **2005**, *9*, 65–76. [CrossRef] [PubMed]
18. Catalan-Moreno, A.; Caballero, C.J.; Irurzun, N.; Cuesta, S.; López-Sagaseta, J.; Toledo-Arana, A. One evolutionary selected amino acid variation is sufficient to provide functional specificity in the cold shock protein paralogs of *Staphylococcus aureus*. *Mol. Microbiol.* **2020**, *113*, 826–840. [CrossRef] [PubMed]
19. Catalan-Moreno, A.; Cela, M.; Mendenez-Gil, P.; Irurzun, N.; Caballero, C.J.; Caldelari, I.; Toledo-Arana, A. RNA thermoswitches modulate *Staphylococcus aureus* adaptation to ambient temperatures. *Nucleic Acids Res.* **2021**, *49*, 3409–3426. [CrossRef]
20. Kunst, F.; Rapoport, G. Salt stress is an environmental signal affecting degradative enzyme synthesis in *Bacillus subtilis*. *J. Bacteriol.* **1995**, *177*, 2403–2407. [CrossRef]
21. Sambrook, J.; Fritsch, E.F.; Maniatis, T. *Molecular Cloning: A Laboratory Manual*, 2nd ed.; Cold Spring Harbor Laboratory: Cold Spring Harbor, NY, USA, 1989.
22. Lamm-Schmidt, V.; Fuchs, M.; Sulzer, J.; Gerovac, M.; Hör, J.; Dersch, P.; Vogel, J.; Faber, F. Grad-seq identifies KhpB as a global RNA-binding protein in *Clostrioides difficile* that regulates toxin production. *microLife* **2021**, *2*, uqab004. [CrossRef]
23. Barbe, V.; Cruveiller, S.; Kunst, F.; Lenoble, P.; Meurice, G.; Sekowska, A.; Vallenet, D.; Wang, T.; Moszer, I.; Médigue, C.; et al. From a consortium sequence to a unified sequence: The *Bacillus subtilis* 168 reference genome a decade later. *Microbiology* **2009**, *155*, 1758–1775. [CrossRef]

24. Kearse, M.; Moir, R.; Wilson, A.; Stones-Havas, S.; Cheung, M.; Sturrock, S.; Buxton, S.; Cooper, A.; Markowitz, S.; Duran, C.; et al. Geneious basic: An integrated and extendable desktop software platform for the organization and analysis of sequence data. *Bioinformatics* **2012**, *28*, 1647–1649. [CrossRef] [PubMed]

25. Martin-Verstraete, I.; Débarbouillé, M.; Klier, A.; Rapoport, G. Mutagenesis of the *Bacillus subtilis* "-12, -24" promoter of the levanase operon and evidence for the existence of an upstream activating sequence. *J. Mol. Biol.* **1992**, *226*, 85–99. [CrossRef]

26. Lehnik-Habrink, M.; Pförtner, H.; Rempeters, L.; Pietack, N.; Herzberg, C.; Stülke, J. The RNA degradosome in *Bacillus subtilis*: Identification of CshA as the major helicase in the multiprotein complex. *Mol. Microbiol.* **2010**, *77*, 958–971. [CrossRef] [PubMed]

27. Guérout-Fleury, A.M.; Shazand, K.; Frandsen, N.; Stragier, P. Antibiotic-resistance cassettes for *Bacillus subtilis*. *Gene* **1995**, *167*, 335–336. [CrossRef]

28. Hames, C.; Halbedel, S.; Schilling, O.; Stülke, J. MMR: A method for the simultaneous introduction of multiple mutations into the *glpK* gene of *Mycoplasma pneumoniae*. *Appl. Environ. Microbiol.* **2005**, *71*, 4097–4100. [CrossRef] [PubMed]

29. Langmead, B.; Trapnell, C.; Pop, M.; Salzberg, S.L. Ultrafast and memory-efficient alignment of short DNA sequences to the human genome. *Genome Biol.* **2009**, *10*, R25. [CrossRef] [PubMed]

30. Love, M.I.; Huber, W.; Anders, S. Moderated estimation of fold change and dispersion for RNA-seq data with DESeq2. *Genome Biol.* **2014**, *15*, 550. [CrossRef] [PubMed]

31. Hilker, R.; Stadermann, K.B.; Schwengers, O.; Anisiforov, E.; Jaenicke, S.; Weisshaar, B.; Zimmermann, T.; Goesmann, A. ReadXplorer 2—Detailed read mapping analysis and visualization from one single source. *Bioinformatics* **2016**, *32*, 3702–3708. [CrossRef] [PubMed]

32. Schilling, O.; Frick, O.; Herzberg, C.; Ehrenreich, A.; Heinzle, E.; Wittmann, C.; Stülke, J. Transcriptional and metabolic responses of *Bacillus subtilis* to the availability of organic acids: Transcription regulation is important but not sufficient to account for metabolic adaptation. *Appl. Environ. Microbiol.* **2007**, *73*, 499–507. [CrossRef] [PubMed]

33. Diethmaier, C.; Pietack, N.; Gunka, K.; Wrede, C.; Lehnik-Habrink, M.; Herzberg, C.; Hübner, S.; Stülke, J. A novel factor controlling bistability in *Bacillus subtilis*: The YmdB protein affects flagellin expression and biofilm formation. *J. Bacteriol.* **2011**, *193*, 5997–6007. [CrossRef] [PubMed]

34. Mehne, F.M.P.; Gunka, K.; Eilers, H.; Herzberg, C.; Kaever, V.; Stülke, J. Cyclic-di-AMP homeostasis in *Bacillus subtilis*: Both lack and high-level accumulation of the nucleotide are detrimental for cell growth. *J. Biol. Chem.* **2013**, *288*, 2004–2017. [CrossRef] [PubMed]

35. Zhao, H.; Sun, Y.; Peters, J.M.; Gross, C.A.; Garner, E.C.; Helmann, J.D. Depletion of undecaprenyl phosphatases disrupts cell envelope biogenesis in *Bacillus subtilis*. *J. Bacteriol.* **2016**, *198*, 2925–2935. [CrossRef]

36. Bae, W.; Xia, B.; Inouye, M.; Severinov, K. *Escherichia coli* CspA-family RNA chaperones are transcription antiterminators. *Proc. Natl. Acad. Sci. USA* **2000**, *97*, 7784–7789. [CrossRef] [PubMed]

37. Switzer, R.L.; Turner, R.J.; Lu, Y. Regulation of the *Bacillus subtilis* pyrimidine biosynthetic operon by transcriptional attenuation: Control of gene expression by an mRNA-binding protein. *Prog. Nucleic Acid Res. Mol. Biol.* **1999**, *62*, 329–367. [PubMed]

38. Galperin, M.Y.; Wolf, Y.I.; Makarova, K.S.; Vera Alvarez, R.; Landsman, D.; Koonin, E.V. COG database update: Focus on microbial diversity, model organisms, and widespread pathogens. *Nucleic Acids Res.* **2021**, *49*, D274–D281. [CrossRef]

39. Commichau, F.M.; Pietack, N.; Stülke, J. Essential genes in *Bacillus subtilis*: A re-evaluation after ten years. *Mol. BioSyst.* **2013**, *9*, 1068–1075. [CrossRef]

40. Reuß, D.R.; Commichau, F.M.; Gundlach, J.; Zhu, B.; Stülke, J. The blueprint of a minimal cell: *MiniBacillus*. *Microbiol. Mol. Biol. Rev.* **2016**, *80*, 955–987. [CrossRef]

41. Reuß, D.R.; Faßhauer, P.; Mroch, P.J.; Ul-Haq, I.; Koo, B.M.; Poehlein, A.; Gross, C.A.; Daniel, R.; Brantl, S.; Stülke, J. Topoisomerase IV can functionally replace all type 1A topoisomerases in *Bacillus subtilis*. *Nucleic Acids Res.* **2019**, *47*, 5231–5242. [CrossRef] [PubMed]

42. Lei, Y.; Oshima, T.; Ogasawara, N.; Ishikawa, S. Functional analysis of the protein Veg, which stimulates biofilm formation in *Bacillus subtilis*. *J. Bacteriol.* **2013**, *195*, 1697–1705. [CrossRef] [PubMed]

43. Giangrossi, M.; Brandi, A.; Gioliodori, A.M.; Gualerzi, C.O.; Pon, C.L. Cold-shock-induced de novo transcription and translation of infA and role of IF1 during cold adaptation. *Mol. Microbiol.* **2007**, *64*, 807–821. [CrossRef] [PubMed]

44. Benda, M.; Woelfel, S.; Fasshauer, P.; Gunka, K.; Klumpp, S.; Poehlein, A.; Kálalová, D.; Šanderová, H.; Daniel, R.; Krásný, L.; et al. Quasi-essentiality of RNase Y in *Bacillus subtilis* is caused by its critical role in the control of mRNA homeostasis. *Nucleic Acids Res.* **2021**. [CrossRef] [PubMed]

*microorganisms*

Article

# Comparison of the Genetic Features Involved in *Bacillus subtilis* Biofilm Formation Using Multi-Culturing Approaches

Yasmine Dergham [1,2], Pilar Sanchez-Vizuete [1], Dominique Le Coq [1,3], Julien Deschamps [1], Arnaud Bridier [4], Kassem Hamze [2] and Romain Briandet [1,*]

[1] Micalis Institute, INRAE, AgroParisTech, Université Paris-Saclay, 78350 Jouy-en-Josas, France; yasmin.dorghamova-dergham@inrae.fr (Y.D.); pilarsanviz@gmail.com (P.S.-V.); dominique.le-coq@inrae.fr (D.L.C.); julien.deschamps@inrae.fr (J.D.)

[2] Faculty of Science, Lebanese University, 1003 Beirut, Lebanon; kassem.hamze@ul.edu.lb

[3] Centre National de la Recherche Scientifique (CNRS), Micalis Institute, INRAE, AgroParisTech, Université Paris-Saclay, 78350 Jouy-en-Josas, France

[4] Fougères Laboratory, Antibiotics, Biocides, Residues and Resistance Unit, Anses, 35300 Fougères, France; arnaud.bridier@anses.fr

* Correspondence: romain.briandet@inrae.fr

**Citation:** Dergham, Y.; Sanchez-Vizuete, P.; Le Coq, D.; Deschamps, J.; Bridier, A.; Hamze, K.; Briandet, R. Comparison of the Genetic Features Involved in *Bacillus subtilis* Biofilm Formation Using Multi-Culturing Approaches. *Microorganisms* **2021**, *9*, 633. https://doi.org/10.3390/microorganisms9030633

Academic Editor: Imrich Barák

Received: 19 February 2021
Accepted: 16 March 2021
Published: 18 March 2021

**Publisher's Note:** MDPI stays neutral with regard to jurisdictional claims in published maps and institutional affiliations.



**Abstract:** Surface-associated multicellular assemblage is an important bacterial trait to withstand harsh environmental conditions. *Bacillus subtilis* is one of the most studied Gram-positive bacteria, serving as a model for the study of genetic pathways involved in the different steps of 3D biofilm formation. *B. subtilis* biofilm studies have mainly focused on pellicle formation at the air-liquid interface or complex macrocolonies formed on nutritive agar. However, only few studies focus on the genetic features of *B. subtilis* submerged biofilm formation and their link with other multicellular models at the air interface. NDmed, an undomesticated *B. subtilis* strain isolated from a hospital, has demonstrated the ability to produce highly structured immersed biofilms when compared to strains classically used for studying *B. subtilis* biofilms. In this contribution, we have conducted a multi-culturing comparison (between macrocolony, swarming, pellicle, and submerged biofilm) of *B. subtilis* multicellular communities using the NDmed strain and mutated derivatives for genes shown to be required for motility and biofilm formation in pellicle and macrocolony models. For the 15 mutated NDmed strains studied, all showed an altered phenotype for at least one of the different culture laboratory assays. Mutation of genes involved in matrix production (i.e., *tasA*, *epsA-O*, *cap*, *ypqP*) caused a negative impact on all biofilm phenotypes but favored swarming motility on semi-solid surfaces. Mutation of *bslA*, a gene coding for an amphiphilic protein, affected the stability of the pellicle at the air-liquid interface with no impact on the submerged biofilm model. Moreover, mutation of *lytF*, an autolysin gene required for cell separation, had a greater effect on the submerged biofilm model than that formed at aerial level, opposite to the observation for *lytABC* mutant. In addition, *B. subtilis* NDmed with *sinR* mutation formed wrinkled macrocolony, less than that formed by the wild type, but was unable to form neither thick pellicle nor structured submerged biofilm. The results are discussed in terms of the relevancy to determine whether genes involved in colony and pellicle formation also govern submerged biofilm formation, by regarding the specificities in each model.

**Keywords:** *Bacillus subtilis*; NDmed; biofilm; pellicle; complex macrocolonies; swarming; confocal laser scanning microscopy (CLSM)

## 1. Introduction

Bacteria in nature frequently exist in communities that display complex social behavior, which involves intercellular signaling to permit survival and dissemination in a wide variety of habitats [1]. Even within a pure culture biofilm, where cells are genetically identical, different patterns of gene expression co-exist and therefore produce subpopulations of functionally distinct cell types [2]. Surface-associated biofilm develops in a sequential process in which sessile bacterial cells secrete extracellular matrix and aggregate as structured multicellular groups [3,4]. In nature, microbial biofilms participate in many biogeochemical cycling processes for most elements in water, soil, sediments, and subsurface environments [5]. In addition, utilization of microbial antagonists as biological control agents is a promising biotechnological alternative to the use of pesticides, which often accumulate in plants and end up by affecting humans in a direct or indirect way [6]. However, in terms of public health and with the medical science progress, more and more medical devices and/or artificial organs are being applied in the treatment of human diseases. As a consequence, biofilm-associated infections has become also frequent. It has been estimated that many bacterial infections in human are correlated with biofilm formation and are associated with the indwelling medical devices (such as catheters or needles) [7].

Over the last decades, *Bacillus subtilis*, a Gram-positive, motile, spore-forming bacterium has served as a model organism for molecular studies on biofilm formation [5]. These studies were mainly based on the development of complex macrocolonies on the agar-air interface, or floating pellicle at the air-liquid interface, and only few on submerged biofilms [8–14]. These models allowed highlighting that the transition from motile to sessile biofilm lifestyle, and vice versa, is controlled by complex genes regulatory networks. Four pairs of global regulators—the Spo0A/AbrB, SinI/SinR, SlrR/SlrA, and DegS/DegU—have been shown to play major roles, directly and indirectly, on both the formation and development of complex multicellular communities and on expression of the motility-involved genes [8,12,15–20]. Flagella required for motility are partly encoded by the *fla/che* operon, which, in addition to flagellar genes, includes chemotaxis genes and the *sigD* gene. In turn, the sigma D factor has been shown to direct transcription of other flagellar genes outside the *fla/che* operon (i.e., *hag* gene and other SigD-dependent motility genes) and genes involved in autolytic enzymes synthesis (*lytC, lytD*, and *lytF*) that mediate the separation of sister cells after cell division [21–24]. On the other hand, Spo0A phosphorylation represses two negative biofilm formation regulators, AbrB and SinR, therefore leading to expression of genes involved in the synthesis of biofilm matrix (polysaccharide synthesis by *epsA-O*, amyloid like fiber TasA encoded by the *tapA-sipW-tasA* operon, and the amphiphilic matrix protein produced by *bslA*) [2,25].

In specific conditions, cells from a bacterial colony can become highly motile and migrate over the substrate with specific collective patterns, a process known as swarming [4]. Swarming—a remarkable example of cooperative behavior in bacteria—is a mass, coordinated, and rapid migration (2 to 10 mm/hr) of cells on a surface [26]. In *B. subtilis*, this developmental process is observed on semi-solid agar (0.6%–1% agar) and has been shown to be completely dependent on flagella and surfactin production [26–29].

In 2001, Hamon and Lazazzera have shown that *B. subtilis* has the ability to adhere to abiotic surfaces and form structured biofilms [8], which have grabbed biofilm researches to reconsider the importance of the immersed surface-associated biofilm model for this species. In this context, architectural comparative submerged biofilm studies performed on various *B. subtilis* strains from different origins, including NCIB3610 and 168 reference strains, have revealed an undomesticated *B. subtilis* NDmed strain as able to form the highest submerged biofilm biovolume [11,13].

The NDmed strain, isolated from a hospital endoscope washer-disinfector was found to resist to the action of peracetic acid (an oxidizing agent commonly used in formulations used for the endoscope disinfection) and to have the ability to protect the pathogen *Staphylococcus aureus* in mixed biofilms [30,31]. By the use of confocal and electronic microscopy techniques, it has been shown that the hyper-resistant phenotype was related to the complex architectural biofilm formed and to the large amount of extracellular matrix produced that could prevent the diffusion-reaction of oxidizing agents [30]. Moreover, further genetic comparison between NDmed and other *B. subtilis* reference strains pinpointed that the *ypqP* gene (renamed *spsM* [32]), potentially involved in the synthesis of polysaccharide, was involved indirectly in this resistance by participating to the strong spatial organization of the *B. subtilis* NDmed biofilms, both at air and liquid interfaces [13]. This gene is disrupted by the SPβ prophage in both *B. subtilis* NCIB3610 and 168 strains [13]. These new observations suggested that interfaces between surfaces and liquids could, as for most other bacteria, be a relevant biotope for *B. subtilis* biofilm.

Our knowledge for the molecular mechanisms controlling the formation and the behavior of *B. subtilis* 3D communities is still limited. In this contribution, *B. subtilis* NCIB3610 and 168 strains were compared to NDmed in different laboratory culture conditions. Moreover, 15 mutants derived from the NDmed strain and defective in genes previously described as triggering biofilm formation were compared through a multi-culturing approach using four multicellular models, at the interface with air (solid agar, semi-solid agar, liquid medium) or at the interface between a solid surface (polystyrene) and a liquid medium, submerged model. Thus, this provided a global view over the different biofilm laboratory assays used to study the effect of gene mutation on both motility and biofilm formation in *B. subtilis* wild type.

## 2. Materials and Methods

### 2.1. Bacterial Strains and Growth Conditions

All bacterial strains and mutants used in this study are listed in Table 1. The *B. subtilis* NDmed derivatives mutated in various genes were obtained by transformation with chromosomal DNA extracted from strains carrying the corresponding different alleles of interest marked with a suitable antibiotic resistance cassette. Transforming chromosomal DNA was extracted according to the method of Marmur [33], and transformation of *B. subtilis* was performed according to the method of Anagnostopoulos and Spizizen [34], including the use of the MGI and MGII media of Borenstein and Ephrati-Elizur [35]. Transformants were selected on Lysogeny Broth (LB) plates supplemented with the relevant antibiotic at the following concentrations: spectinomycin, 100 µg ml$^{-1}$; chloramphenicol, 4 µg ml$^{-1}$; erythromycin, 0.5 µg ml$^{-1}$; tetracycline, 10 µg ml$^{-1}$; neomycin and kanamycin, 8 µg ml$^{-1}$. Before each experiment, cells were subcultured in Tryptone Soya Broth (TSB, BioMérieux, France; pH 7.2) and supplemented with antibiotics when necessary. For biofilm formation, bacteria were grown in TSB at 30 °C for 8 h with agitation, then diluted 1/100 in 10 mL TSB incubated overnight at 30 °C. This culture was then used to grow biofilms on different assays. Bacteria for swarming experiments were grown with agitation at 37 °C in synthetic B-medium composed of (all final concentrations; pH 7.2) 15 mM $(NH_4)_2SO_4$, 8 mM $MgSO_4.7H_2O$, 27 mM KCl, 7 mM sodium citrate.$2H_2O$, 50 mM Tris/HCl (pH 7.5), and 2 mM $CaCl_2.2H_2O$, 1 µM $FeSO_4.7H_2O$, 10 µM $MnSO_4.4H_2O$, 0.6 mM $KH_2PO_4$, 4.5 mM glutamic acid (pH 8), 862 µM lysine, 784 µM tryptophan, 1 mM threonine and 0.5% glucose were added before use [36]. Antibiotics were added to bacterial cultures when needed.

**Table 1.** *Bacillus subtilis* strains used in this study.

| Strain | Genotype or Isolation Source | Construction [a] or Reference |
|---|---|---|
| NDmed | Undomesticated, isolated from endoscope washer-disinfectors | [31] |
| NCIB3610 | Natural isolate, less domesticated | [37] |
| 168 | *trpC2* (domesticated strain) | [37] |
| GM3248 | NDmed Δ*ypqP*:: kan | [13] |
| GM3533 | NDmed Δ*sinR*:: cm | Tf NDmed/DNA ABS840 [38] |
| GM3535 | NDmed Δ*epsA-O*:: tet | Tf NDmed/DNA GM3532 [NCIB3610, Δ*tasA*:: kan, Δ*epsA-O*:: tet] ( our lab collection) |
| GM3539 | NDmed Δ*sinI*:: kan | Tf NDmed/DNA ABS803 [39] |
| GM3545 | NDmed Δ*cap*:: pKPSd/ cm | Tf NDmed/DNA GM3543 [NCIB3610 Δ*cap*:: pKPSd/ cm] (our lab collection) |

**Table 1.** *Cont.*

| Strain | Genotype or Isolation Source | Construction [a] or Reference |
|---|---|---|
| GM3555 | NDmed Δ*abrB*:: cm | Tf NDmed/DNA MM1717 [40] |
| GM3559 | NDmed Δ*degU*:: neo | Tf NDmed/DNA GM719 [41] |
| GM3561 | NDmed Δ*bslA*:: cm | Tf NDmed/DNA NRS2097 [20] |
| GM3602 | NDmed Δ*lytF*:: spec | Tf NDmed/DNA NRS3295 [42] |
| GM3611 | NDmed Δ*lytABC*:: kan | Tf NDmed/DNA NRS3295 [42] |
| GM3614 | NDmed Δ*tasA*:: kan | Tf NDmed/DNA GM3532 [NCIB3610, Δ*tasA*:: kan, Δ*epsA-O*:: tet] ( our lab collection) |
| GM3618 | NDmed Δ*slrR*:: spec | Tf NDmed/DNA GM3598 [NCIB3610 Δ*slrR*:: spec] (our lab collection) |
| GM3619 | NDmed Δ*srfAA*:: ery | Tf NDmed/DNA GM3599 [NCIB3610 Δ*srfAA*:: ery] (our lab collection) |
| GM3652 | NDmed *amyE*:: Phyperspank-GFP/spec, Δ*hag*::cm | Tf NDmedGFP [30]/DNA OMG954 [29] |
| GM3671 | NDmed *amyE*:: Phyperspank-GFP/spec, Δ*spo0A*:: Kan | Tf NDmedGFP [30]/DNA FBT2 [43] |

[a] TF NDmed/DNA stands for transformation of NDmed by chromosomal DNA of indicated strains.

### 2.2. Submerged Biofilm Developmental Assays

Submerged biofilms were grown on the surface of polystyrene 96-well microtiter plates with a µclear® base (Greiner Bio-one, France) enabling high-resolution fluorescence imaging as previously described [44]. An amount of 200 µL of an overnight culture in TSB (adjusted to an OD 600 nm of 0.02) was added in each well. The microtiter plate was then incubated at 30 °C for 90 min to allow the bacteria to adhere to the bottom of the wells. Wells were then rinsed with TSB to eliminate non-adherent bacteria and refilled with 200 µL of sterile TSB. The plates were incubated at 30 °C for 24 h, and 5 µM of the cell permeant nucleic acid dye SYTO 9 (diluted 1:1000 in TSB from a SYTO 9 stock solution at 5 mM in DMSO; Invitrogen, France) were added to the 200 µL culture, obtain green fluorescent bacteria. For each strain, at least 9 to 15 wells were analyzed independently.

### 2.3. Macrocolony Experimental Conditions

For colony architectural formation, 3 µL of an overnight culture in TSB were inoculated on 1.5% Tryptone Soya Agar (TSA) with 40 µg/mL Congo Red (Sigma-Aldrich, St. Quentin Fallavier, France) and 20 µg/mL Coomassie Brilliant Blue (Sigma-Aldrich, St. Quentin Fallavier, France). Congo Red has been shown to bind extracellular matrix components and allows to compare the ability of different bacterial strains, including *B. subtilis*, which binds to amyloidic proteins [45,46]. The Coomassie Blue has a high affinity to bind proteins and is commonly used to detect, visualize, and quantify proteins separated on polyacrylamide gels [47,48]. The plates were then incubated at 30 °C for 6 days. Digital images of the colonies on the plates were taken using a Canon EOS 80D with 24 MP (6000 × 4000 pixels). Macrocolony experiments were performed three to five times independently.

### 2.4. Swarming Experiment Conditions

Cultures for the swarm inoculum were prepared in 10 mL B-medium inoculated with a single colony and shaken overnight at 37 °C. The culture was then diluted to an $OD_{600nm}$ of approximately 0.1 and grown until it reached an $OD_{600nm}$ of approximately 0.2. This procedure was repeated twice and finally the culture was grown to T4 (4 h after the transition from exponential growth). The $OD_{600nm}$ was measured and the culture was diluted, and 2 µL of diluted bacterial culture ($10^4$ CFU) were inoculated at the center of B-medium agar plate and incubated for 24 h at 30 °C with 50% relative humidity. Plates (9 cm diameter) containing 25 mL agar medium (0.7% agar) were prepared 1 h before inoculation and dried with lids open for 5 min before inoculation. Pictures were taken by a digital Nikon Coolpix P100 (10 MP) camera. Swarming experiments were repeated three to five times independently.

## 2.5. Pellicle Experiments

After an overnight culture in TSB at 30 °C, 10 µL of the bacterial suspension were used to inoculate 2 mL of TSB in 24-well plates (TPP, Trasadingen, Switzerland). Plates were then incubated at 30 °C for 24 h. Digital images of the pellicles were taken using a digital Nikon Coolpix P100 (10 MP) camera. This experiment was repeated three up to five times independently for each condition.

## 2.6. Non-Invasive Confocal Laser Scanning Microscopy (CLSM) of Submerged Biofilms

Immersed biofilms were observed using a Leica SP8 AOBS inverter confocal laser scanning microscope (CLSM, LEICA Microsystems, Wetzlar, Germany) at the INRAE MIMA2 platform (www6.jouy.inra.fr/mima2_eng/ accessed on 1 December 2020). For observation, strains were tagged fluorescently in green with SYTO 9 (1:1000 dilution in TSB), a nucleic acid marker. After 20 min of incubation in the dark at 30 °C to enable fluorescent labeling of the bacteria, plates were then mounted on the motorized stage of the confocal microscope. Biofilms on the bottom of the wells were scanned using a HC PL APO CS2 63x/1.2 water immersion objective lens. SYTO 9 excitation was performed at 488 nm with an argon laser, and the emitted fluorescence was recorded within the range 500–600 nm on hybrids detectors. The 3D (xyz) acquisitions were performed ($512 \times 512$ pixels, pixel size 0.361 µm, 1 image every z = 1 µm with a scan speed of 600 Hz). Easy 3D projections were constructed from Z-series images using IMARIS v9.0 software (Bitplane AG, Zurich, Switzerland). Biofilms biomass was estimated through extraction of the biofilm biovolume (in $\mu m^3/\mu m^2$) after isosurfaces automatic detection using the IMARIS quantification module from a minimum of twenty confocal image z-series.

## 2.7. Statistical Analysis

One-way ANOVA was performed using GraphPad Prism 8 software (GraphPad, CA, USA). Significance was defined as a $p$ value associated with a Fisher test value lower than 0.05.

## 3. Results and Discussion

### 3.1. Bacillus Subtilis NDmed forms Highly Structured Biofilms Compared to the NCIB3610 and 168 Strains

In the last decades, NCIB3610 has been widely used as a model for the "wild type" of *B. subtilis*. This strain has been shown to form more elaborate and robust biofilm communities when compared to the domesticated laboratory stain 168 [49,50]. However, in both the NCIB3610 and 168 strains, the *ypqP* gene is disrupted by the SPβ prophage, contrary to several sequenced natural isolates of *B. subtilis* [13]. This gene has been shown to be involved in the strong spatial organization of biofilms of the undomesticated *B. subtilis* NDmed strain, both at air and liquid interfaces [13]. In this study, a phenotypical characterization of NDmed grown under different laboratory culture conditions was performed, in comparison with the classical reference strains NCIB3610 and 168.

Macrocolonies of these strains were observed after being grown for 6 days on indicator plates containing both Congo Red (labeling amyloidic proteinaceous compounds in *B. subtilis* biofilm matrix) and Coomassie blue (proteinaceous matrix counterstain) [46,47]. As shown in Figure 1, the NDmed strain formed a highly structured and more compact macrocolony, contrary to the NCIB3610 and 168 strains that formed flat macrocolonies without or with only very fine wrinkles. In addition, the NDmed macrocolony was more intensely stained by the Congo Red, indicating a higher amount of exopolymeric substances and proteins produced compared to the two other strains.

As the biofilms formed by these three strains had such profound architectural differences, we wondered whether they might also present marked differences in another structured multicellular behavior i.e., swarming. Hence, to better visualize differences between them, semi-solid plates (swarming plates) were used as a 2D model to view bacterial surface colonization initiating from a macrocolony. Dendritic swarming pattern of *B. subtilis* was previously best characterized on a synthetic fully defined medium (B-medium) with optimized nutrient and temperature conditions [28]. Figure 1 shows the swarming patterns obtained on the synthetic B-medium (0.7% agar) after 24 h of incubation at 30 °C for the studied *B. subtilis* strains. Obviously, both NDmed and NCIB3610 showed swarming on B-medium but with varied dendritic patterns. NCIB3610 displayed a thin highly complex dendritic swarming pattern that spread all over the plate within 24 h of incubation, whereas NDmed swarmed with a thick countable less spread dendritic pattern. The mother colony of the NDmed appears to be highly structured with slimy texture when disrupted mechanically by a loop. On the other hand, a less structured widely spread mother colony was formed by NCIB3610, suggesting that less extracellular polymeric substances are produced in this strain compared to the NDmed strain. The mother colony in a swarm for both NDmed and NCIB3610 closely resembles the structural architecture of the macrocolonies formed. Consistent with previous observation, the 168 *B. subtilis* strain failed to swarm on this synthetic medium, essentially because of a frameshift mutation in the *sfp* gene, required for the surfactin biosynthesis that facilitates the migration over the surface by reducing the surface tension [27].

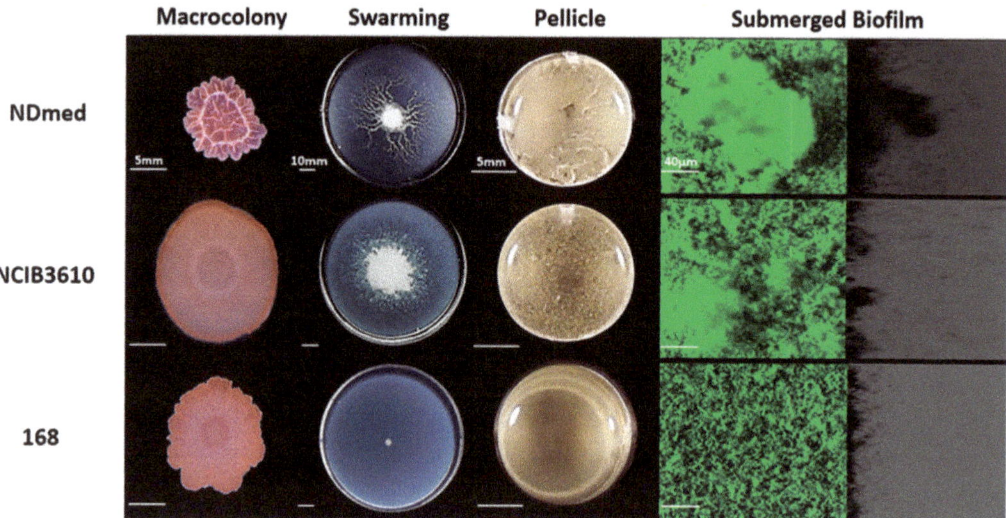


**Figure 1.** Comparative phenotype for *B. subtilis* strains on different laboratory assays. Macrocolonies grown on 1.5% TSA for 6 days at 30 °C after a central spot of 3 µL of an overnight bacterial culture in TSB (scale bars 5 mm). 0.7% of B-synthetic medium is used for swarming plates (9 cm diameter) that are incubated for 24 h at 30 °C (scale bars represent 10 mm). For pellicles, bacterial cells have been cultured in a 24-well plate with TSB for 24 h at 30 °C (scale bars 5 mm). Macrocolony, swarming, and pellicle images are representative for the majority of the phenotype from at least three replicates for each strain, which reveal variation for the surface architecture. In a 96 well microplate system, immersed biofilms are labeled by SYTO 9 after 24 h of incubation at 30 °C. The shadow on the right represents the virtual lateral shadow projection of the submerged biofilm (scale bars represent 40 µm).

Other models of biofilm are formed in liquid cultures, either at the air-liquid interface (pellicle) or as submerged biofilms at the solid-liquid interface [8,10–12]. To characterize the ability of *B. subtilis* to adhere and to form submerged multicellular communities on surface, CLSM has been used to acquire confocal stack images for the submerged biofilms, from which an Easy-3D reconstruction by the IMARIS software could reveal the three-dimensional structure with a lateral virtual shadow projection. As shown in Figure 1, and in accordance with previous reports, *B. subtilis* NDmed formed well-structured air-liquid biofilm (pellicle) and highly spatially organized submerged biofilm at the solid-liquid interface [11,13,30].

NCIB3610 strain did not form a thick pellicle within 24 h of incubation at 30 °C but produced well-structured biofilms (with a biovolume of 11 $\mu m^3/\mu m^2$, significantly smaller than the 14 $\mu m^3/\mu m^2$ biovolume formed by the NDmed, $p < 0.05$). The 168 strain, as previously been observed [11], was unable to form a pellicle in these conditions and displayed only a much less dense submerged biofilm (with a 6 $\mu m^3/\mu m^2$ biovolume) in comparison with the two other *B. subtilis* strains ($p < 0.05$).

In comparison between the three *B. subtilis* strains studied here, NDmed displayed complex architectural biofilm formation on/in both solid and liquid medium, and had the ability to swarm rather efficiently.

*3.2. Mutants Affected in Matrix-Producing Components Fail to Form Well-Firmed Surface Cohesive Biofilms*

In order to determine whether the genes involved in *B. subtilis* colony and pellicle formation also govern submerged biofilm formation, we constructed a set of derivative mutants of the NDmed strain and analyzed the corresponding phenotypes in the different biofilm models.

Extracellular matrix, mainly composed of polymeric substances, is essential for the biofilm structural formation. In *B. subtilis*, the amyloid-like fiber TasA encoded by the *tapA-sipW-tasA* operon, and the polysaccharides synthesized by the products of the *epsA-O* operon are mainly responsible for the synthesis of biofilm matrix, which bundles cells together and maintains their stability [2,46,49,51,52]. In addition, the BslA protein exhibits amphiphilic properties by forming a hydrophobic layer at the air interface [53] and activates the formation of complex colony development and pellicle formation [20,54]. Poly-γ-glutamate (γ-PGA), a secreted polymeric substance that accumulates in the culture media like the biofilm matrix [9] and in the capsule, is synthesized by the enzymes encoded by the *cap* operon. Recently it has been shown that in many tested environmental *B. subtilis* isolates γ-PGA production contributed to the complex morphology and robustness by enhancing cell-surface interactions of the colony biofilms [55]. The *ypqP* gene in both *B. subtilis* strains 168 and NCIB3610 is disrupted by the SPβ prophage, whose excision during sporulation phase reconstitutes a functional *ypqP* gene allowing addition of polysaccharides to the spore envelope [32]. In the undomesticated NDmed strain, *ypqP* non-disrupted by the SPβ prophage, has been identified as a requirement for the spatial biofilm organization [13].

Figure 2 shows the effect of matrix gene mutation on different laboratory culture assays. Macrocolonies formed by *tasA*, *epsA-0*, *bslA*, *cap*, and *ypqP* mutants on TSA agar medium were flat contrary to the highly structured and wrinkled wild type NDmed colony (Figure 2). Interestingly, the *tasA* mutant was the least stained, by proteinaceous dyes, indicating a drastic negative effect of the corresponding mutation on extracellular matrix production.

Effects of matrix gene mutations on surface motility were visualized through swarming plates. All mutants affected in matrix synthesis tested were observed to swarm better than the wild type NDmed strain after 24 h of incubation on minimal B-medium. The mother colony (place of bacterial inoculation) for the *tasA*, *epsA-O*, and *ypqP* mutants was producing a very viscous and loose matrix. This suggests that all together the TasA (amyloid-like fibers) with the exopolysaccharide synthesized (through the products of *epsA-O* and *ypqP*) are important for the cell interlock and the structural stability in a biofilm.

However, it is difficult to differentiate the importance of each gene individually on the biofilm structural formation on agar. Hence, submerged biofilms revealed how in the *tasA* and *epsA-O* mutants biofilm cells were clearly unbundled and unable to form structured biofilms (Figure 2). Submerged biofilm formed by the *bslA* mutant was not affected at all, and those formed by the *cap* and the *ypqP* mutants were quite less affected after 24 h of incubation. Such observation has been numerically confirmed by an estimation of the biovolume and the thickness of submerged biofilms formed for all NDmed mutants studied here, which are represented in Figure 5. Indeed, in this study the *ypqP* mutation had a less effect on submerged biofilm biovolume and thickness after 24 h of incubation, however, the effect was more drastic when compared to the wild type NDmed after 48 h of incubation [13]. Moreover, *ypqP* was slightly expressed after 24 h and strongly transcribed only after 48 h (our unpublished data). This could suggest that *ypqP* is involved in the late structural biofilm spatial organization.

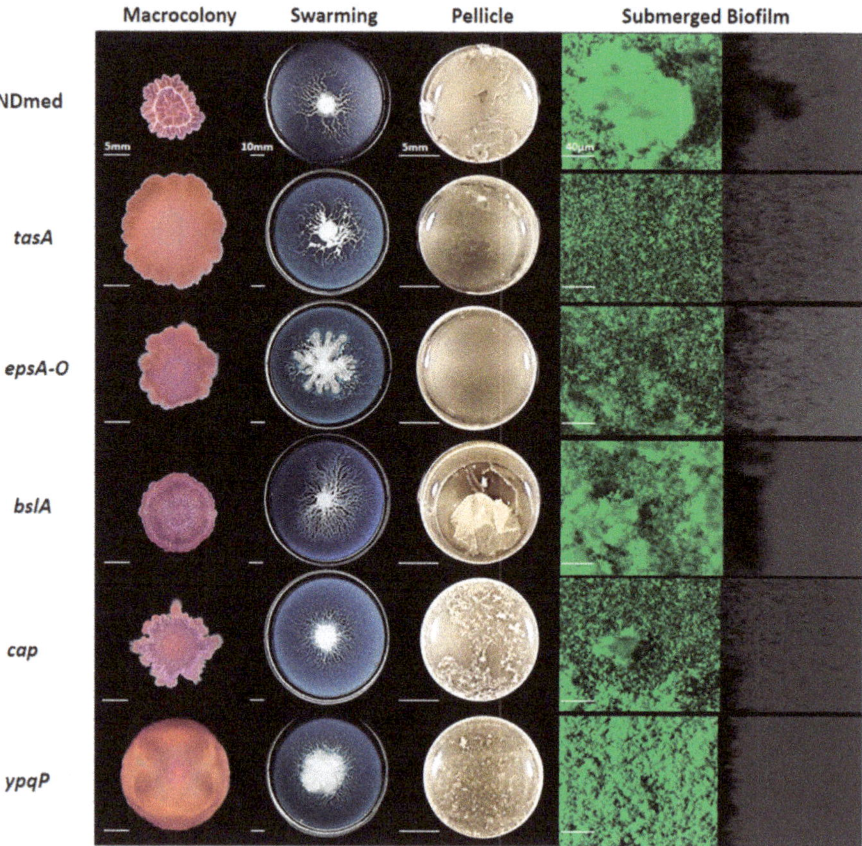



**Figure 2.** Different *B. subtilis* NDmed mutants of genes involved in extracellular matrix production on different culture assays. On 1.5% TSA, macrocolonies grown for 6 days at 30 °C after a central spot of 3 µL of an overnight bacterial culture in TSB. For swarming model, 2 µL of bacterial culture ($10^4$ bacterial dilution) have been inoculated on the middle of 0.7% B-medium plates and cultured for 24 h at 30 °C. In a 24-well plate, bacteria in TSB are cultured at 30 °C and pellicles were obtained after 24 h. Macrocolony, swarming, and pellicle images are representative for the majority of the phenotype from at least three replicates for each strain revealing the effect of mutations on the biofilm formation. In a microplate system, immersed biofilms are labeled by SYTO 9 after 24 h on incubation at 30 °C. The shadow on the right represents the vertical projection of the submerged biofilm (scale bars represent 40 µm).

Regarding biofilms formed on liquid-air interface, our observations also highlight the importance of amyloid fibers and exopolysaccharides in the biofilm formation. In rich medium after 24 h of incubation the *tasA* and *epsA-O* mutants could form only very thin delicate pellicle (Figure 2), similar to what has been shown by previous studies on *B. subtilis* NCIB3610 [46,52]. As for the *ypqP* and *cap* mutants a less structured pellicle was formed. On the other hand, a delicate pellicle formed by the *bslA* mutant was very fragile and sensitive to any small plate movement, and sank to the bottom of the well due to cells lacking the hydrophobic layer that allows the pellicle to be stable at the air-liquid interface. These results suggest that *tasA* and *epsA-O* are crucial matrix genes, required in architectural settlement of *B. subtilis* multicellular communities in the different biofilm models. The genes *cap*, *ypqP*, and *bslA* also play an important role in formation of a highly structured and stable biofilm but in a more model-dependent way.

### 3.3. Motility and Autolysins are Essentially Required for Architectural Submerged Biofilm Formation of B. subtilis NDmed

In the mid-exponential growth phase of *B. subtilis*, two populations of cells were described: individual motile cells, and long chains of sessile cells [56]. Motility is a way for bacteria to colonize more favorable niches. Bacterial motility has also a positive role in nascent biofilm maturation and spreading, as it has been shown that motile cells can create transient pores that increase the nutrient flow in the matrix of mature biofilms [57]. In *B. subtilis*, flagellar motility studies have focused on both swarming over semi-solid agar plates and swimming in liquid culture [27,28,56,58]. As previously shown, *B. subtilis hag* mutants, affected in a gene encoding flagellin protein for flagellum formation, fail to swarm over different media tested including the B-medium [27,29]. In liquid culture, *B. subtilis hag* mutant was shown to have a delayed flagellar formation [10,58].

In Figure 3, the NDmed *hag* mutant formed a slightly wrinkled macrocolony on agar plate, while it failed to swarm on an optimal semi-solid plate. In static liquid culture after 24 h of incubation, this *hag* mutant was able to produce non-structured submerged compact biofilm with diminished thickness unaffecting the biovolume at the solid-liquid interface (Figure 5). Nevertheless, the *hag* mutant did not form pellicle at the air-liquid interface after 24 h of incubation in a rich medium (TSB). This suggested that the inability to swim prevented the cells to reach the air-liquid interface and thus inhibited or caused a delay in the formation of a pellicle, as previously observed [10].

For efficient growth and motility, bacteria need to continuously divide and adapt the cell wall composition (peptidoglycan), thanks to the autolysin system in *B. subtilis*. Expression of two major autolysin genes, *lytF* and *lytC* involved in cell separation is controlled by sigma factor D that also directs the transcription of motility and chemotaxis genes [24,59]. We have studied the effect of *lytF* and *lytABC* mutation on the different assays of biofilm formation (Figure 3). The NDmed *lytF* mutants showed better aerial (macrocolony and pellicle) biofilm formation than the *lytABC* autolysin mutant, that formed flat and pale color macrocolony (due to the reduced autolytic enzymes produced). However, in submerged biofilm, the *lytF* mutant was more affected and showed reduced biovolume (Figure 5A; $p < 0.05$) while the biofilm biovolume of the *lytABC* mutant was only slightly decreased. To look at the effect on motility, we have tested these mutant strains on swarming plates. Similarly to previous observation with *B. subtilis* NCIB3610 strain [59] the *lytF* mutant was able to swarm better than the *lytABC* mutant, which led to the proposition that *lytF* is principally dedicated in cell separation and *lytC* is more involved in the proper flagellar function [59]. Hence, among the different autolysins, encoded by more than 35 genes encoding peptidoglycan hydrolases, inactivation of only one gene will have an impact on one of the biofilm models studied. However, absolute long chain cells phenotype could not be always seen, since different autolysins could replace each other [24,60].

Interestingly, the *srfAA* mutation, affecting surfactin production and competence, has no effect on the structural biofilms developed as macrocolonies, pellicle, and submerged one (Figure 3) when compared to the wild type *B. subtilis* NDmed ($p > 0.05$). On swarming plates, surfactin production reduces surface tension during bacterial surface translocation. The 168 strain, carrying a frame-shift mutation in *sfp*, fails to produce surfactin and is thus unable to migrate over the B-medium swarming plate [27,29]. Moreover, studies with the NCIB3610 *sfrAA* mutant have also shown its inability to swarm [61]. However, either 168 or NCIB3610 *srfAA* mutants, have been shown to regain the ability of swarming, when provided with exogenous surfactin [27,61]. Interestingly, in our study, the NDmed *srfAA* mutant, which lacks a surfactin ring, displayed a monolayer dendritic swarming pattern having migrated from a more viscous mother colony, suggesting that other extracellular proteases have been secreted to facilitate the translocation.

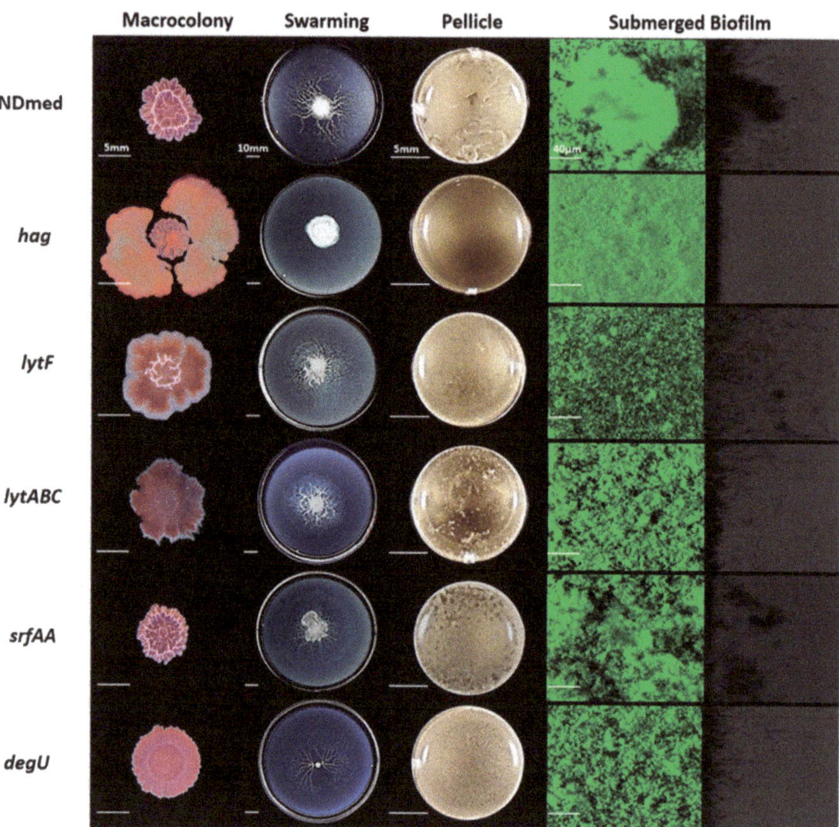



**Figure 3.** Motility and autolysin genes mutants of *B. subtilis* NDmed strain on different laboratory assays. Macrocolonies for mutated regulator genes are cultured on TSA for 6 days at 30 °C. Swarming plates are formed on B-synthetic medium (0.7% agar) that are cultured for 24 h at 30 °C. Pellicle and submerged biofilms were formed after 24 h of incubation at 30 °C in TSB medium. For submerged images the scale bars represent 40 µm. Macrocolony, swarming, and pellicle images are representative for the majority of the phenotype from at least three replicates for each strain revealing the effect of mutations on the biofilm formation.

Previous studies have shown that mutation of *degU* affects transcription of more than 200 genes, which intervene in the genetic network activation for both flagellum and biofilm formation [54]. It has been demonstrated that different levels of DegU~P co-ordinates *B. subtilis* multicellular behavior i.e., low level of DegU~P activates swarming motility and complex architectural colony formation whereas high level of DegU~P inhibits swarming and complex colony formation and is mainly required for the activation of exoprotease production [54,62]. In *B. subtilis* NCIB3610, DegU targets two proteins that have been shown to be involved in biofilm formation, YuaB (BslA) and YvcA (a putative membrane-bound lipoprotein). However, for the *B. subtilis* ATCC6051 strain, highly genetically similar to NCIB3610 (they are both descending from the original Marburg strain [37]), YvcA has been shown to be required only for complex colony formation but not for pellicle formation [20,54,62]. Hence, multicellular communities differ from strain to strain, which highlights the interest to test *degU* mutation affecting the undomesticated strain NDmed and observe its effect over the different laboratory assays (Figure 3). Such *degU* mutation has a negative impact on the complex architectural macrocolony formed on agar surface and only slightly affects the biovolume formed by the submerged biofilm (Figure 5A,

$p > 0.05$). A slight delay was observed in the swarming motility as well as for the pellicle formation indicating that a complex regulatory network, like phosphorylated Spo0A [20], intervenes to ensure a comparable biofilm formation.

### 3.4. Mutation of B. subtilis NDmed Biofilm Regulators do Not Have the Same Impact on All Biofilm Models

Spo0A, a key regulator of biofilm formation, is driven by exogenous and endogenous signals [63]. Activated Spo0A governs the genetic pathway controlling the matrix production gene expression by inducing SinI which binds and inhibits SinR, a repressor of the *eps* and *tapA-sipW-tasA* operons. Another role for Spo0A is to repress the expression of AbrB, a negative regulator for the initiation of biofilm formation [8,64]. Hence, the transition from surface-attached cells to three-dimensional biofilm structure is dependent on the activated Spo0A regulator [8]. To determine and clearly visualize the effect of these regulators on biofilm formation, *spo0A, abrB, sinR, sinI*, and *slrR* mutants of NDmed were tested under different biofilm culture conditions (Figure 4).

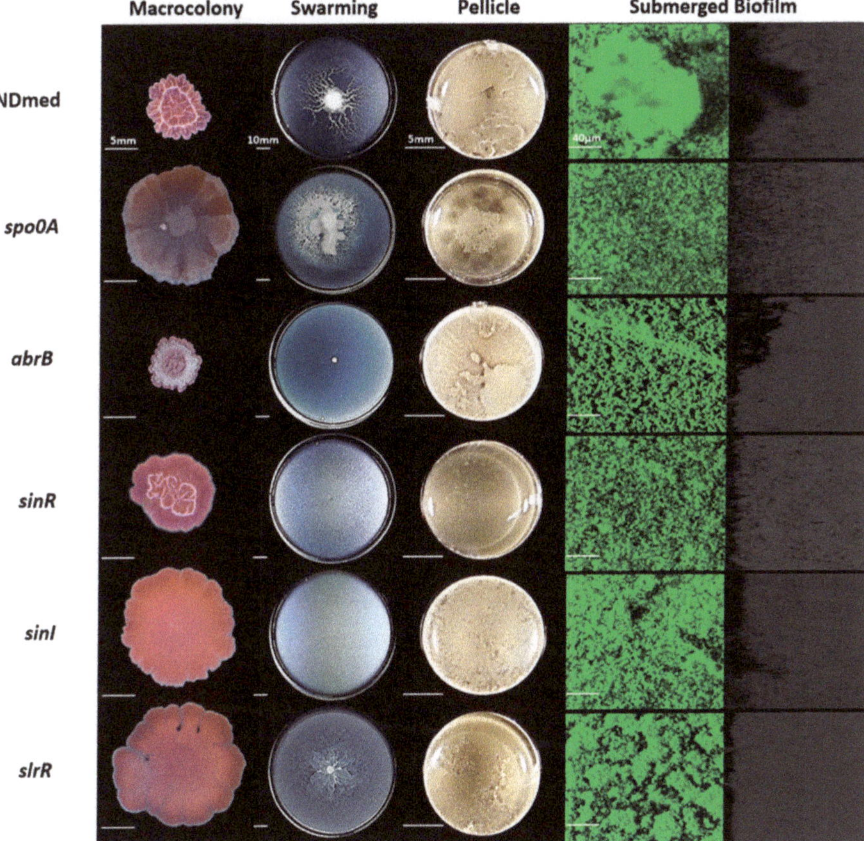


**Figure 4.** Mutational effect of global regulators required for biofilm development. Macrocolonies for mutants of regulator genes have been cultured on TSA for 6 days at 30 °C after a central spot of 3µl of an overnight bacterial culture in TSB. Swarming plates are formed by B-synthetic medium (0.7% agar) incubated for 24 h at 30 °C. Pellicle formed after 24 h of incubation at 30 °C in TSB medium. Macrocolony, swarming, and pellicle images are representative for the majority of the phenotype from at least three replicates for each strain revealing the effect of mutations on the biofilm formation. In a microplate system, immersed biofilms are labeled by SYTO 9 after 24 h on incubation at 30 °C. The shadow on the right represents the vertical projection of the submerged biofilm (scale bars represent 40 µm).

The *spo0A* mutant grew as a structureless spread macrocolony, while the *abrB* mutant showed a very vigorous and structured macrocolony on solid agar medium (Figure 4). In liquid culture, previous studies have shown that *B. subtilis spo0A* mutant cells were able to adhere to a surface and attach only as a monolayer form, suggesting that these mutants lack cell-cell interactions necessary for multicellular biofilm formation [8]. By using the CLSM, we could observe that the *spo0A* mutant cells did not form any thick submerged biofilm and rather remained essentially dispersed in the medium (Figure 4). These dispersed cells seemed to reach the surface of the liquid-air interface and form a highly disconnected pellicle-like structure in the middle of the well. On the other hand, the *abrB* mutant could form an extremely firm and highly structured pellicle, even more than that formed by the wild type NDmed strain, as well as thick highly structured architectural submerged biofilm (Figures 4 and 5B). Quantification of the submerged biofilm biovolumes (Figure 5A) formed by the *spo0A* and *abrB* mutants assures the role of Spo0A/AbrB pair as the main regulator for biofilm formation. On swarming plates, the *abrB* mutant was strongly affected, where even though producing an extensive surfactin zone, it was only able to form few small bud-like structures that emerged from the mother colony and then failed to proceed further. A similar behavior was observed for the *abrB* mutant of *B. subtilis* (168 *sfp+*) whose cells within the bud accumulate as long-chain forms [29]. Besides this, we could observe that the *spo0A* mutant on the swarming plates (Figure 4) showed extensive motility that filled all the plate rapidly with viscous multicellular biofilm formation in the middle of the plate. This could indicate that this viscous layer is due to an extensive secretion of surfactin or of extracellular proteases from a huge number of bacterial cells that lack cell-cell interaction, facilitating the movement over the surface.

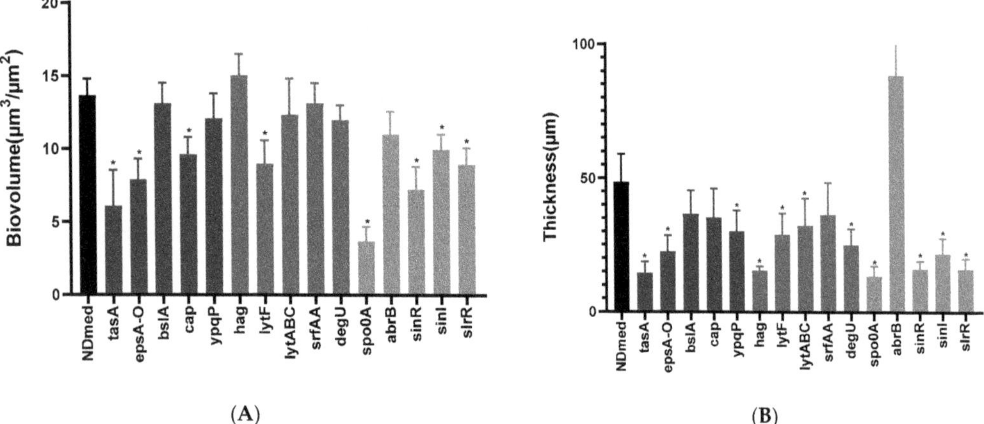

(**A**)  (**B**)

**Figure 5.** Submerged biofilms biovolumes and maximum thickness of the *B. subtilis* NDmed mutant strains studied. Biovolume (**A**) and maximum thickness (**B**) obtained were calculated from twenty confocal image series each. The color of the bars indicate gene categories (black for wild type NDmed, dark grey for matrix genes, grey for motility and autolytic genes, and light grey for global regulators). The error bars indicate the 95% confidence level, and the asterisk indicates the statistically significant differences (* is for $p < 0.05$) with the NDmed wild-type strain.

Biofilm formation, under appropriate conditions, is initiated by motile *B. subtilis* cells that adhere to the surface become sessile and form long chains of non-motile cells, held together by extracellular matrix. The transcription factor SinR, a central regulator in the assembly of *B. subtilis* cells into multicellular communities [17], controls both motility and biofilm formation by directly repressing the *eps* and *tapA-sipW-tasA* operons [65]. SinI, induced by phosphorylated Spo0A, binds directly to SinR and causes its inhibition. Moreover, SinI derepresses the action of SlrR [18,66]. SlrR, an additional regulatory protein, binds to and antagonizes SlrA, and thus constitutes a negative regulatory double loop with

SinR, in which the *slrR* gene is repressed by SinR and in turn SlrR prevents SinR from repressing *slrR* [16,67]. SlrA could play only a minor role in biofilm formation; however, it can be substituted functionally by SinI, its equivalent paralog [16,18]. Hence, SinR is inhibited by two paralogous antirepressors, SinI and SlrA [16].

A *sinR* mutation, in the NCIB3610 strain has been shown to lead to the formation of extremely thick colony when compared to the wild type, while *sinI* or *slrR* mutants formed flat structureless colonies on agar surface [17,65]. We have investigated the role of these major gene regulators on submerged biofilm formation and motility in the *B. subtilis* NDmed strain. Figure 4 shows similar phenotype for both *sinI* and *slrR* mutants with flat structureless macrocolonies on agar surface; however, the *sinR* mutant formed wrinkled macrocolony less structured than that formed by the wild type.

Swarming is a phenomenon taking place in two consecutive stages, migration over the surface of highly motile cells followed by their differentiation to less motile matrix producing cells that become stacked in a three-dimensional structure [26,68]. On swarming plates and after 24 hr of incubation, the NDmed *sinR* mutant swarmed all over the plate with a multilayered biofilm dendritic pattern, which could indicate that swarming cells are unable to separate. In contrast, the NDmed *sinI* mutant eventually swarmed all over the plate in a monolayer form (Figure 4) similar to what has been described for *sinI* mutant in the NCIB3610 context [17]. This suggests that when matrix production genes are blocked, mutant bacterial strains were only able to reach the first stage of swarming. SlrR stimulates transcription of the *tapA-sipW-tasA* operon but not of the *eps* operon and represses genes that mediate cell separation [10,18]. Thus, *slrR* mutation affects the expression of TasA but not Eps production and promotes cell separation. On swarming plates, the NDmed *slrR* mutant was able to swarm rapidly in a monolayer form all over the plate with less structured biofilm in the mother colony (place of inoculation) when compared to the wild type (Figure 4).

In liquid culture, a NDmed *sinR* mutant cultivated in TSB medium for 24 h of incubation, formed very thin pellicle (Figure 4). This could be due to cells unable to reach easily the surface. The NDmed *sinI* mutant was able to form a rather good pellicle, suggesting that the motile swimmer cells were able to reach the surface. These phenotypes are similar to what has been observed previously for ATCC6051 *sinR* and *sinI* mutant strains [10,18]. A defect in flagellar formation in the *sinR* mutant [10,18] and a functional complementation between SinI and SlrA [16] in the *sinI* mutant could account for these phenotypes. Another hypothesis could be the occurrence of natural frameshift mutations within the *sinR* open reading frame, which suppress the blocking biofilm formation effects of a *sinI* mutation, as shown by Kearns *et al.* [17]. A NDmed *slrR* mutant could form a thin pellicle at the air-liquid interface, similarly to what has been observed in the NCIB3610 context [65].

The submerged biofilm biovolume of the NDmed *sinR* mutant (Figure 5A) was more negatively affected than that of the *sinI* or the *slrR* mutants when compared to the wild type NDmed (with a $p < 0.05$ for these three mutated genes compared to the wild type). This could stress the importance of motility and autolysin in the formation of biofilm and suggest that mutation in one gene could be overcome and controlled by other regulatory pathways. Thus, these results further indicate that the SinI/SinR pair are the main regulators controlling the mode of bacterial life, motile or sessile, cells.

## 4. Conclusions

Overall, this study highlights the value of the NDmed strain as an undomesticated, naturally competent *B. subtilis* isolate, to point out the effect of gene mutation on the different structural biofilm communities formed. Gene mutation could exhibit a similar impact on all the different biofilm models formed on different culturing conditions. For instance, the *tasA* and *epsA-O* gene mutation affected all the surface associated communities formed but improved surface translocation. However, the *bslA* gene mutation has a negative effect just on the aerial biofilm models, structural microcolonies, and the pellicle stability, and no effect on the submerged biofilm formation. Our results emphasize the importance of the submerged model to further understand the molecular mechanisms during biofilm formation. Biofilm development throughout different environmental culturing conditions could have similar genetic profile, but these multicellular communities can also display considerable differences on the structural, chemical, and biological heterogeneity levels across different biofilm models. A whole transcriptional analysis could be done for the differently localized heterogeneous compartments of a biofilm to further understand the core of the transcriptional network that takes place between and during the biofilm development.

**Author Contributions:** Conceptualization, R.B., D.L.C., P.S.-V., Y.D. and K.H.; methodology, Y.D., P.S.-V., and J.D.; formal analysis, Y.D.; writing—original draft preparation, Y.D.; writing—review and editing, R.B., K.H., D.L.C., A.B., P.S.-V. and J.D.; supervision, R.B. and K.H.; funding acquisition, K.H. and R.B. All authors have read and agreed to the published version of the manuscript.

**Funding:** Yasmine Dergham is the recipient of fundings from the Union of Southern Suburbs Municipalities of Beirut, INRAE, Campus France PHC CEDRE 42280PF and Fondation AgroParisTech.

**Data Availability Statement:** Not applicable.

**Acknowledgments:** We thank Arnaud Chastanet, Etienne Dervyn, Michal Obuchowski, Nicola R. Stanley-Wall, and Roberto Kolter for providing strains, and the MIMA2 INRAE imaging platform for microscopic analysis.

**Conflicts of Interest:** The authors declare no conflict of interest.

## References


1. Shapiro, J.A. Thinking about bacterial populations as multicellular organisms. *Annu. Rev. Microbiol.* **1998**, *52*, 81–104. [CrossRef]
2. Vlamakis, H.; Chai, Y.; Beauregard, P.B.; Losick, R.; Kolter, R. Sticking together: Building a biofilm the *Bacillus subtilis* way. *Nat. Rev. Microbiol.* **2013**, *11*, 157–168. [CrossRef]
3. Kearns, D.B. A field guide to bacterial swarming motility. *Nat. Rev. Microbiol.* **2010**, *8*, 634–644. [CrossRef] [PubMed]
4. Verstraeten, N.; Braeken, K.; Debkumari, B.; Fauvart, M.; Fransaer, J.; Vermant, J.; Michiels, J. Living on a surface: Swarming and biofilm formation. *Trends Microbiol.* **2008**, *16*, 496–506. [CrossRef] [PubMed]
5. Lemon, K.P.; Earl, A.M.; Vlamakis, H.C.; Aguilar, C.; Kolter, R. Biofilm Development with an Emphasis on *Bacillus subtilis*. *Curr. Top. Microbiol. Immunol.* **2008**, *322*, 1–16. [CrossRef] [PubMed]
6. Nagórska, K.; Bikowski, M.; Obuchowski, M. Multicellular behaviour and production of a wide variety of toxic substances support usage of *Bacillus subtilis* as a powerful biocontrol agent. *Acta Biochim. Pol.* **2007**, *54*, 495–508. [CrossRef]
7. Jamal, M.; Ahmad, W.; Andleeb, S.; Jalil, F.; Imran, M.; Nawaz, M.A.; Hussain, T.; Ali, M.; Rafiq, M.; Kamil, M.A. Bacterial biofilm and associated infections. *J. Chin. Med. Assoc.* **2018**, *81*, 7–11. [CrossRef]
8. Hamon, M.A.; Lazazzera, B.A. The sporulation transcription factor Spo0A is required for biofilm development in *Bacillus subtilis*. *Mol. Microbiol.* **2002**, *42*, 1199–1209. [CrossRef]
9. Stanley, N.R.; Lazazzera, B.A. Defining the genetic differences between wild and domestic strains of *Bacillus subtilis* that affect poly-γ-dl-glutamic acid production and biofilm formation. *Mol. Microbiol.* **2005**, *57*, 1143–1158. [CrossRef] [PubMed]
10. Kobayashi, K. *Bacillus subtilis* Pellicle Formation Proceeds through Genetically Defined Morphological Changes. *J. Bacteriol.* **2007**, *189*, 4920–4931. [CrossRef]
11. Bridier, A.; Le Coq, D.; Dubois-Brissonnet, F.; Thomas, V.; Aymerich, S.; Briandet, R. The Spatial Architecture of *Bacillus subtilis* Biofilms Deciphered Using a Surface-Associated Model and In Situ Imaging. *PLoS ONE* **2011**, *6*, e16177. [CrossRef] [PubMed]
12. Hamon, M.A.; Stanley, N.R.; Britton, R.A.; Grossman, A.D.; Lazazzera, B.A. Identification of AbrB-regulated genes involved in biofilmformation by *Bacillus subtilis*. *Mol. Microbiol.* **2004**, *52*, 847–860. [CrossRef] [PubMed]
13. Sanchez-Vizuete, P.; Le Coq, D.; Bridier, A.; Herry, J.M.; Aymerich, S.; Briandet, R. Identification of *ypqP* as a new *Bacillus subtilis* biofilm determinant that mediates the protection of *Staphylococcus aureus* against antimicrobial agents in mixed-species communities. *Appl. Environ. Microbiol.* **2015**, *81*, 109–118. [CrossRef]
14. Ostrov, I.; Sela, N.; Belausov, E.; Steinberg, D.; Shemesh, M. Adaptation of Bacillus species to dairy associated environment facilitates their biofilm forming ability. *Food Microbiol.* **2019**, *82*, 316–324. [CrossRef] [PubMed]
15. Chu, F.; Kearns, D.B.; Branda, S.S.; Kolter, R.; Losick, R. Targets of the master regulator of biofilm formation in *Bacillus subtilis*. *Mol. Microbiol.* **2006**, *59*, 1216–1228. [CrossRef] [PubMed]
16. Chai, Y.; Kolter, R.; Losick, R. Paralogous antirepressors acting on the master regulator for biofilm formation in *Bacillus subtilis*. *Mol. Microbiol.* **2009**, *74*, 876–887. [CrossRef]
17. Kearns, D.B.; Chu, F.; Branda, S.S.; Kolter, R.; Losick, R. A master regulator for biofilm formation by *Bacillus subtilis*. *Mol. Microbiol.* **2004**, *55*, 739–749. [CrossRef]
18. Kobayashi, K. SlrR/SlrA controls the initiation of biofilm formation in *Bacillus subtilis*. *Mol. Microbiol.* **2008**, *69*, 1399–1410. [CrossRef]

19. Murray, E.J.; Kiley, T.B.; Stanley-Wall, N.R. A pivotal role for the response regulator DegU in controlling multicellular behaviour. *Microbiology* **2009**, *155*, 1–8. [CrossRef]

20. Verhamme, D.T.; Murray, E.J.; Stanley-Wall, N.R. DegU and Spo0A Jointly Control Transcription of Two Loci Required for Complex Colony Development by *Bacillus subtilis*. *J. Bacteriol.* **2008**, *191*, 100–108. [CrossRef] [PubMed]

21. Márquez-Magaña, L.M.; Chamberlin, M.J. Characterization of the *sigD* transcription unit of *Bacillus subtilis*. *J. Bacteriol.* **1994**, *176*, 2427–2434. [CrossRef] [PubMed]

22. Marquez, L.M.; Helmann, J.D.; Ferrari, E.; Parker, H.M.; Ordal, G.W.; Chamberlin, M.J. Studies of σ(D)-dependent functions in *Bacillus subtilis*. *J. Bacteriol.* **1990**, *172*, 3435–3443. [CrossRef]

23. Serizawa, M.; Yamamoto, H.; Yamaguchi, H.; Fujita, Y.; Kobayashi, K.; Ogasawara, N.; Sekiguchi, J. Systematic analysis of SigD-regulated genes in *Bacillus subtilis* by DNA microarray and Northern blotting analyses. *Gene* **2004**, *329*, 125–136. [CrossRef]

24. Margot, P.; Pagni, M.; Karamata, D. *Bacillus subtilis* 168 gene *lytF* encodes a γ-D-glutamate-meso-diaminopimelate muropeptidase expressed by the alternative vegetative sigma factor, σD. *Microbiology* **1999**, *145*, 57–65. [CrossRef] [PubMed]

25. Ostrowski, A.; Mehert, A.; Prescott, A.; Kiley, T.B.; Stanley-Wall, N.R. YuaB Functions Synergistically with the Exopolysaccharide and TasA Amyloid Fibers to Allow Biofilm Formation by *Bacillus subtilis*. *J. Bacteriol.* **2011**, *193*, 4821–4831. [CrossRef] [PubMed]

26. Harshey, R.M. Bacterial Motility on a Surface: Many Ways to a Common Goal. *Annu. Rev. Microbiol.* **2003**, *57*, 249–273. [CrossRef] [PubMed]

27. Julkowska, D.; Obuchowski, M.; Holland, I.B.; Séror, S.J. Comparative Analysis of the Development of Swarming Communities of *Bacillus subtilis* 168 and a Natural Wild Type: Critical Effects of Surfactin and the Composition of the Medium. *Society* **2005**, *187*, 65–76. [CrossRef]

28. Julkowska, D.; Obuchowski, M.; Holland, I.B.; Séror, S.J. Branched swarming patterns on a synthetic medium formed by wild-type *Bacillus subtilis* strain 3610: Detection of different cellular morphologies and constellations of cells as the complex architecture develops. *Microbiology* **2004**, *150*, 1839–1849. [CrossRef]

29. Hamze, K.; Julkowska, D.; Autret, S.; Hinc, K.; Nagorska, K.; Sekowska, A.; Holland, I.B.; Séror, S.J. Identification of genes required for different stages of dendritic swarming in *Bacillus subtilis*, with a novel role for *phrC*. *Microbiology* **2009**, *155*, 398–412. [CrossRef]

30. Bridier, A.; Sanchez-Vizuete, M.d.P.; Le Coq, D.; Aymerich, S.; Meylheuc, T.; Maillard, J.-Y.; Thomas, V.; Dubois-Brissonnet, F.; Briandet, R. Biofilms of a *Bacillus subtilis* Hospital Isolate Protect *Staphylococcus aureus* from Biocide Action. *PLoS ONE* **2012**, *7*, e44506. [CrossRef]

31. Martin, D.; Denyer, S.; McDonnell, G.; Maillard, J.-Y. Resistance and cross-resistance to oxidising agents of bacterial isolates from endoscope washer disinfectors. *J. Hosp. Infect.* **2008**, *69*, 377–383. [CrossRef]

32. Abe, K.; Kawano, Y.; Iwamoto, K.; Arai, K.; Maruyama, Y.; Eichenberger, P.; Sato, T. Developmentally-Regulated Excision of the SPβ Prophage Reconstitutes a Gene Required for Spore Envelope Maturation in *Bacillus subtilis*. *PLoS Genet.* **2014**, *10*, e1004636. [CrossRef] [PubMed]

33. Marmur, J. A procedure for the isolation of deoxyribonucleic acid from micro-organisms. *J. Mol. Biol.* **1961**, *3*, 208–218. [CrossRef]

34. Anagnostopoulos, C.; Spizizen, J. Requirements for Transformation in *Bacillus subtilis*. *J. Bacteriol.* **1961**, *81*, 741–746. [CrossRef]

35. Borenstein, S.; Ephrati-Elizur, E. Spontaneous release of DNA in sequential genetic order by *Bacillus subtilis*. *J. Mol. Biol.* **1969**, *45*, 137–152. [CrossRef]

36. Antelmann, H.; Engelmann, S.; Schmid, R.; Sorokin, A.; Lapidus, A.; Hecker, M. Expression of a stress- and starvation-induced *dps/pexB*-homologous gene is controlled by the alternative sigma factor σ(B) in *Bacillus subtilis*. *J. Bacteriol.* **1997**, *179*, 7251–7256. [CrossRef] [PubMed]

37. Zeigler, D.R.; Prágai, Z.; Rodriguez, S.; Chevreux, B.; Muffler, A.; Albert, T.; Bai, R.; Wyss, M.; Perkins, J.B. The Origins of 168, W23, and Other *Bacillus subtilis* Legacy Strains. *J. Bacteriol.* **2008**, *190*, 6983–6995. [CrossRef]

38. Bidnenko, V.; Nicolas, P.; Grylak-Mielnicka, A.; Delumeau, O.; Auger, S.; Aucouturier, A.; Guérin, C.; Repoila, F.; Bardowski, J.; Aymerich, S.; et al. Termination factor Rho: From the control of pervasive transcription to cell fate determination in *Bacillus subtilis*. *PLoS Genet.* **2017**, *13*, e1006909. [CrossRef]

39. Monteferrante, C.G.; MacKichan, C.; Marchadier, E.; Prejean, M.-V.; Van Dijl, J.M.; Carballido-López, R. Mapping the twin-arginine protein translocation network of *Bacillus subtilis*. *Proteomics* **2013**, *13*, 800–811. [CrossRef]

40. Nagórska, K.; Hinc, K.; Strauch, M.A.; Obuchowski, M. Influence of the σB Stress Factor and *yxaB*, the Gene for a Putative Exopolysaccharide Synthase under σB Control, on Biofilm Formation. *J. Bacteriol.* **2008**, *190*, 3546–3556. [CrossRef] [PubMed]

41. Crutz, A.M.; Steinmetz, M. Transcription of the *Bacillus subtilis sacX* and *sacY* genes, encoding regulators of sucrose metabolism, is both inducible by sucrose and controlled by the DegS-DegU signalling system. *J. Bacteriol.* **1992**, *174*, 6087–6095. [CrossRef] [PubMed]

42. Murray, E.J.; Stanley-Wall, N.R. The sensitivity of *Bacillus subtilis* to diverse antimicrobial compounds is influenced by Abh. *Arch. Microbiol.* **2010**, *192*, 1059–1067. [CrossRef]

43. Dervyn, E.; Poncet, S.; Klier, A.; Rapoport, G. Transcriptional regulation of the cryIVD gene operon from *Bacillus thuringiensis* subsp. israelensis. *J. Bacteriol.* **1995**, *177*, 2283–2291. [CrossRef] [PubMed]

44. Bridier, A.; Dubois-Brissonnet, F.; Boubetra, A.; Thomas, V.; Briandet, R. The biofilm architecture of sixty opportunistic pathogens deciphered using a high throughput CLSM method. *J. Microbiol. Methods* **2010**, *82*, 64–70. [CrossRef] [PubMed]

45. Friedman, L.; Kolter, R. Genes involved in matrix formation in *Pseudomonas aeruginosa* PA14 biofilms. *Mol. Microbiol.* **2003**, *51*, 675–690. [CrossRef]

46. Romero, D.; Aguilar, C.; Losick, R.; Kolter, R. Amyloid fibers provide structural integrity to *Bacillus subtilis* biofilms. *Proc. Natl. Acad. Sci. USA* **2010**, *107*, 2230–2234. [CrossRef]

47. Kiersztyn, B.; Siuda, W.; Chróst, R. Coomassie Blue G250 for Visualization of Active Bacteria from Lake Environment and Culture. *Pol. J. Microbiol.* **2017**, *66*, 365–373. [CrossRef]

48. Neumann, U.; Khalaf, H.; Rimpler, M. Quantitation of electrophoretically separated proteins in the submicrogram range by dye elution. *Electrophoresis* **1994**, *15*, 916–921. [CrossRef]

49. Branda, S.S.; González-Pastor, J.E.; Ben-Yehuda, S.; Losick, R.; Kolter, R. Fruiting body formation by *Bacillus subtilis*. *Proc. Natl. Acad. Sci. USA* **2001**, *98*, 11621–11626. [CrossRef] [PubMed]

50. McLoon, A.L.; Guttenplan, S.B.; Kearns, D.B.; Kolter, R.; Losick, R. Tracing the Domestication of a Biofilm-Forming Bacterium. *J. Bacteriol.* **2011**, *193*, 2027–2034. [CrossRef]

51. Branda, S.S.; González-Pastor, J.E.; Dervyn, E.; Ehrlich, S.D.; Losick, R.; Kolter, R. Genes Involved in Formation of Structured Multicellular Communities by *Bacillus subtilis*. *J. Bacteriol.* **2004**, *186*, 3970–3979. [CrossRef] [PubMed]

52. Branda, S.S.; Chu, F.; Kearns, D.B.; Losick, R.; Kolter, R. A major protein component of the *Bacillus subtilis* biofilm matrix. *Mol. Microbiol.* **2006**, *59*, 1229–1238. [CrossRef] [PubMed]

53. Kobayashi, K.; Iwano, M. BslA(YuaB) forms a hydrophobic layer on the surface of *Bacillus subtilis* biofilms. *Mol. Microbiol.* **2012**, *85*, 51–66. [CrossRef]

54. Kobayashi, K. Gradual activation of the response regulator DegU controls serial expression of genes for flagellum formation and biofilm formation in *Bacillus subtilis*. *Mol. Microbiol.* **2007**, *66*, 395–409. [CrossRef] [PubMed]

55. Yu, Y.; Yan, F.; Chen, Y.; Jin, C.; Guo, J.-H.; Chai, Y. Poly-γ-Glutamic Acids Contribute to Biofilm Formation and Plant Root Colonization in Selected Environmental Isolates of *Bacillus subtilis*. *Front. Microbiol.* **2016**, *7*, 1811. [CrossRef]

56. Kearns, D.B.; Losick, R. Cell population heterogeneity during growth of *Bacillus subtilis*. *Genes Dev.* **2005**, *19*, 3083–3094. [CrossRef] [PubMed]

57. Houry, A.; Gohar, M.; Deschamps, J.; Tischenko, E.; Aymerich, S.; Gruss, A.; Briandet, R. Bacterial swimmers that infiltrate and take over the biofilm matrix. *Proc. Natl. Acad. Sci. USA* **2012**, *109*, 13088–13093. [CrossRef]

58. Hölscher, T.; Bartels, B.; Lin, Y.-C.; Gallegos-Monterrosa, R.; Price-Whelan, A.; Kolter, R.; Dietrich, L.E.P.; Kovács, Á.T. Motility, Chemotaxis and Aerotaxis Contribute to Competitiveness during Bacterial Pellicle Biofilm Development. *J. Mol. Biol.* **2015**, *427*, 3695–3708. [CrossRef]

59. Chen, R.; Guttenplan, S.B.; Blair, K.M.; Kearns, D.B. Role of the σD-dependent autolysins in *Bacillus subtilis* population heterogeneity. *J. Bacteriol.* **2009**, *191*, 5775–5784. [CrossRef]

60. Smith, T.J.; Foster, S.J. Characterization of the involvement of two compensatory autolysins in mother cell lysis during sporulation of *Bacillus subtilis* 168. *J. Bacteriol.* **1995**, *177*, 3855–3862. [CrossRef]

61. Kearns, D.B.; Chu, F.; Rudner, R.; Losick, R. Genes governing swarming in *Bacillus subtilis* and evidence for a phase variation mechanism controlling surface motility. *Mol. Microbiol.* **2004**, *52*, 357–369. [CrossRef] [PubMed]

62. Verhamme, D.T.; Kiley, T.B.; Stanley-Wall, N.R. DegU co-ordinates multicellular behaviour exhibited by *Bacillus subtilis*. *Mol. Microbiol.* **2007**, *65*, 554–568. [CrossRef] [PubMed]

63. Mielich-süss, B.; Lopez, D. Molecular mechanisms involved in *Bacillus subtilis* biofilm formation. *Environ. Microbiol.* **2015**, *17*, 555–565. [CrossRef] [PubMed]

64. Strauch, M.; Webb, V.; Spiegelman, G.; Hoch, J.A. The Spo0A protein of *Bacillus subtilis* is a repressor of the *abrB* gene. *Proc. Natl. Acad. Sci. USA* **1990**, *87*, 1801–1805. [CrossRef] [PubMed]

65. Takada, H.; Morita, M.; Shiwa, Y.; Sugimoto, R.; Suzuki, S.; Kawamura, F.; Yoshikawa, H. Cell motility and biofilm formation in *Bacillus subtilis* are affected by the ribosomal proteins, S11 and S21. *Biosci. Biotechnol. Biochem.* **2014**, *78*, 898–907. [CrossRef]

66. Chu, F.; Kearns, D.B.; McLoon, A.; Chai, Y.; Kolter, R.; Losick, R. A novel regulatory protein governing biofilm formation in *Bacillus subtilis*. *Mol. Microbiol.* **2008**, *68*, 1117–1127. [CrossRef]

67. Chai, Y.; Kolter, R.; Losick, R. Reversal of an epigenetic switch governing cell chaining in *Bacillus subtilis* by protein instability. *Mol. Microbiol.* **2010**, *78*, 218–229. [CrossRef]

68. Copeland, M.F.; Weibel, D.B. Bacterial swarming: A model system for studying dynamic self-assembly. *Soft Matter* **2009**, *5*, 1174–1187. [CrossRef] [PubMed]

 *microorganisms*

*Article*

# The ComX Quorum Sensing Peptide of *Bacillus subtilis* Affects Biofilm Formation Negatively and Sporulation Positively

Mihael Špacapan [1], Tjaša Danevčič [1], Polonca Štefanic [1], Michael Porter [2], Nicola R. Stanley-Wall [2] and Ines Mandic-Mulec [1,*]

[1] Chair of Microbiology, Department of Food Science and Technology, Biotechnical Faculty, University of Ljubljana, Vecna pot 111, 1000 Ljubljana, Slovenia; mihael.spacapan@bf.uni-lj.si (M.Š.); tjasa.danevcic@bf.uni-lj.si (T.D.); polonca.stefanic@bf.uni-lj.si (P.Š.)

[2] Division of Molecular Microbiology, School of Life Sciences, University of Dundee, Dundee DD1 5EH, UK; m.porter@dundee.ac.uk (M.P.); N.R.Stanleywall@dundee.ac.uk (N.R.S.-W.)

* Correspondence: ines.mandicmulec@bf.uni-lj.si; Tel.: +386-1-3203-409

Received: 10 June 2020; Accepted: 22 July 2020; Published: 27 July 2020

**Abstract:** Quorum sensing (QS) is often required for the formation of bacterial biofilms and is a popular target of biofilm control strategies. Previous studies implicate the ComQXPA quorum sensing system of *Bacillus subtilis* as a promoter of biofilm formation. Here, we report that ComX signaling peptide deficient mutants form thicker and more robust pellicle biofilms that contain chains of cells. We confirm that ComX positively affects the transcriptional activity of the $P_{epsA}$ promoter, which controls the synthesis of the major matrix polysaccharide. In contrast, ComX negatively controls the $P_{tapA}$ promoter, which drives the production of TasA, a fibrous matrix protein. Overall, the biomass of the mutant biofilm lacking ComX accumulates more monosaccharide and protein content than the wild type. We conclude that this QS phenotype might be due to extended investment into growth rather than spore development. Consistent with this, the ComX deficient mutant shows a delayed activation of the pre-spore specific promoter, $P_{spoIIQ}$, and a delayed, more synchronous commitment to sporulation. We conclude that ComX mediated early commitment to sporulation of the wild type slows down biofilm formation and modulates the coexistence of multiple biological states during the early stages of biofilm development.

**Keywords:** biofilm; pellicle; quorum; sporulation; bacillus; signal; communication; heterogeneity; matrix; surfactin

---

## 1. Introduction

Biofilms from multicellular collectives encased in the extracellular matrix and represent the default mode of microbial growth in nature [1–3]. Quorum sensing (QS) [4], is a microbial communications system that coordinates the cell density-dependent bacterial gene expression, which often regulates biofilm development [5]. *Bacillus subtilis* is the most studied species in the genus Bacillus [6], and has served over the years as an excellent model to study the development of metabolically dormant and heat resistant spores [7] and more recently as a model of biofilms [8–11], division of labor [12–15], and a variety of social interactions (reviewed in [16]). This Gram-positive spore former relies on several peptide-based signaling systems and encodes many Phr-Rap signaling peptide-phosphatases pairs and the ComQXPA QS system, which influence sporulation, competence, and biofilm development in *B. subtilis* [16]. The ComQXPA QS system is widespread among Firmicutes [17]. The ComQ isoprenoid transferase processes and activates the ComX signaling peptide. The peptide binds to the membrane-bound histidine kinase, ComP, which phosphorylates ComA [18]. ComA-P directly activates the transcription of its target genes [19–21] including the *srfA* operon, responsible for the synthesis of the lipopeptide antibiotic surfactin [22]. Surfactin is a positive regulator of biofilm development and sporulation [23,24]. PhrC, the CSF competence stimulating factor [25], promotes the phosphorylation state of ComA by inhibiting its phosphatase RapC, but its effect on surfactin synthesis is less prominent [26]. Once phosphorylated, ComA-P also increases the transcription of the pleiotropic regulatory gene *degQ* [20,26,27], which increases the DegU phosphorylation rate [28]. DegU-P is ultimately required for the activation of extracellular degradative enzyme production and regulates the biofilm and spore development.

In biofilms, an extracellular matrix composed of polysaccharides (Eps), proteins, and extracellular DNA [2] glues cells together, but the ratios of each matrix constituent differ depending on the specific strain, media, and growth conditions [29]. In *B. subtilis*, the *epsA-O* operon is involved in the production of a critical polysaccharide component of the biofilm matrix [30], which is essential for the development of floating biofilm (pellicle) [31]. TasA, the major matrix protein, encoded by the *tapA-tasA-sipW* operon (hereafter referred to as the *tapA* operon) forms long fibers and gives structural support to floating biofilm [32]. Although TasA is not essential for the formation of floating biofilms, the *tasA* mutant forms less prominent biofilms [33,34]. The molecular regulation of the operons involved in the synthesis of the biofilm matrix components is very complex [35]. Briefly, Spo0A [31], which is controlled by the phosphorelay of multiple histidine kinases [36], initiates biofilm development. Phosphorylated Spo0A (Spo0A-P) activates transcription of the *sinI* operon [37,38]. SinI inhibits SinR [39], which acts as the central transcriptional repressor of the *epsA-O* and *tapA* operons [31,40]. A moderate amount of phosphorylated Spo0A suffices for the activation of the biofilm matrix production operons. However, as the Spo0A-P levels increase, cells commit to sporulation. DegU-P, which is also gradually phosphorylated, also contributes to biofilm formation [41]. DegU indirectly affects the phosphorylation state of Spo0A, shortening the time window of intermediate Spo0A phosphorylation, required to trigger the synthesis of the extracellular matrix [37]. The Spo0A-P levels also positively affects transcription of *bslA* [42], which contributes to biofilm hydrophobicity and influences transcription from a poly-γ-glutamate (*pgs*) operon that only plays a role during the formation of surface adhered biofilms [43].

The ComX QS system affects the transcription of the biofilm matrix operons by increasing surfactin synthesis. Surfactin triggers potassium ion leakage, which positively affects the activity of KinC [23]. This histidine kinase then increases the phosphorylation state of Spo0A [24]. Thus, it has been hypothesized that the ComX QS system increases the phosphorylation of Spo0A indirectly via the lipopeptide surfactin [23]. Consequently, ComX should promote pellicle biofilm formation and may also influence sporulation.

Cells resident in biofilms simultaneously express multiple biological states and this phenotypic heterogeneity results, for example, in the coexistence of matrix producers and highly resistant spores in one biofilm [9,24,44]. QS often regulates adaptive traits that exhibit phenotypic heterogeneity in many different species [45]. However, how ComX density-dependent communication affects the distribution of biological states devoted to the production of matrix components or sporulation in the pellicle biofilm, is less understood.

This work investigates the role of the ComX dependent signaling in pellicle biofilm development and sporulation. It describes the ComX dependent biofilm phenotype and temporal heterogeneity of matrix producers and spores in the pellicle biofilm. Our results provide evidence that ComX positively affects sporulation, but not biofilm development and offers a new outlook on the phenotypic heterogeneity of the phenotypes associated with ComX deficiency.

## 2. Materials and Methods

### 2.1. Bacterial Strains and Strain Construction

All *Bacillus subtilis* strains used in this study are listed in Table 1. The recombinant strains were constructed by transforming specific markers into *B. subtilis* recipients. The recipient strains were grown to competence in competence medium (CM) at 37 °C [46] for 6 h and then approximately 1 µg of DNA added to 500 µL of competent cells. Transformants were selected on LB agar supplemented with a specific antibiotic (chloramphenicol (Cm) 10 µg/mL, kanamycin (Kan) 50 µg/mL, or spectinomycin (Spec) 100 µg/mL) at 37 °C. Competence in strains with *comQ* deletion was induced by the addition of exogenous ComX, as previously described [47]. Integration of P$_{epsA}$-*gfp* reporter fusion into different strains was performed by transformation with YC164 genomic DNA [37] (Table 1). The *srfA::Tn917* (*mls*; lincomycin 12.5 µg/mL and erythromycin 0.5 µg/mL) mutant was prepared by transformation with BM1044 genomic DNA [48] (Table 1). Strains with P$_{tapA}$-*yfp* reporter fusion were constructed by transformation with BM1115 genomic DNA [49]. To prepare strains carrying the P$_{spoIIQ}$-*yfp* reporter fusion, the plasmid pKM3 [50] was transformed into indicated recipients (Table 1). We transformed strains with the plasmid pMS17 with the kanamycin resistance marker, and the plasmid pMS7 with the chloramphenicol resistance marker to construct P$_{43}$-*mKate2* reporter fusion strains (Table 2). To prepare strains with the P$_{srfAA}$-*yfp* reporter fusion, DL722 chromosomal DNA was transformed into different strains [24]. To construct the NCIB 3610 QS mutant, the plasmid pMiniMAD2-updowncomQ (Table 2) was used [47]. The plasmid was transformed into *B. subtilis* NCIB 3610 *comI*$^{Q12L}$ as previously described [47,51].

The P$_{43}$-*yfp* construct from the Pkm3-p43-*yfp* plasmid [49] was digested with EcoRI and BamHI and ligated into the pSac-Cm [52] plasmid to form the pEM1071 plasmid carrying a P$_{43}$-*yfp* fusion inside the *sacA* integration site. P$_{43}$ is a constitutively expressed promoter in *Bacillus subtilis* [53] located upstream from the *yfp* gene in the pEM1071 plasmid. This vector was then digested with HindIII and BamHI restriction enzymes to remove the *yfp* gene from the original vector and then simultaneously ligated with the digested *mKate2* fragment to construct the plasmid pMS7. The *mKate2* sequence was PCR amplified using the P3F and P3R primer pair (Table 3) and the genomic DNA of BM1097 as the template [54]. To construct the plasmid pMS17, the plasmid pMS7 was digested with EcoRI and BamHI restriction enzymes, and the digested P$_{43}$-*mKate2* fragment was ligated into pSac-Kan EcoRI/BamHI restriction sites [52].

**Table 1.** Bacterial strains used in this study.

| Strain Name | Background | Genome Description | Reference |
|---|---|---|---|
| *Bacillus subtilis* **strains** | | | |
| PS-216 | PS-216 | wt | [55] |
| BM1127 | PS-216 | Δ*comQ* | [47] |
| BM1128 | PS-216 | *amyE*::P$_{spoIIQ}$-*yfp* (Sp) | this work |
| BM1129 | PS-216 | Δ*comQ*<br>*amyE*::P$_{spoIIQ}$-*yfp* (Sp) | this work |
| BM1625 | PS-216 | *amyE*::P$_{spoIIQ}$-*yfp* (Sp)<br>*sacA*::P$_{43}$-*mKate2* (Cm) | this work |
| BM1626 | PS-216 | Δ*comQ*<br>*amyE*::P$_{spoIIQ}$-*yfp* (Sp)<br>*sacA*::P$_{43}$-*mKate2* (Cm) | this work |
| BM1613 | PS-216 | *amyE*::P$_{tapA}$-*yfp* (Sp)<br>*sacA*::P$_{43}$-*mKate2* (Cm) | this work |

**Table 1.** *Cont.*

| Strain Name | Background | Genome Description | Reference |
|---|---|---|---|
| BM1614 | PS-216 | $\Delta comQ$<br>$amyE::P_{tapA}$-*yfp* (Sp)<br>$sacA::P_{43}$-*mKate2* (Cm) | this work |
| BM1629 | PS-216 | $sacA::P_{43}$-*mKate2* (Kan) | this work |
| BM1630 | PS-216 | $\Delta comQ$<br>$sacA::P_{43}$-*mKate2* (Kan) | this work |
| BM1631 | PS-216 | $amyE::P_{epsA}$-*gfp* (Cm)<br>$sacA::P_{43}$-*mKate2* (Kan) | this work |
| BM1622 | PS-216 | $\Delta comQ$<br>$amyE::P_{epsA}$-*gfp* (Cm)<br>$sacA::P_{43}$-*mKate2* (Kan) | this work |
| DK1042 | NCIB 3610 | $comI^{Q12L}$ | [56] |
| BM1667 | NCIB 3610 | $\Delta comQ$<br>$comI^{Q12L}$ | this work |
| BM1623 | NCIB 3610 | $comI^{Q12L}$<br>$amyE::P_{epsA}$-*gfp* (Cm) | this work |
| BM1624 | NCIB 3610 | $comI^{Q12L}$<br>$\Delta comQ$<br>$amyE::P_{epsA}$-*gfp* (Cm) | this work |
| YC164 | NCIB 3610 | $amyE::P_{epsA}$-*gfp* (Cm) | [37] |
| BM1103 | PS-216 | $amyE::P_{epsA}$-*gfp* (Cm) | this work |
| BM1458 | PS-216 | $\Delta comQ$<br>$amyE::P_{epsA}$-*gfp* (Cm) | this work |
| BM1115 | PS-216 | $amyE::P_{tapA}$-*yfp* (Sp) | [49] |
| BM1126 | PS-216 | $\Delta comQ$<br>$amyE::P_{tapA}$-*yfp* (Sp) | this work |
| BM1097 | PS-216 | $amyE::P_{hyperspank}$-*mKate2* (Cm) | [51] |
| BM1044 | PS-216 | $srfA::Tn917$ (mls) | [48] |
| BM1673 | PS-216 | $\Delta comQ$<br>$srfA::Tn917$ (mls) | this work |
| DL722 | 3610 | $amyE::P_{srfAA}$-*yfp* (Sp) | [24] |
| BM1454 | PS-216 | $amyE::P_{srfAA}$-*yfp* (Sp) | this work |
| BM1455 | PS-216 | $\Delta comQ$<br>$amyE::P_{srfAA}$-*yfp* (Sp) | this work |
| *Escherichia coli strain* | | | |
| ED367 | BL-21 (DE3) | pET22(b)—*comQ comX* from *B. subtilis* 168 (Amp) | [57] |

**Table 2.** Plasmids used in this study.

| Plasmid Name | Description | Reference |
|---|---|---|
| pKM3 | $amyE::P_{spoIIQ}$-*yfp* (Sp, Amp) | [50] |
| Pkm3-p43-yfp | $amyE::P_{43}$-*yfp* (Sp, Amp) | [49] |
| pSac-Kan | $sacA::kan$ (Amp) | [52] |
| pSac-Cm | $sacA::cat$ (Amp) | [52] |
| pEM1071 | $sacA::P_{43}$-*yfp* (Cm, Amp) | this work |
| pMS7 | $sacA::P_{43}$-*mKate2* (Cm, Amp) | this work |
| pMS17 | $sacA::P_{43}$-*mKate2* (Kan, Amp) | this work |
| pMiniMAD2-updowncomQ | pMiniMAD2 with updown *comQ*<br>between EcoRI and SalI sites (Mls, Amp) | [47] |

**Table 3.** Oligonucleotides used in this study.

| Oligonucleotide Name | Sequence 5′→3′ | Reference |
| --- | --- | --- |
| P3F | GTACAAGCTTAAGGAGGAACTACTATGGATTCAATAGAAAAGGTAAG | [54] |
| P3R | GTACGGATCCTTATCTGTGCCCCAGTTTGCT | [54] |

## 2.2. Growth Conditions

Bacterial overnight cultures were inoculated directly from glycerol stocks at −80 °C grown in LB medium with the appropriate antibiotics at 37 °C with shaking at 200 rpm. Pellicle biofilms were grown by re-suspending an overnight culture (1% *v/v*) in liquid MSgg medium, (5 mM potassium phosphate tribasic (pH 7), 100 mM MOPS (pH 7), 2 mM $MgCl_2$, 700 mM $CaCl_2$, 50 mM $MnCl_2$, 50 mM $FeCl_3$, 1 mM $ZnCl_2$, 2 mM thiamine, 0.5% glycerol, 0.5% glutamate, 50 mg/L tryptophan, 50 mg/L phenylalanine) [31], and incubating the culture in static conditions at 37 °C for up to 40 h. The heterologously expressed ComX in M9 spent medium was prepared as described previously [47]. Briefly, *E. coli* ED367 [57] was used as a heterologous expression strain where the transcription of the *comQ* and *comX* genes was induced with isopropyl β-D-1-thiogalactopyranoside (IPTG). After induction with IPTG, the ED367 filtered spent medium with ComX was added to the liquid MSgg medium (20% *v/v*). *E. coli* ED367 spent medium, where *comQX* expression was not induced by IPTG, was used as a control.

## 2.3. Biochemical Composition of Extracellular Polymers and CFU Counts Determination in Pellicle Biofilms

Pellicle biofilms were grown in 20 mL of liquid MSgg medium at 37 °C for up to 40 h in Petri dishes (90 mm diameter). At indicated times, pellicle biofilms were collected and transferred into two centrifuge tubes containing 1 mL of physiological saline solution. Pellicle biofilms were kept on ice during sonication with the MSE 150 Watt Ultrasonic Disintegrator Mk2 at three 5-s bursts and amplitude of 15 μm. After disintegration, cell counts were determined in the pellicle biofilm by colony forming units (CFU) on LB agar after 24 h of incubation at 37 °C. Spores were enumerated as CFUs/mL after heating cell suspensions to 80 °C for 30 min. The spore fraction was determined by dividing heat resistant CFUs with total CFUs. Extracellular polymers were extracted from pellicle biofilms, as described previously [29]. The total sugar content in the extracted polymer fraction was determined by the phenol–sulfuric acid method as described before [58] and the total protein content by the Bradford protein assay [59].

## 2.4. Spent Media Droplet Surface Wetting Assay

Pellicle biofilm spent media were collected after 40 h of static growth in MSgg medium at 37 °C, two-fold serially diluted in distilled water, and surfactin concentrations were estimated visually by comparing 20 μL droplets placed on a parafilm strip. As the increase in the wetting surface area of the droplet increases with surfactant concentration, we estimated relative surfactin concentration in the spent medium based on the surface area. Distilled water droplets served as a control.

## 2.5. Pellicle Biofilm Bulk Fluorescence Measurements

Pellicle biofilms were grown in 200 μL MSgg medium in sterile 96-well black transparent bottom microtiter plate. If needed, and as indicated, strains were supplemented with ComX. Microtiter plates were incubated in the Cytation 3 imaging reader (BioTek Instruments, Inc., Winooski, VT, USA) at 37 °C without shaking. Optical density at 650 nm and fluorescence intensity were measured in half hour intervals for up to 60 h. The gain settings, excitation, and emission wavelengths for every transcriptional reporter are shown in Table S1. Each strain was tested in three independent replicates at specified times. Results represent the averages of all the wells with 95% confidence intervals.

Isogenic strains without the fluorescent reporter (background strains) were always cultured in the same microtiter plate as the experimental strains to estimate the background fluorescence. To calculate the final fluorescence of the experimental strain, the average fluorescence intensity of its background strain (grown in eight wells) was deducted from the fluorescence intensity of the experimental strain at each time point.

### 2.6. Pellicle Biofilm Morphology, Hydrophobicity Estimation and Confocal Laser Scanning Microscopy

Pellicle biofilms were grown in 2 mL of MSgg medium in 35-mm diameter glass-bottom Petri dishes and imaged with the Leica WILD M10 stereomicroscope. Pellicle biofilm hydrophobicity was evaluated by spotting a 50 μL droplet of 0.5% (*w/v*) methylene blue solution in the center of the pellicle biofilm and the diffusion of the dye recorded over time by a video camera and then edited by Lightworks v. 14.5 trial video editing software. Cells in pellicle biofilms were visualized by confocal laser scanning microscope after removing the spent medium from underneath the pellicle biofilm by careful pipetting. We used the 100× immersion objective (NA 0.4) and roughly determined the bottom of the pellicle to set up a Z-stack. Pellicle biofilms carrying transcriptional reporters fused to specific promoters or/and constitutive promoters were then visualized using an LSM 800 microscope (Zeiss, Göttingen, Germany). Excitation laser wavelength, emission filter setup, and other laser settings are shown in Table S2. Averaging was set at eight, scanning was unidirectional, pixel time equaled 0.58 μs, and the frame time was 9.11 s.

Fluorescence images were taken with an AxioCam MRm Rev.3 camera. The captured images were analyzed using Zen Blue v. 2.3 lite software (Zeiss, Göttingen, Germany). Pixel intensity values were normalized using the best fit function. Images are presented as orthogonal displays with the Z-stack image that exhibited the highest fluorescence intensity of either the $P_{epsA}$-*gfp* $P_{tapA}$-*yfp* or $P_{spoIIQ}$-*yfp* channels.

### 2.7. Flow Cytometry

Cultures were inoculated with 1% (*v/v*) of an overnight LB culture and grown at 37 °C in 4 mL of MSgg medium in 12-well microtiter plates without shaking. Pellicles were partially homogenized with a 5 mL pipette tip and transferred into a new container. The pellicle was partially disrupted by sonication with a 30-s burst at an amplitude of 15 μm. Afterward, the disrupted pellicle was treated with 4% (*w/v*) paraformaldehyde for 7 min at room temperature and washed in GTE buffer (glucose 50 mM, ethylenediaminetetraacetic acid (EDTA) 10 mM, Tris-HCl 20 mM, pH 8 ± 0.2). Cells were stored at room temperature until further analysis. The disrupted pellicle biofilms were suspended in 1 mL phosphate-buffered saline (PBS) containing 0.5% (*w/v*) bovine serum albumin. The cell suspensions were diluted 1:1000 into filtered GTE buffer containing 1 μg/mL 4′,6-diamidino-2-phenylindole (DAPI) and analyzed using a LSR Fortessa cytometer (BD Biosciences, Franklin Lakes, NJ, USA). Forward and side scatter, DAPI (355 nm laser, 450 nm/50 nm bandpass filter), and green fluorescent protein/yellow fluorescent protein (GFP/YFP) (488 nm laser, 530 nm/30 nm bandpass filter) were detected with photomultiplier voltages between 300 and 500 V. GFP and YFP signals were measured for 30,000 DAPI-gated events per sample. Data were analyzed using FlowJo 10 with samples gated on the upper limit of signal from fluorophore-free cells to define the GFP/YFP negative/positive cells in the population analyzed.

### 2.8. Statistical Analysis

All experiments were performed in at least three time-independent biological replicates. In the experiments concerning bulk pellicle fluorescence measurements in the microplate reader, we subtracted the estimate for background fluorescence intensity (averaged from all the wells with strains without fluorescent transcriptional reporter fusions) from every microtiter plate as well as fluorescence measurements. Afterward, fluorescence intensity in each well was averaged and a 95% confidence interval calculated from all of the wells. For biochemical tests, flow cytometry tests, and CFUs, the averages of three independent biological replicates are shown with the standard error of means (SEM).

In all experiments, two groups of samples were compared by calculating a Student's *t*-test and a one-way non-parametric Mann–Whitney U test. We treated the two groups of samples as being statistically different if both tests showed a *p*-value < 0.05.

## 3. Results

### 3.1. The Quorum Sensing ComQ Mutant Forms Thicker Pellicles than the Wild Type

To investigate the role of ComX in pellicle biofilm formation, we grew the wild type and the ComX deficient PS-216 mutant ($\Delta comQ$) in MSgg medium [30]. We found that the $\Delta comQ$ mutant formed a prominent pellicle biofilm, which appeared thicker than the parental strain (Figure 1).

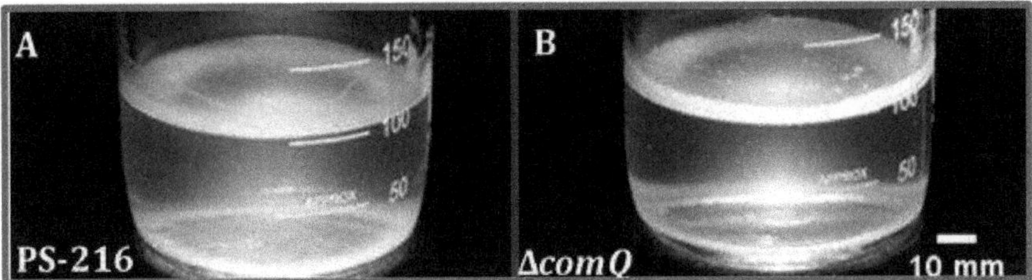


**Figure 1.** Pellicle biofilms of the *Bacillus subtilis* wild type PS-216 (**A**) and the ComX deficient PS-216 $\Delta comQ$ strain (BM1127) (**B**) grown statically in MSgg medium at 37 °C and in 250 mL bottles for 16 h.

This observation contrasted with the generally held view that the ComQXPA QS system positively regulates biofilm formation due to its requirement for surfactin synthesis [23]. Therefore, we tested surfactin synthesis in both the wild type PS-216 [55] (hereafter PS-216) and the $\Delta comQ$ mutant (BM1127) strains. As expected, surfactin synthesis was approximately eight-fold lower in the $\Delta comQ$ mutant compared to PS-216 (Figure S1). The BM1044 mutant with an inactivated *srf* operon was used as a negative control. These results suggest that either surfactin is not essential for pellicle biofilm development in the PS-216 strain or other unknown factors may contribute to the proficient biofilm phenotype of the ComX deficient mutant.

Since *Bacillus* species can form highly hydrophobic biofilms [60,61], we assessed if the altered pellicle of the ComX mutants preserved this property. To estimate pellicle biofilm hydrophobicity, we spotted a 0.5% (*m/v*) methylene blue solution on the upper surface of the $\Delta comQ$ mutant and PS-216 pellicle biofilms. We observed that the droplet spread more slowly on the $\Delta comQ$ (BM1127) pellicle biofilm (Video S1), compared to PS-216, which suggests that the mutant pellicle is more hydrophobic and therefore structurally different to the PS-216 pellicle.

### 3.2. Expression of $P_{epsA}$ and $P_{tapA}$ Promoters Per Pellicle Is Higher in the QS Mutant than in the Wild Type Strain

The thicker pellicles formed by the peptide deficient mutant led us to question the postulated role of ComX in promoting biofilm formation. Therefore, we analyzed the activity of the $P_{epsA}$ promoter, which controls the expression of the *epsA-O* operon responsible for the synthesis of the major biofilm matrix polysaccharide. We also tested the $P_{tapA}$ promoter activity, which controls the synthesis of the TasA fibers [62] in the pellicles. We first monitored the activity of both promoters at the population level in the PS-216 strain (wt; BM1631) and $\Delta comQ$ (BM1622) strains carrying the $P_{epsA}$-*gfp* reporter or in the BM1613 (wt) and BM1314 ($\Delta comQ$) strains carrying the $P_{tapA}$-*yfp* reporter, respectively. Both reporters are transcriptional fusions. To estimate fluorescence from both reporters in the entire pellicle biofilm, we measured the bulk pellicle fluorescence by fluorimetry. Our analysis revealed that both reporters exhibited a higher pellicle fluorescence intensity in the $\Delta comQ$ compared to the wild type (Figure 2A,B), which is inconsistent with ComX promoting overall biofilm formation.

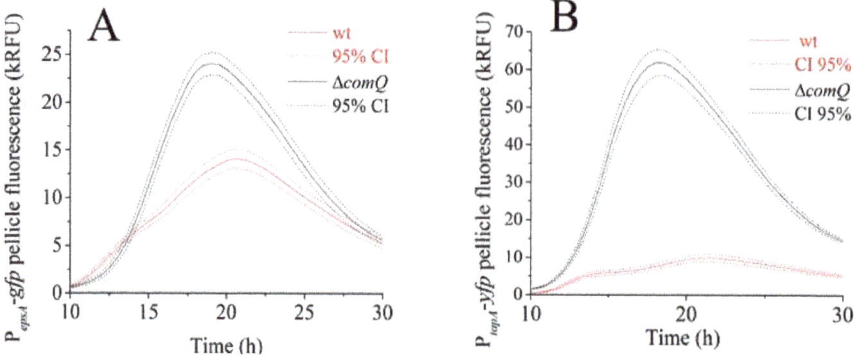

**Figure 2.** Bulk pellicle biofilm fluorescence measured in a microplate reader in *B. subtilis* PS-216 during static growth in MSgg medium at 37 °C. (**A**) Pellicle biofilm fluorescence intensity of the wild type phenotype $P_{epsA}$-*gfp* reporter (wt; BM1631) strain and the $\Delta comQ$ mutant phenotype (BM1622) reporter strains. (**B**) Pellicle biofilm fluorescence intensity of the $P_{tapA}$-*yfp* wild type phenotype (wt; BM1613) and $\Delta comQ$ phenotype (BM1614) reporter constructs. The line labelled CI shows the 95% confidence interval of the measurements.

The impact of the *comQ* mutation on pellicles formed by the *B. subtilis* NCIB 3610 strain was also tested. The PS-216 isolate, used here as the primary model strain, is very similar at the genomic sequence level to NCIB 3610, which is frequently used in *B. subtilis* biofilm research [55,63]. The NCIB 3610 isolate responded to ComX deficiency in a similar manner to the PS-216 isolate (Figure S2), albeit with some minor differences. The NCIB 3610 isolate produced comparatively less surfactin than PS-216, which further decreased in $\Delta comQ$ (Figure S1). It also responded to ComX deficiency similarly albeit less dramatically at the level of bulk $P_{epsA}$-*gfp* expression than PS-216 derivative (Figure 2A, Figure S2). Although subtle strain specific differences between PS-216 and NCIB 3610 in biofilm formation exist, both strains respond comparably to ComX deficiency. This suggests that biofilm associated phenotypes observed in PS-216 may be more generally applicable.

*3.3. The Quorum Sensing Mutant Pellicles Exhibit a Different Pellicle Morphology and Distributions of Cells with Active $P_{epsA}$ and $P_{tapA}$ Dependent Expression*

In a biofilm, *B. subtilis* cells are subjected to spatiotemporal regulation of gene expression, which results in the coexistence of multiple cell types and a complex macroscale architecture in the mature community [64,65]. We lack an understanding of how ComX influences the spatiotemporal distribution of matrix producing cells in pellicle biofilms. To shed light on the influence exerted by ComX, we monitored the distribution of cells expressing the matrix genes using $P_{epsA}$-*gfp* or $P_{tapA}$-*yfp* reporters as markers, while detecting metabolically active cells using a $P_{43}$-*mKate2* fluorescence reporter [53] carried by each strain. After removing the liquid media from the well at indicated time points, we directly visualized the pellicles by confocal microscopy. The $\Delta comQ$ pellicle biofilms lacking ComX had a different morphology from wild type, and the pellicles were 2.5-fold thicker with a depth of up to 30 μm compared with the wild type biofilms that had a depth of up to 12 μm (Figures 3 and 4).

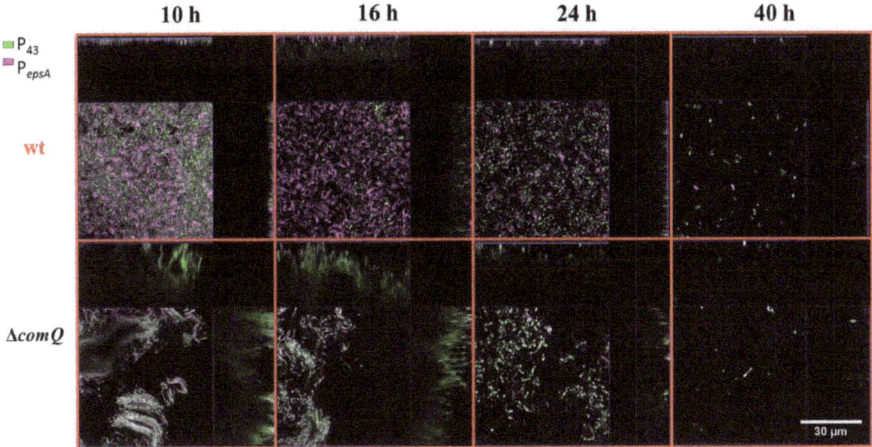



**Figure 3.** Pellicle biofilms of the PS-216 *Bacillus subtilis* wild type (wt; BM1631) and the QS mutant (Δ*comQ*; BM1622) phenotypes carrying P$_{43}$-*mKate2* (false-colored green) and P$_{epsA}$-*gfp* (false-colored magenta) transcriptional reporters during static growth in MSgg medium at 37 °C and visualized under a confocal microscope at the indicated time points.

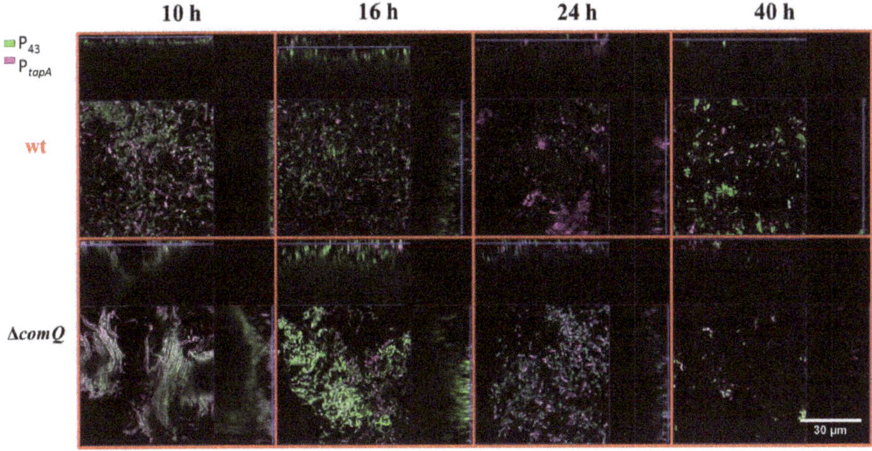



**Figure 4.** Pellicle biofilms of the PS-216 *Bacillus subtilis* wild type (wt; BM1613) and the QS mutant (Δ*comQ*; BM1614) phenotypes carrying P$_{43}$-*mKate2* (false-colored green) and P$_{tapA}$-*yfp* (false-colored magenta) transcriptional reporters during static growth in MSgg medium at 37 °C and visualized under a confocal microscope at the indicated time points.

This analysis confirms our initial visual assessment of the pellicle biofilm (Figure 1). We further noted that the wild type cells in the pellicle were densely packed and evenly distributed at the 10 h time point. In contrast, the cells in the Δ*comQ* biofilm were less densely packed than those within the wild type pellicles. The Δ*comQ* mutant formed bundles of long chains of cells at early time points (10 h): cell chaining is one of the hallmarks of the biofilm phenotype [66].

Confocal microscopy images (taken at 10 h and 16 h time points) suggest that the wild type biofilms have a higher proportion of cells with the active P$_{epsA}$ promoter than the mutant biofilms, which is consistent with the published results [23]. Mostly, the P$_{epsA}$ activated promoter cells occupied the bottom of the pellicle (Figure 3). In contrast, we did not observe a substantial difference in the activity of the $P_{tapA}$ promoter between the two strains, the wild type (BM1631), and $\Delta comQ$ (BM1622) for any of the examined time points (Figure 4). By 40 h, we detected only a few $P_{43}$ fluorescing (false-colored green) cells in both the wild type and *comQ* biofilms, which now appeared similar in morphology. Although the weak fluorescence from the $P_{43}$-*mKate2* fluorescence reporter suggested their low metabolic activity, the biofilms were still visible by the naked eye, demonstrating that the biomass was not dispersed. In total, the confocal microscopy confirms that ComX during early stages of biofilm development shapes the thickness and morphology of this multicellular structure. It also supports a positive effect of ComX on the $P_{epsA}$, but not on $P_{tapA}$ promoter.

*3.4. $P_{tapA}$ But Not $P_{epsA}$ Dependent Expression Is More Active and in a More Substantial Portion of Cells in the $\Delta comQ$ Mutant than in the Wild Type Strain*

The microscopic imaging of the whole pellicle indicated heterogeneous activation of $P_{epsA}$ and $P_{tapA}$. However, the microscopy of the undisturbed pellicle only provided qualitative results of the distribution of cell types. To obtain quantitative information at a single-cell resolution, we opted to follow the activity of these two promoters by flow cytometry. The wild type and ComX deficient ($\Delta comQ$) strains were grown at 37 °C in 4 mL of MSgg medium in 12-well microtiter plates without shaking for the stated times. We disrupted the pellicle at 16 and 24 h time points and employed flow cytometry on the cell suspension to determine the percentage of cells that express the *epsA* and *tapA* operons. These experiments calculate the relative percentage of cells expressing each operon and the level of transcription at a single cell level.

The results presented in Figure 5A show that only 10.4% ± 1.3% of cells express the $P_{epsA}$-*gfp* reporter and the percentage (relative to the total cell count) of cells with $P_{epsA}$ promoter guided transcription being higher in the wild type (BM1103) than in the mutant (4.4% ± 0.7%; BM1458) at the 16 h time point. Although still detectable, this difference is not statistically significant at the 24 h time-point, where 12.7% ± 3.1% of wild type and only 6.5% ± 2.0% of $\Delta comQ$ cells express the $P_{epsA}$ regulated fluorescent reporter (Figure 5A). The percentage of cells that express the $P_{tapA}$-*yfp* reporter in the wild type phenotype (BM1115) equaled to 35.0% ± 1.5% of cells at 16 h and 23.5% ± 3.6% of cells at 24 h. At both time points, the *comQ* pellicle (BM1126) harbored a significantly higher percentage of fluorescent cells (59.8% ± 3.9% at 16 h and 47.2% ± 2.1% at 24 h) (Figure 5B). There were no statistically significant differences detected in terms of average single cell $P_{epsA}$ expression (Figure 5B). The average intensity of $P_{epsA}$ expressing cells in the wild type reporter (BM1103) equaled 144 RFU ± 8 RFU at 16 h and 148 RFU ± 5 RFU at 24 h. The average intensity of $P_{epsA}$ expressing cells in the $\Delta comQ$ reporter (BM1458) was similar, and equaled 138 RFU ± 8 RFU at 16 h and 138 RFU ± 8 RFU at 24 h (Figure 5C). We did, however, detect a statistically significant difference between the wild type and the mutant expression profiles when we measured the average $P_{tapA}$ activity per cell. The average intensity of $P_{tapA}$ expressing cells in the wild type reporter (BM1115) equaled 223 RFU ± 16 RFU at 16 h and 200 RFU ± 10 RFU at 24 h. The average intensity of $P_{tapA}$ expressing cells in the $\Delta comQ$ reporter (BM1126) on the other hand, equaled 329 RFU ± 32 RFU at 16 h and 329 RFU ± 15 RFU at 24 h (Figure 5D).

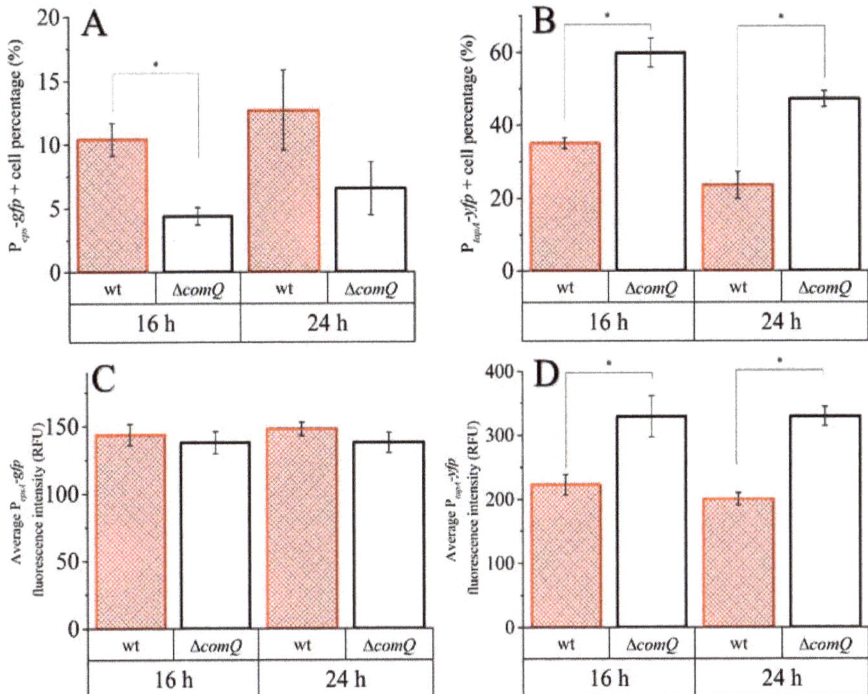

**Figure 5.** Comparisons of single-cell $P_{epsA}$ (wt; BM1103 and $\Delta comQ$; BM1458) and $P_{tapA}$ (wt; BM1115 and $\Delta comQ$ BM1126) promoter activity in *B. subtilis* PS-216 during static growth in MSgg medium at 37 °C. (**A**) Percentage (relative to total cell count) of wild type phenotype (wt) and QS mutant ($\Delta comQ$) cells expressing from $P_{epsA}$. (**B**) Percentage (relative to total cell count) of cells expressing from $P_{tapA}$. (**C**) Average fluorescence intensity of cells expressing from $P_{epsA}$. (**D**) Average fluorescence intensity of cells expressing from $P_{tapA}$. Data points represent averages with the standard error of means (SEM) of three biological replicates. Statistical significance (marked with *) was determined using a one-way Mann–Whitney test and a Student's *t*-test ($p < 0.05$).

In conclusion, the ComQX QS system increases the fraction of cells expressing the *epsA* operon during early biofilm development without changing the average intensity of $P_{epsA}$ guided transcription. In contrast, ComX decreases the average activity and fraction of cells with $P_{tapA}$ activity.

### 3.5. Extracellular Polymer Extracts of QS Mutant Pellicles Contain More Sugar and Protein

Our results show that ComX negatively affected $P_{tapA}$ transcription, but positively guided $P_{epsA}$ transcription. Therefore, it was essential to verify whether ComX deficient mutants preserve this pattern at the level of production of matrix components. To estimate the overall monosaccharide and protein content in the biofilm matrix, we extracted the extracellular polymers from the wild type (PS-216) and $\Delta comQ$ (BM1127) pellicles using a simplified method of Dogsa et al. [29]. We then quantified the amount of monosaccharide and protein in the extracellular polymer extracts by using the reducing sugar assay [58] and the Bradford protein assay [59].

Despite our finding that a lower percentage of cells actively expressed the $P_{epsA}$ promoter in the *comQ* strain using flow cytometry, the overall monosaccharide content did not follow this pattern. Polymer extracts of the $\Delta comQ$ biofilms harbored a higher amount of monosaccharide per pellicle than the wild type extracts (Student's *t*-test and one-way Mann–Whitney U test; $p < 0.05$) at 16 h (10-fold more) and 24 h (1.6-fold more) (Figure 6A).

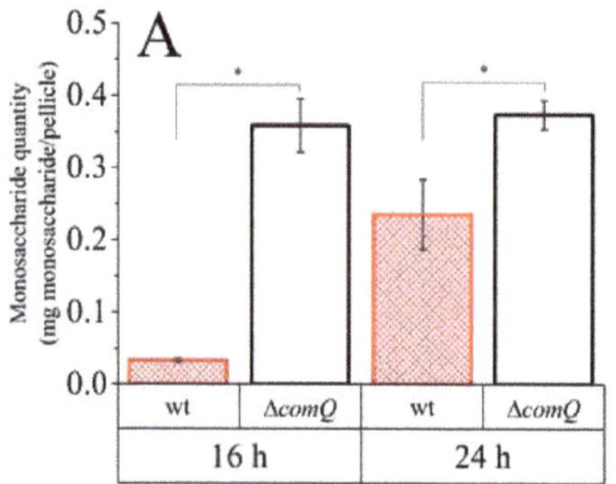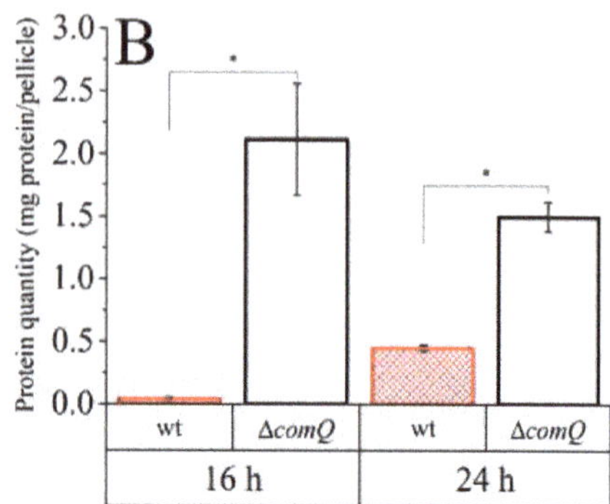

**Figure 6.** Monosaccharide and protein quantity in *B. subtilis* PS-216 (wt; PS-216) and QS mutant phenotype (ΔcomQ; BM1127) pellicles during static growth in the MSgg medium at 37 °C. (**A**) Monosaccharide content was determined with the phenol–sulfuric acid method. (**B**) Protein content was determined with the Bradford protein assay. Averages with the standard error of means (SEM) of three biological replicates are shown. Statistical significance (marked with *) was determined using a one-way Mann–Whitney test and a Student's *t*-test ($p < 0.05$).

Therefore, the synthesis of matrix components followed the activity pattern of the $P_{tapA}$, but not of the $P_{epsA}$ promoter.

### 3.6. The QS Deficient Mutant Has Lower Spore Counts during the Early Stages of Biofilm Development

Matrix transcribing cells transition into spore forming cells during the time course of biofilm formation [67]. Thus, we next tested whether ComX influences the dynamics of sporulation during biofilm development. To do this, we examined the impact of ComX on the expression of the sporulation marker gene, *spoIIQ*. First, we measured the $P_{spoIIQ}$-*yfp* dependent bulk fluorescence by fluorometry over time and normalized it by $P_{43}$ (constitutive promoter) dependent bulk fluorescence to roughly account for culture growth (Figure 7A). Normalized fluorescence was found to be higher in the wild type (BM1625) than in the mutant (BM1626) during the first 10–14 h of incubation. Around 14 h of incubation, the normalized $P_{spoIIQ}$-*yfp* fluorescence increased rapidly, reaching the fluorescence of the wild type. When we added heterologous ComX to the ΔcomQ mutant (BM1626) at the beginning of the incubation, its expression profile became comparable to when heterologous ComX was added to the wild type strain (BM1625) (Figure 7A). The sole addition of heterologous ComX to the wt strain, however, also slightly modified its normalized expression profile. This is most likely because ComX was already added at the onset of growth. We next quantified the number of cells in the pellicle that express the *spoIIQ* gene by flow cytometry (Figure 7B) using the $P_{spoIIQ}$-*yfp* labelled wild type strain (BM1128) and the ΔcomQ strain (BM1129). These results reveal that the $P_{spoIIQ}$ promoter was active in 42.9% ± 0.4% (SEM; standard error of means) of the cells in wild type (BM1128) at 16 h and 49% ± 3.0% of cells at 24 h. In the ComX deficient mutant (BM1129), 7.2% ± 0.8% of cells activated this gene at 16 h and 15.2% ± 4.5% at 24 h.

We finally compared the number of colony-forming units (CFU) in the wild type (PS-216) and ΔcomQ (BM1127) pellicles (Figure 7C). The total CFU counts were comparable at 24 h and 40 h, however, the ΔcomQ mutant exhibited a significantly higher CFU count in the pellicle at the 16 h time-point (Figure 7C).

We also measured the quantity of heat resistant spores in the population (Figure 7D). The wild type pellicle contained 9.9% ± 3.3% (SEM) at 16 h and 45.3% ± 5.9% at 24 h heat resistant CFU per total pellicle CFU. The QS mutant pellicle contained 1.1% ± 0.4% and 14.5% ± 3% of heat resistant CFU per total CFU at 16 h and 24 h, respectively (Figure 7B). At 16 h, the wild type pellicle therefore contained approximately 5-fold and at 24 h 3-fold more heat resistant CFUs than the Δ*comQ* mutant. After 40 h, all the pellicle CFUs were heat resistant in both strains (Figure 7D).

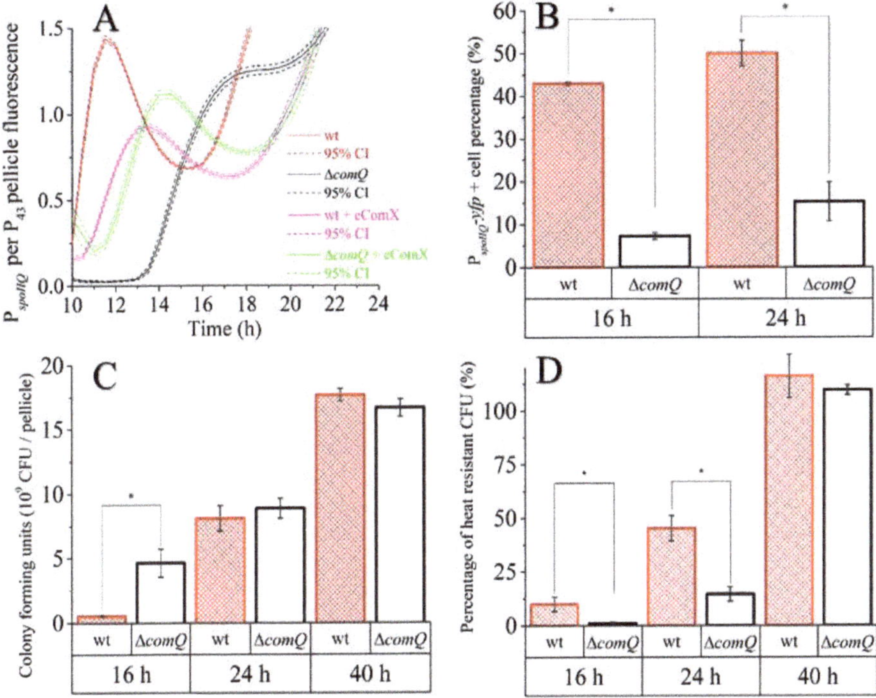

**Figure 7.** Comparisons of *B. subtilis* PS-216 wild type (wt) and QS mutant (Δ*comQ*) during static growth in MSgg medium at 37 °C. (**A**) Pellicle biofilm $P_{spoIIQ}$-*yfp* fluorescence normalized per $P_{43}$-mKate2 fluorescence with and without exogenous heterologously expressed ComX. (**B**) The fraction of cells with an active $P_{spoIIQ}$-*yfp* promoter per all cells analyzed by flow cytometry. (**C**) Total pellicle biofilm CFU counts. (**D**) The fraction of the heat resistant CFU per the total non-heat-treated pellicle CFU. Averages with the standard error of means (SEM) of three biological replicates are shown. Statistical significance (marked with *) was determined using a one-way Mann–Whitney test and a Student's *t*-test ($p < 0.05$), and by calculating the 95% confidence interval on panel B (dashed line).

Taken in combination, the difference in timing of the $P_{spoIIQ}$ promoter activity between the wild type and Δ*comQ* pellicle biofilm visualized by confocal microscopy (Figure 8) further supports that the wild type enters sporulation after only 10 h of incubation. In contrast, the Δ*comQ* pellicle biofilms did not show detectable $P_{spoIIQ}$-*yfp* activity at this time.

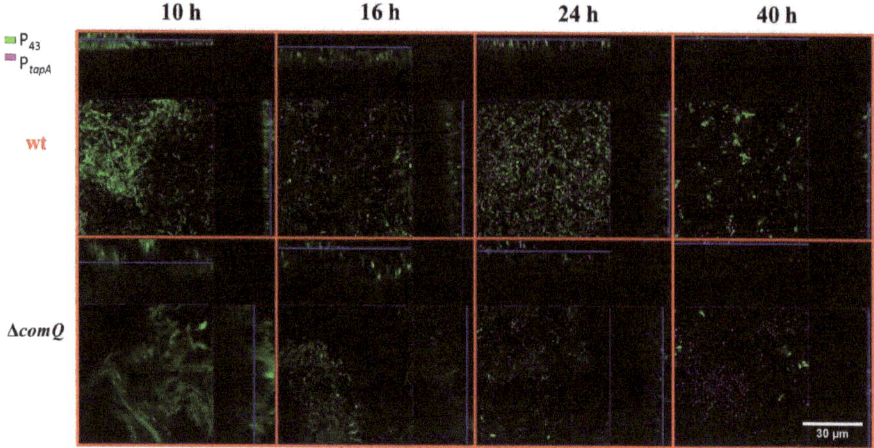


**Figure 8.** Confocal visualization of P$_{spoIIQ}$-*yfp* (false-colored magenta) and P$_{43}$-*mKate2* (false-colored green) expression in wild type (wt; BM1625) and QS mutant (Δ*comQ*; BM1626) pellicle biofilms during static growth in MSgg medium at 37 °C at the indicated time points.

Collectively, these results indicate that ComX deficiency decreases sporulation during the early stages of pellicle formation, a decrease that is most measurable at the 10 h time point where P$_{spoIIQ}$ activated cells are not detected by confocal microscopy in the Δ*comQ* mutant, but are visible in the wild type. We therefore conclude that ComX signals to the cell population to halt cell growth and production of matrix components. It has been recently shown that the main protein matrix component TasA also serves as an extracellular signal, which can trigger cell motility [68], therefore potentially enabling cells to migrate and colonize new ecological niches. However, since ComX mediated quorum sensing appears to decrease TasA production, it appears that QS promotes an increased investment into spore development instead of cell motility.

## 4. Discussion

The ComQXPA QS system upregulates surfactin production [22,57,69], which promotes transcription of the *epsA-O* operon [23]. Hence, a generally accepted view is that ComX promotes biofilm formation. However, our results showed that the ComX deficient mutant still developed prominent and structurally even more prominent biofilms than the wild type strain. The QS mutant biofilms expressed the P$_{tapA}$ promoter in a higher proportion of cells and accumulated a greater quantity of proteins and sugars than the wild type biofilms. In contrast, ComX deficient mutants produced fewer spores during early stages of biofilm development. These results indicate that ComX dependent QS negatively affects investment into multicellularity and positively to commitment to spore formation.

The average expression level from the *tapA* promoter and the percentage of cells expressing *tapA* were both higher in the QS mutant cells (Figure 5B,D), which indicates that ComX negatively regulates the synthesis of matrix proteins, but not the P$_{epsA}$ activity. Although, earlier work suggests that *epsA-O* and *tapA* operons are co-regulated [67,70], recently published results [71] and our observations do not support their tight co-regulation. The effect on P$_{tapA}$ and the absence of an effect on P$_{epsA}$ may be related to differential affinities of these promoters for SinR or AbrB as the concentration of Spo0A-P gradually increases due to the surfactin effect on phosphorelay. We also found ComX deficient mutants formed more elongated and chained cells (Figure 3, Figure 4, and Figure 8) compared to the wild type strain. Chained cells, a hallmark of *B. subtilis* biofilm formation, express P$_{tapA}$ regulated genes, while properly segmented cells have increased *epsA-O* transcription [71]. Therefore, we speculate that a yet unknown but ComX dependent mechanism inhibits the P$_{tapA}$ activity in pellicles and decreases the synthesis of biofilm matrix protein. Although ComX does not negatively influence $P_{epsA}$ activity (Figure 5A,C), the mutant biofilm matrix still accumulates more monosaccharides, which are indicative of a higher concentration of polysaccharides (Figure 6A,B). At this point, we cannot explain the molecular basis of this observation, but only speculate that the synthesis of matrix polysaccharides might be regulated by additional mechanisms that are under the influence of ComX and are independent of $P_{epsA}$ promoter activity.

Moreover, other signaling pathways like PhrC (CSF) and its cognate RapC phosphatase, which are involved in surfactin production through ComA-P [26], albeit with a lesser effect than ComX [72], may contribute to the synthesis of matrix components and biofilm development. The phosphorelay, which regulates accumulation of Spo0A-P, integrates a variety of signaling pathways including kinases, phosphatase, and additional Phr-Rap pairs [65]. However, future work will show whether ComX dependent regulation of biofilm development acts primarily through surfactin, the DegQ-DegS-DegU regulatory system [42], or the Spo0A phosphorelay [67].

Our results are in accordance with recent findings, indicating that surfactin is not generally essential for the formation of pellicle biofilms [73] with surfactin mutants exhibiting similar shifts in the expression of *epsA* and *tapA* operons, as we show here for the QS mutant. However, the results presented here suggest that the main regulatory path of biofilm development may include ComX and only subsequently surfactin.

By analyzing single-cell expression, we confirmed that ComX positively regulates the *epsA-O* operon [23]. We also quantified heat resistant spores, the percentage of the $P_{SpoIIQ}$-*yfp* active cells by flow cytometry, and the $P_{SpoIIQ}$-*yfp* bulk fluorescence during biofilm growth. We found that ComX promotes sporulation in a subpopulation of cells during the early stages of biofilm growth. Bacterial spores stop growing and are metabolically inactive [74]. Although we determined the lower proportion of cells with the $P_{epsA}$ active promoter, the absence of spores early in the ComX mutant may contribute to higher absolute numbers of matrix producers and higher sugar content due to prolonged growth. This result is in accordance with the idea that the phosphorylation of the Spo0A gradually increases [75]. An intermediate level of Spo0A-P triggers the transcription of the *epsA-O* operon, while a high level of Spo0A-P triggers sporulation [36]. This is evidenced by the delayed and more synchronous sporulation of the QS mutant (Figure 7). We speculate that low levels of Spo0A-P persist longer in the ComX deficient mutant due to low levels of surfactin, which adds to the activation of Spo0A-P in the wild type [23]. It is possible that other Phr-Rap pairs may substitute for surfactin signaling, but these act a bit late during the stationary phase [65,76]. Hence, without ComX to detect the increase in cell density and approaching starvation early on, cells produce polysaccharides rather than spores.

We here show that ComX deficient mutants have delayed sporulation in biofilms compared to the wild type, so ComX acts as a positive signal for sporulation. Recent work demonstrates that the timing of sporulation entry has consequences for spore quality, with early spore formers showing more efficient germination than late spore formers [77]. In addition to this delay, we also observed that the ComX defective mutant induced *spoIIQ* more synchronously than the wild type, which shows prominent heterogeneity in the expression of this marker gene. Phenotypic heterogeneity arises from several different intrinsic and extrinsic factors [78]. Sporulation in *B. subtilis* is subject to phenotypic heterogeneity due to the noise in the expression of transcriptional regulators, the phosphate transfer in the phosphorylation regulation cascades, and positive and negative feedback loop mechanisms [79]. There are several potential input signals, which may contribute to heterogeneous sporulation [80] and the division of labor in biofilms [81]. How a QS system in general and ComX in particular mechanistically contribute to the heterogeneity of gene expression is not yet understood [45]. We speculate that ComX informs the population on its growth rate and may act as a gatekeeper, which allows investments into late growth adaptations only when growth slows down. Given this, ComX deficiency will enable the population to skip this safety valve and overinvest into growth and matrix production rather than prepare for dormancy. However, ComX omission extends growth only for a short period. Soon, other signaling systems may attain their thresholds and inform the population of the famine and stress and override ComX deficiency.

*B. subtilis* is known for its bet hedging behavior in terms of spore formation, where a sub-population of cells initiates sporulation stochastically, regardless of the sensed external environmental stimuli (e.g., nutrient starvation) [13]. In a wild type population, the endospore forming process already start at the onset of floating biofilm formation. The bacterial population with an inactivated QS, however, did not exhibit such behavior, but invested into spore formation later. This suggests that the ComQXPA QS system serves as a signal that down modulates investment into growth and assures early investment into sporulation. This, on one hand, may enable bet hedging behavior, but on the other hand, may burden cells with the metabolic costs of early sporulation initiation.

## 5. Conclusions

This work shows that ComX inhibits matrix production, especially the TasA protein, and promotes sporulation during early stages of biofilm development in the wild type. Hence, we propose that this QS signaling system fine-tunes commitment to different biological states and therefore contributes to the phenotypic heterogeneity of *B. subtilis* biofilms.

**Supplementary Materials:** The following are available online at http://www.mdpi.com/2076-2607/8/8/1131/s1, Figure S1: Semi quantification of surfactant concentrations in the pellicle biofilm spent media after 40 h of static growth in MSgg medium at 37 °C using a droplet surface wetting assay, Figure S2: Bulk pellicle fluorescence measurements of the P$_{epsA}$-*gfp* reporter activity of *Bacillus subtilis* NCIB 3610 wild type (wt) and QS mutant ($\Delta comQ$) phenotypes during static growth in MSgg medium at 37 °C. Averages of three biological replicates are shown. Statistical significance was determined by calculating the 95% confidence interval (dashed line); Table S1: Emission and excitation wavelengths and gain settings for transcriptional reporters used in this study, Table S2: Confocal laser scanning microscope settings for different transcriptional reporters used in this study; Video S1: Fast forward video showing diffusion of the 50 μL methylene blue droplet placed on the PS-216 wild type (wt) or QS mutant ($\Delta comQ$) pellicle biofilms grown statically in MSgg medium at 37 °C for the time periods indicated on the video, https://www.youtube.com/watch?v=1FRfrhzIPKY&feature=youtu.be. Upload date: 17.9.2018.

**Author Contributions:** Conceptualization, M.Ć., T.D., I.M.-M., and N.R.S.-W. Methodology, M.Ć., M.P., T.D., P.Ć., I.M.-M., and N.R.S.-W.; Formal analysis, M.Ć., M.P.; Investigation, M.Ć., T.D., and M.P.; Writing—original draft preparation, M.Ć., T.D., and I.M.-M.; Writing—review and editing, I.M.-M. and N.R.S.-W.; Visualization, M.Ć.; Supervision, T.D., I.M.-M., and N.R.S.-W.; Project administration, I.M.-M.; Funding acquisition, I.M.-M. and N.R.S.-W. All authors have read and agreed to the published version of the manuscript.

**Funding:** Work in the I.M.-M. lab was supported by the ARRS (Slovenian Research Agency) P4-0116 program grant, the J4-9302 research grant awarded to I.M.-M., and the Young Researcher grant awarded to M.S. Work in the N.R.S.-W. lab was supported by the Biotechnology and Biological Sciences Research Council [BB/M013774/1; BB/N022254/1; BB/R012415/1].

**Acknowledgments:** We would like to acknowledge the support of the infrastructural center "Microscopy of biological sample" of the Biotechnical Faculty, University of Ljubljana. We thank D. Rudner for the kind gift of the pKM3 plasmid. We would also like to thank Á.T. Kovács for constructive discussions.

**Conflicts of Interest:** The authors declare no conflicts of interest.

## References

1. Hall-Stoodley, L.; Costerton, J.W.; Stoodley, P. Bacterial biofilms: from the Natural environment to infectious diseases. *Nat. Rev. Genet.* **2004**, *2*, 95–108. [CrossRef]
2. Flemming, H.-C.; Wingender, J. The biofilm matrix. *Nat. Rev. Genet.* **2010**, *8*, 623–633. [CrossRef] [PubMed]
3. Jefferson, K.K. What drives bacteria to produce a biofilm? *FEMS Microbiol. Lett.* **2004**, *236*, 163–173. [CrossRef] [PubMed]
4. Bloom-Ackermann, Z.; Ganin, H.; Kolodkin-Gal, I. Quorum-sensing Cascades Governing Bacterial Multicellular Communities. *Isr. J. Chem.* **2015**, *56*, 302–309. [CrossRef]
5. Platt, T.G.; Fuqua, C. What's in a name? The semantics of quorum sensing. *Trends Microbiol.* **2010**, *18*, 383–387. [CrossRef]
6. Mandic-Mulec, I.; Stefanic, P.; Van Elsas, J.D. Ecology of Bacillaceae. *Microbiol. Spectr.* **2015**, *3*, 59–85. [CrossRef]

7.    Piggot, P.J.; Coote, J.G. Genetic aspects of bacterial endospore formation. *Bacteriol. Rev.* **1976**, *40*, 908–962. [CrossRef]

8.    Cairns, L.; Hobley, L.; Stanley-Wall, N.R. Biofilm formation by *Bacillus subtilis*: new insights into regulatory strategies and assembly mechanisms. *Mol. Microbiol.* **2014**, *93*, 587–598. [CrossRef]

9.    Dragos, A.; Kiesewalter, H.; Martin, M.; Hsu, C.-Y.; Hartmann, R.; Wechsler, T.; Eriksen, C.; Brix, S.; Drescher, K.; Stanley-Wall, N.; et al. Division of Labor during Biofilm Matrix Production. *Curr. Boil.* **2018**, *28*, 1903–1913. [CrossRef]

10.   Abee, T.; Kovács, Á.T.; Kuipers, O.P.; Van Der Veen, S. Biofilm formation and dispersal in Gram-positive bacteria. *Curr. Opin. Biotechnol.* **2011**, *22*, 172–179. [CrossRef]

11.   Kobayashi, K. Gradual activation of the response regulator DegU controls serial expression of genes for flagellum formation and biofilm formation in *Bacillus subtilis*. *Mol. Microbiol.* **2007**, *66*, 395–409. [CrossRef] [PubMed]

12.   López, D.; Kolter, R. Extracellular signals that define distinct and coexisting cell fates in *Bacillus subtilis*. *FEMS Microbiol. Rev.* **2010**, *34*, 134–149. [CrossRef] [PubMed]

13.   Veening, J.-W.; Smits, W.K.; Kuipers, O.P. Bistability, Epigenetics, and Bet-Hedging in Bacteria. *Annu. Rev. Microbiol.* **2008**, *62*, 193–210. [CrossRef] [PubMed]

14.   Schultz, D.; Wolynes, P.G.; Ben-Jacob, E.; Onuchic, J.N. Deciding fate in adverse times: Sporulation and competence in *Bacillus subtilis*. *Proc. Natl. Acad. Sci. USA* **2009**, *106*, 21027–21034. [CrossRef] [PubMed]

15.   Lopez, D.; Vlamakis, H.; Kolter, R. Biofilms. *Cold Spring Harb. Perspect. Boil.* **2010**, *2*, a000398. [CrossRef]

16.   Kalamara, M.; Spacapan, M.; Mandic-Mulec, I.; Stanley-Wall, N.R. Social behaviours by *Bacillus subtilis*: quorum sensing, kin discrimination and beyond. *Mol. Microbiol.* **2018**, *110*, 863–878. [CrossRef] [PubMed]

17.   Dogsa, I.; Choudhary, K.S.; Marsetic, Z.; Hudaiberdiev, S.; Vera, R.; Pongor, S.; Mandic-Mulec, I. ComQXPA Quorum Sensing Systems May Not Be Unique to *Bacillus subtilis*: A Census in Prokaryotic Genomes. *PLoS ONE* **2014**, *9*, e96122. [CrossRef] [PubMed]

18.   Roggiani, M.; Dubnau, D. ComA, a phosphorylated response regulator protein of *Bacillus subtilis*, binds to the promoter region of *srfA*. *J. Bacteriol.* **1993**, *175*, 3182–3187. [CrossRef]

19.   Ogura, M. DNA microarray analysis of *Bacillus subtilis* DegU, ComA and PhoP regulons: an approach to comprehensive analysis of *B.subtilis* two-component regulatory systems. *Nucleic Acids Res.* **2001**, *29*, 3804–3813. [CrossRef]

20.   Comella, N.; Grossman, A.D. Conservation of genes and processes controlled by the quorum response in bacteria: characterization of genes controlled by the quorum-sensing transcription factor ComA in *Bacillus subtilis*. *Mol. Microbiol.* **2005**, *57*, 1159–1174. [CrossRef]

21.   Wolf, D.; Rippa, V.; Mobarec, J.C.; Sauer, P.; Adlung, L.; Kolb, P.; Bischofs, I.B. The quorum-sensing regulator ComA from *Bacillus subtilis* activates transcription using topologically distinct DNA motifs. *Nucleic Acids Res.* **2015**, *44*, 2160–2172. [CrossRef] [PubMed]

22.   Nakano, M.M.A.; Marahiel, M.; Zuber, P. Identification of a genetic locus required for biosynthesis of the lipopeptide antibiotic surfactin in *Bacillus subtilis*. *J. Bacteriol.* **1988**, *170*, 5662–5668. [CrossRef] [PubMed]

23.   López, D.; Fischbach, M.A.; Chu, F.; Losick, R.; Kolter, R. Structurally diverse natural products that cause potassium leakage trigger multicellularity in *Bacillus subtilis*. *Proc. Natl. Acad. Sci. USA* **2008**, *106*, 280–285. [CrossRef] [PubMed]

24.   López, D.; Vlamakis, H.; Losick, R.; Kolter, R. Paracrine signaling in a bacterium. *Genes Dev.* **2009**, *23*, 1631–1638. [CrossRef]

25.   Lazazzera, B.A.; Kurtser, I.G.; McQuade, R.S.; Grossman, A.D. An Autoregulatory Circuit Affecting Peptide Signaling in *Bacillus subtilis*. *J. Bacteriol.* **1999**, *181*, 5193–5200. [CrossRef] [PubMed]

26.   Auchtung, J.; Lee, C.A.; Grossman, A.D. Modulation of the ComA-Dependent Quorum Response in *Bacillus subtilis* by Multiple Rap Proteins and Phr Peptides. *J. Bacteriol.* **2006**, *188*, 5273–5285. [CrossRef] [PubMed]

27.   Ishii, H.; Tanaka, T.; Ogura, M. The *Bacillus subtilis* Response Regulator Gene *degU* Is Positively Regulated by CcpA and by Catabolite-Repressed Synthesis of ClpC. *J. Bacteriol.* **2012**, *195*, 193–201. [CrossRef] [PubMed]

28.   Msadek, T.; Kunst, F.; Klier, A.; Rapoport, G. DegS-DegU and ComP-ComA modulator-effector pairs control expression of the *Bacillus subtilis* pleiotropic regulatory gene *degQ*. *J. Bacteriol.* **1991**, *173*, 2366–2377. [CrossRef]

29.   Dogsa, I.; Brloznik, M.; Stopar, D.; Mandic-Mulec, I. Exopolymer Diversity and the Role of Levan in *Bacillus subtilis* Biofilms. *PLoS ONE* **2013**, *8*, e62044. [CrossRef]

30. Branda, S.S.; González-Pastor, J.E.; Ben-Yehuda, S.; Losick, R.; Kolter, R. Fruiting body formation by *Bacillus subtilis*. *Proc. Natl. Acad. Sci. USA* **2001**, *98*, 11621–11626. [CrossRef]

31. Kearns, D.B.; Chu, F.; Branda, S.S.; Kolter, R.; Losick, R. A master regulator for biofilm formation by *Bacillus subtilis*. *Mol. Microbiol.* **2004**, *55*, 739–749. [CrossRef]

32. Romero, D.; Aguilar, C.; Losick, R.; Kolter, R. Amyloid fibers provide structural integrity to *Bacillus subtilis* biofilms. *Proc. Natl. Acad. Sci. USA* **2010**, *107*, 2230–2234. [CrossRef] [PubMed]

33. Terra, R.; Stanley-Wall, N.R.; Cao, G.; Lazazzera, B.A. Identification of *Bacillus subtilis* SipW as a Bifunctional Signal Peptidase That Controls Surface-Adhered Biofilm Formation. *J. Bacteriol.* **2012**, *194*, 2781–2790. [CrossRef] [PubMed]

34. Hamon, M.A.; Stanley, N.R.; Britton, R.A.; Grossman, A.D.; Lazazzera, B.A. Identification of AbrB-regulated genes involved in biofilm formation by *Bacillus subtilis*. *Mol. Microbiol.* **2004**, *52*, 847–860. [CrossRef] [PubMed]

35. Vlamakis, H.; Chai, Y.; Beauregard, P.; Losick, R.; Kolter, R. Sticking together: building a biofilm the *Bacillus subtilis* way. *Nat. Rev. Genet.* **2013**, *11*, 157–168. [CrossRef]

36. Jiang, M.; Shao, W.; Perego, M.; Hoch, J.A. Multiple histidine kinases regulate entry into stationary phase and sporulation in *Bacillus subtilis*. *Mol. Microbiol.* **2000**, *38*, 535–542. [CrossRef]

37. Chai, Y.; Chu, F.; Kolter, R.; Losick, R. Bistability and biofilm formation in *Bacillus subtilis*. *Mol. Microbiol.* **2007**, *67*, 254–263. [CrossRef] [PubMed]

38. Shafikhani, S.H.; Mandic-Mulec, I.; Strauch, M.A.; Smith, I.; Leighton, T. Postexponential Regulation of *sin* Operon Expression in *Bacillus subtilis*. *J. Bacteriol.* **2002**, *184*, 564–571. [CrossRef] [PubMed]

39. Bai, U.; Mandić-Mulec, I.; Smith, I. SinI modulates the activity of SinR, a developmental switch protein of *Bacillus subtilis*, by protein-protein interaction. *Genes Dev.* **1993**, *7*, 139–148. [CrossRef] [PubMed]

40. Chu, F.; Kearns, D.B.; Branda, S.S.; Kolter, R.; Losick, R. Targets of the master regulator of biofilm formation in *Bacillus subtilis*. *Mol. Microbiol.* **2006**, *59*, 1216–1228. [CrossRef]

41. Kobayashi, K.; Iwano, M. BslA(YuaB) forms a hydrophobic layer on the surface of *Bacillus subtilis* biofilms. *Mol. Microbiol.* **2012**, *85*, 51–66. [CrossRef] [PubMed]

42. Verhamme, D.T.; Kiley, T.B.; Stanley-Wall, N.R. DegU co-ordinates multicellular behaviour exhibited by *Bacillus subtilis*. *Mol. Microbiol.* **2007**, *65*, 554–568. [CrossRef] [PubMed]

43. Stanley, N.R.; Lazazzera, B.A. Defining the genetic differences between wild and domestic strains of *Bacillus subtilis* that affect poly-γ-dl-glutamic acid production and biofilm formation. *Mol. Microbiol.* **2005**, *57*, 1143–1158. [CrossRef] [PubMed]

44. Garde, R.; Ibrahim, B.; Kovács, Á.T.; Schuster, S. Differential equation-based minimal model describing metabolic oscillations in *Bacillus subtilis* biofilms. *R. Soc. Open Sci.* **2020**, *7*, 190810. [CrossRef] [PubMed]

45. Bettenworth, V.; Steinfeld, B.; Duin, H.; Petersen, K.; Streit, W.R.; Bischofs, I.B.; Becker, A. Phenotypic Heterogeneity in Bacterial Quorum Sensing Systems. *J. Mol. Boil.* **2019**, *431*, 4530–4546. [CrossRef] [PubMed]

46. Albano, M.; Hahn, J.; Dubnau, D. Expression of competence genes in *Bacillus subtilis*. *J. Bacteriol.* **1987**, *169*, 3110–3117. [CrossRef]

47. Spacapan, M.; Danevčič, T.; Mandic-Mulec, I. ComX-Induced Exoproteases Degrade ComX in *Bacillus subtilis* PS-216. *Front. Microbiol.* **2018**, *9*. [CrossRef]

48. Oslizlo, A.; Stefanic, P.; Vatovec, S.; Glaser, S.B.; Rupnik, M.; Mandic-Mulec, I. Exploring ComQXPA quorum-sensing diversity and biocontrol potential of *Bacillus* spp. isolates from tomato rhizoplane. *Microb. Biotechnol.* **2015**, *8*, 527–540. [CrossRef]

49. Stefanic, P.; Kraigher, B.; Lyons, N.A.; Kolter, R.; Mandic-Mulec, I. Kin discrimination between sympatric *Bacillus subtilis* isolates. *Proc. Natl. Acad. Sci. USA* **2015**, *112*, 14042–14047. [CrossRef]

50. Doan, T.; Marquis, K.A.; Rudner, D.Z. Subcellular localization of a sporulation membrane protein is achieved through a network of interactions along and across the septum. *Mol. Microbiol.* **2005**, *55*, 1767–1781. [CrossRef]

51. Patrick, J.E.; Kearns, D.B. MinJ (YvjD) is a topological determinant of cell division in *Bacillus subtilis*. *Mol. Microbiol.* **2008**, *70*, 1166–1179. [CrossRef] [PubMed]

52. Middleton, R.; Hofmeister, A. New shuttle vectors for ectopic insertion of genes into *Bacillus subtilis*. *Plasmid* **2004**, *51*, 238–245. [CrossRef]

53. Song, Y.; Nikoloff, J.M.; Fu, G.; Chen, J.; Li, Q.; Xie, N.; Zheng, P.; Sun, J.; Zhang, D. Promoter Screening from Bacillus subtilis in Various Conditions Hunting for Synthetic Biology and Industrial Applications. *PLoS ONE* **2016**, *11*, e0158447. [CrossRef] [PubMed]

54. Norman, T.M.; Lord, N.D.; Paulsson, J.; Losick, R. Stochastic Switching of Cell Fate in Microbes. *Annu. Rev. Microbiol.* **2015**, *69*, 381–403. [CrossRef] [PubMed]

55. Stefanic, P.; Mandic-Mulec, I. Social Interactions and Distribution of *Bacillus subtilis* Pherotypes at Microscale. *J. Bacteriol.* **2009**, *191*, 1756–1764. [CrossRef]

56. Parashar, V.; Konkol, M.A.; Kearns, D.B.; Neiditch, M.B. A Plasmid-Encoded Phosphatase Regulates *Bacillus subtilis* Biofilm Architecture, Sporulation, and Genetic Competence. *J. Bacteriol.* **2013**, *195*, 2437–2448. [CrossRef]

57. Ansaldi, M.; Marolt, D.; Stebe, T.; Mandic-Mulec, I.; Dubnau, D. Specific activation of the *Bacillus* quorum-sensing systems by isoprenylated pheromone variants. *Mol. Microbiol.* **2002**, *44*, 1561–1573. [CrossRef]

58. Cuesta, G.; Suarez, N.I.; Bessio, M.; Ferreira, F.; Massaldi, H. Quantitative determination of pneumococcal capsular polysaccharide serotype 14 using a modification of phenol-sulfuric acid method. *J. Microbiol. Methods* **2003**, *52*, 69–73. [CrossRef]

59. Bradford, M.M. A rapid and sensitive method for the quantitation of microgram quantities of protein utilizing the principle of protein-dye binding. *Anal. Biochem.* **1976**, *72*, 248–254. [CrossRef]

60. Arnaouteli, S.; Macphee, C.E.; Stanley-Wall, N.R. Just in case it rains: building a hydrophobic biofilm the *Bacillus subtilis* way. *Curr. Opin. Microbiol.* **2016**, *34*, 7–12. [CrossRef]

61. Hobley, L.; Ostrowski, A.; Rao, F.V.; Bromley, K.M.; Porter, M.; Prescott, A.R.; Macphee, C.E.; Van Aalten, D.; Stanley-Wall, N.R. BslA is a self-assembling bacterial hydrophobin that coats the *Bacillus subtilis* biofilm. *Proc. Natl. Acad. Sci. USA* **2013**, *110*, 13600–13605. [CrossRef] [PubMed]

62. Branda, S.S.; Chu, F.; Kearns, D.B.; Losick, R.; Kolter, R. A major protein component of the *Bacillus subtilis* biofilm matrix. *Mol. Microbiol.* **2006**, *59*, 1229–1238. [CrossRef]

63. Durrett, R.; Miras, M.; Mirouze, N.; Narechania, A.; Mandic-Mulec, I.; Dubnau, D. Genome Sequence of the *Bacillus subtilis* Biofilm-Forming Transformable Strain PS216. *Genome Announc.* **2013**, *1*, 1. [CrossRef] [PubMed]

64. Dervaux, J.; Magniez, J.C.; Libchaber, A. On growth and form of *Bacillus subtilis* biofilms. *Interface Focus* **2014**, *4*, 20130051. [CrossRef]

65. Bischofs, I.B.; Hug, J.A.; Liu, A.W.; Wolf, D.M.; Arkin, A.P. Complexity in bacterial cell-cell communication: Quorum signal integration and subpopulation signaling in the *Bacillus subtilis* phosphorelay. *Proc. Natl. Acad. Sci. USA* **2009**, *106*, 6459–6464. [CrossRef]

66. Chai, Y.; Kolter, R.; Losick, R. Reversal of an epigenetic switch governing cell chaining in *Bacillus subtilis* by protein instability. *Mol. Microbiol.* **2010**, *78*, 218–229. [CrossRef] [PubMed]

67. Vlamakis, H.; Aguilar, C.; Losick, R.; Kolter, R. Control of cell fate by the formation of an architecturally complex bacterial community. *Genome Res.* **2008**, *22*, 945–953. [CrossRef]

68. Steinberg, N.; Keren-Paz, A.; Hou, Q.; Doron, S.; Yanuka-Golub, K.; Olender, T.; Hadar, R.; Rosenberg, G.; Jain, R.; Cámara-Almirón, J.; et al. The extracellular matrix protein TasA is a developmental cue that maintains a motile subpopulation within *Bacillus subtilis* biofilms. *Sci. Signal.* **2020**, *13*, eaaw8905. [CrossRef]

69. Magnuson, R.; Solomon, J.; Grossman, A.D. Biochemical and genetic characterization of a competence pheromone from *B. subtilis*. *Cell* **1994**, *77*, 207–216. [CrossRef]

70. Beauregard, P.B.; Chai, Y.; Vlamakis, H.; Losick, R.; Kolter, R. *Bacillus subtilis* biofilm induction by plant polysaccharides. *Proc. Natl. Acad. Sci. USA* **2013**, *110*, E1621–E1630. [CrossRef]

71. Van Gestel, J.; Vlamakis, H.; Kolter, R. New Tools for Comparing Microscopy Images: Quantitative Analysis of Cell Types in *Bacillus subtilis*. *J. Bacteriol.* **2014**, *197*, 699–709. [CrossRef] [PubMed]

72. Lazazzera, A.; Solomon, B.; Grossman, J.M.A.D. An Exported Peptide Functions Intracellularly to Contribute to Cell Density Signaling in *B. subtilis*. *Cell* **1997**, *89*, 917–925. [CrossRef]

73. Thérien, M.; Kiesewalter, H.T.; Auria, E.; Charron-Lamoureux, V.; Wibowo, M.; Maróti, G.; Kovács, Á.T.; Beauregard, P.B. Surfactin production is not essential for pellicle and root-associated biofilm development of *Bacillus subtilis*. *Biofilm* **2020**, *2*, 100021. [CrossRef]

74. Higgins, D.; Dworkin, J. Recent progress in *Bacillus subtilis* sporulation. *FEMS Microbiol. Rev.* **2012**, *36*, 131–148. [CrossRef] [PubMed]

75. Fujita, M.; González-Pastor, J.E.; Losick, R. High- and Low-Threshold Genes in the Spo0A Regulon of *Bacillus subtilis*. *J. Bacteriol.* **2005**, *187*, 1357–1368. [CrossRef]

76. Lazazzera, A.B. Quorum Sensing and starvation: signals for entry into stationary phase. *Curr. Opin. Microbiol.* **2000**, *3*, 177–182. [CrossRef]

77. Mutlu, A.; Trauth, S.; Ziesack, M.; Nagler, K.; Bergeest, J.-P.; Rohr, K.; Becker, N.; Höfer, T.; Bischofs, I.B. Phenotypic memory in *Bacillus subtilis* links dormancy entry and exit by a spore quantity-quality tradeoff. *Nat. Commun.* **2018**, *9*, 1–12. [CrossRef]

78. Gasperotti, A.; Brameyer, S.; Fabiani, F.; Jung, K. Phenotypic heterogeneity of microbial populations under nutrient limitation. *Curr. Opin. Biotechnol.* **2020**, *62*, 160–167. [CrossRef]

79. Veening, J.W.; Stewart, E.J.; Berngruber, T.W.; Taddei, F.; Kuipers, O.P.; Hamoen, L.W. Bet-hedging and epigenetic inheritance in bacterial cell development. *Proc. Natl. Acad. Sci. USA* **2008**, *105*, 4393–4398. [CrossRef]

80. Grossman, A.D. Genetic Networks Controlling the Initiation of Sporulation and the Development of Genetic Competence in *Bacillus Subtilis*. *Annu. Rev. Genet.* **1995**, *29*, 477–508. [CrossRef]

81. Shank, E.A.; Kolter, R. Extracellular signaling and multicellularity in *Bacillus subtilis*. *Curr. Opin. Microbiol.* **2011**, *14*, 741–747. [CrossRef] [PubMed]

 *microorganisms*

*Article*

# The Role of Surfactin Production by *Bacillus velezensis* on Colonization, Biofilm Formation on Tomato Root and Leaf Surfaces and Subsequent Protection (ISR) against *Botrytis cinerea*

Alexandra Stoll [1,2,*], Ricardo Salvatierra-Martínez [2], Máximo González [1] and Michael Araya [3]

[1] Laboratorio de Microbiología Aplicada, Centro de Estudios Avanzados de Zonas Áridas (CEAZA), La Serena 1720256, Chile; maximo.gonzalez@ceaza.cl
[2] Programa de Doctorado en Biología y Ecología Aplicada, Departamento de Biología, Universidad de La Serena, La Serena 1720170, Chile; rsmesteban@gmail.com
[3] Centro de Investigación y Desarrollo Tecnológico en Algas CIDTA, Universidad Católica del Norte, Coquimbo 1781421, Chile; mmaraya@ucn.cl
* Correspondence: alexandra.stoll@ceaza.cl

**Abstract:** Many aspects regarding the role of lipopeptides (LPs) in bacterial interaction with plants are not clear yet. Of particular interest is the LP family of surfactin, immunogenic molecules involved in induced systemic resistance (ISR) and the bacterial colonization of plant surfaces. We hypothesize that the concentration of surfactin produced by a strain correlates directly with its ability to colonize and persist on different plant surfaces, which conditions its capacity to trigger ISR. We used two *Bacillus velezensis* strains (BBC023 and BBC047), whose antagonistic potential in vitro is practically identical, but not on plant surfaces. The surfactin production of BBC047 is 1/3 higher than that of BBC023. Population density and SEM images revealed stable biofilms of BBC047 on leaves and roots, activating ISR on both plant surfaces. Despite its lower surfactin production, strain BBC023 assembled stable biofilms on roots and activated ISR. However, on leaves only isolated, unstructured populations were observed, which could not activate ISR. Thus, the ability of a strain to effectively colonize a plant surface is not only determined through its production of surfactin. Multiple aspects, such as environmental stressors or compensation mechanisms may influence the process. Finally, the importance of surfactin lies in its impacts on biofilm formation and stable colonization, which finally enables its activity as an elicitor of ISR.

**Keywords:** colonization pattern; population density; surfactin production; ISR

**Citation:** Stoll, A.; Salvatierra-Martínez, R.; González, M.; Araya, M. The Role of Surfactin Production by *Bacillus velezensis* on Colonization, Biofilm Formation on Tomato Root and Leaf Surfaces and Subsequent Protection (ISR) against *Botrytis cinerea*. *Microorganisms* **2021**, *9*, 2251. https://doi.org/10.3390/microorganisms9112251

Academic Editor: Imrich Barák

Received: 7 September 2021
Accepted: 14 October 2021
Published: 28 October 2021

**Publisher's Note:** MDPI stays neutral with regard to jurisdictional claims in published maps and institutional affiliations.



## 1. Introduction

Currently, several strains of *Bacillus* sp. are used as biocontrol agents against crop disease [1–4]. These strains exhibit different biocontrol mechanisms [5], of which antibiosis and the induction of plant resistance (ISR) are considered the most important. Some of these mechanisms of action are directly linked to the synthesis of lipopeptides (LPs) by *Bacillus* spp. [1,6–8]. LPs are secondary metabolites formed by cyclic peptides linked to a lipid tail or other lipophilic molecules, which self-assemble in diverse configurations and therefore have high structural diversity. Three families, namely iturins, fengycins and surfactins, have been distinguished and characterized to have different capacities, e.g., as antibiotics (bacillomycin and fengycin) or elicitors of ISR [9,10].

In the context of biocontrol, it has been postulated that both direct antagonism and ISR require efficient colonization of plant surfaces [7,11,12]. The colonization of plants is a complex process that involves the attraction to and establishment of microorganisms in different plant compartments. The attraction phase, particularly in the roots, occurs through chemotaxis induced by plant exudates, while establishment involves swarming motility and biofilm production [13–16]. For both mechanisms, a dependence on the production of the immunogenic LP surfactin has been suggested [2,6,17].

According to previous studies, surfactin contributes to both direct and indirect antagonism and is key to plant colonization [2,18,19]. Zeriouh et al. [6] reported that although the LP antibiotics bacillomycin and fengycin, synthesized by *B. velezensis* UMAF6614, generate biocontrol of powdery mildew by direct action, mutations that affect the production of surfactin in this strain decrease its biocontrol capacity despite it continuing to produce the antibiotic LPs. This synergistic effect between the different lipopeptide families was also described for other *Bacillus* spp. strains [2,19], enhancing the dual functionality of surfactin during antibiosis and colonization. Several authors reported a physiological threshold dose of surfactin (10 µm/mL), required to trigger ISR in plants [7,20]. For the colonization of and biofilm formation on plant surfaces, such a threshold has not been published. Recently, Therien et al. [21] suggested that in *B. subtilis* strains biofilm formation could be independent from surfactin production.

In a previous study, we compared the biocontrol effect generated by two strains of *B. velezensis* (strains BBC023 and BBC047) against *Botrytis cinerea* (a necrotrophic fungus of great global importance) in vitro and in vivo [22]. We observed that under in vitro conditions, the antagonistic potential of both strains was practically identical, while the results between the strains differed when they were required to establish on plant surfaces.

Thus, we postulate that these differences could be correlated with the concentration of surfactin produced by a strain, which could determine the colonization ability of the respective strain and its permanence on different plant surfaces. Thereby, our work evaluated the production of surfactin and the colonization in time of two *B. velezensis* strains (BBC023 and BBC047) in a pathosystem comprising a tomato (host) and *B. cinerea*.

## 2. Materials and Methods

### 2.1. Bacterial Strains, Broth and Growth Conditions

The native strains of *B. velezensis* BBC023 and BBBC047 [20], the model strain FZB42 (producer of surfactin A, bacillomycin D and fengycin A and B; [23]) and its CH1 mutant deficient in the production of surfactin [24] were used. All bacterial strains were grown in Luria–Bertani (LB) broth for inoculum production, and 1.5% agar was used for solid plates. For the growth of the CH1 mutant, 1 µg/mL erythromycin was added to the growth medium.

For plant inoculation, bacterial strains were grown in a MOLP medium [25] at 30 °C and 170 rpm for 48 h (~$10^9$ colony-forming units (CFU) per mL). Finally, the bacterial concentration was adjusted to ~$10^7$ CFU/mL.

### 2.2. Quantification of Surfactin Production with HPLC

The LPs were obtained by the acid precipitation method [26]. The different strains were grown in flasks with 100 mL of MOLP medium for 48 h at 30 °C and 170 rpm. Subsequently, the media were centrifuged at 10,000 rpm for 15 min, and the supernatant was recovered for the next steps. LPs were precipitated by adding 3 N HCl to the supernatant until pH 2.0 was reached. Afterwards, the samples were incubated for 30 min at 4 °C, and the LPs were precipitated by centrifugation at 10,000 rpm for 15 min. Finally, the supernatant was discarded, and the LPs were resuspended in 1 mL of methanol (100%). This fraction was injected into a Jasco CO-2065 Plus HPLC system (SpectraLab Scientific Inc., Markham, ON, Canada) equipped with a 5 µm particle size Sunniest C18 250 × 4.6 mm column (CromaNik Technologies Inc., Osaka, Japan) and a UV detector. Surfactin concentrations were determined by calibration curves according to the method of Ali et al. [27]. Quantification was performed in triplicate for each strain using the same protocol.

### 2.3. Biofilm Formation, Motility and Adherence

Biofilm formation was evaluated on Petri dishes with solid MSgg medium (2.5 mM PBS, pH 7, 100 mM MOPS, pH 7, 50 µM $FeCl_3$, 2 mM $MgCl_2$, 50 µM $MnCl_2$, 1 µM $ZnCl_2$,

2 μM thiamine, 50 mg phenylalanine, 0.5% glycerol, 0.5% glutamate and 700 μM CaCl$_2$ [28]). Each inoculated Petri dish was subsequently incubated at 37 °C for 48 h [22]. Motility was evaluated using semisolid LB medium (0.75% agar), while adherence was characterized in 15 × 150 mm tubes with 10 mL of MOLP medium, which were incubated for 48 h at 30 °C [29].

The initial inoculum for each assay was 3 μL of bacterial culture grown in liquid LB for 12 h, adjusted to 10$^7$ CFU/mL.

### 2.4. Characterization of Surfactin Profiles with UHPLC-MS

To identify the LP isoforms, 20 μL of the methanolic fraction was injected into a Dionex UltiMate 3000 UHPLC-MS system coupled to a Thermo Scientific Orbitrap Q Exactive Focus detector (Thermo Fisher Scientific, Waltham, MA, USA) and equipped with a C18 2.1 column × 50 mm, 1.7 μm. A method based on acetonitrile gradients as the mobile phase was used ([27], with modifications). For each run an elution gradient was employed from 95% A:5% B (eluent A: 18 M water/0.1% formic acid; eluent B: 100% methanol/0.1% formic acid) to 5% A:95% B during 12 min. Positive ionization mode data (ESI+) were recorded in full-scan mode for a mass range between 200–1500 *m/z*. The results expressed in an *m/z* ratio corresponding to the surfactin family were used to identify the isoforms. Chromatogram cleaning and data treatment were performed using Excalibur 3.2 software. Chromatograms and mass spectra are expressed as relative abundance and relative percentage against the surfactin standard (Sigma Aldrich, S3523, St. Louis, MO, USA).

### 2.5. Tomato Plant Protection (ISR) against Botrytis Cinerea by Inoculation with Suspension of Bacillus velezensis

#### 2.5.1. Preparation Procedures of the Plants

In this experiment, tomato plants of the hybrid cultivar CalAce (PetoSeed Co. Chile Ltda, Santiago, Chile), 30 days old (second true leaf), were used. Seeds were germinated and grown in Speedling trays until reaching 30 days of age. Then, the roots were completely cleaned of substrate, and the plants were acclimated in a sterile Hoagland nutrient solution for three days. Next, the roots were sterilized, immersed in 70% ethanol for 1 min and then immersed in 5% sodium hypochlorite for 5 min. Finally, the roots were washed 5 times with sterile distilled water and placed in a new nutrient solution. After 5 days of recovery, plants were used in the different experiments.

#### 2.5.2. Experimental Design

Protection (ISR) experiments of tomato plants against disease caused by *B. cinerea* by inoculation of root or leaf surfaces with bacterial suspensions of the antagonist strains and spore suspension of the pathogen. Disease severity indices (DSI) caused by *B. cinerea* were evaluated to express the protection (ISR) provided by the bacterial strains. For each surface, four treatments were established: strain BBC023, strain BBC047, surfactin standard at 5 μg/mL and at 25 μg/mL (Table 1). The treatments were applied physically separate from the fungal infection (e.g., separate leaves or root/leaf), as schematically shown in Supplementary Figure S1.

Finally, a control treatment was established in which only *B. cinerea* was infected.

For foliar application, one leaf per plant was submerged for 30 min in a bacterial suspension of ~10$^7$ CFU/mL or in a surfactin solution (5 or 25 μg/mL), depending on the respective treatment. Root application was carried out by immersing the roots for 2 h in a bacterial suspension of ~10$^7$ CFU/mL or in a surfactin solution (5 or 25 μg/mL).

**Table 1.** Treatments used in protection (ISR) experiment with tomato.

| Treatment | Treatment Application | | B. cinerea Application |
| --- | --- | --- | --- |
| | Root | Leaf 1 | Leaf 2 |
| Control | | | x |
| BBC023 | x | | x |
| BBC047 | x | | x |
| Surf (5 µg/mL) | x | | x |
| Surf (25 µg/mL) | x | | x |
| BBC023 | | x | x |
| BBC047 | | x | x |
| Surf (5 µg/mL) | | x | x |
| Surf (25 µg/mL) | | x | x |

After the establishment of the treatments, the plants were kept in a hydroponic system with a Hoagland nutrient solution and using a photoperiod of 12 h light and 12 h dark for 72 h. Afterwards, infection with *B. cinerea* was carried out according to Mariutto et al. [30] with some modifications. Small wounds were made with a dissecting needle at 6 points on a leaf (a different leaf than for the foliar treatment application, see Figure S1); at each point, 10 µL of a *B. cinerea* conidia suspension ($10^5$ conidia/µL) was placed. Seventy-two hours post-infection, the disease severity around the point of injury was evaluated. A diagrammatic severity scale was used: level 1: 0–10 $mm^2$; level 2: 10–20 $mm^2$; level 3: 20–30 $mm^2$; and level 4: 30–40 $mm^2$. The progression of the disease was calculated using ImageJ software. Finally, the disease severity index was calculated using the following formula: DSI (%) = ($\Sigma$ (f $*$ v)/N $*$ X) $*$ 100, where f = the number of infection points where *B. cinerea* spread beyond the injury point, v = the value on the severity scale, N = the total number of evaluated inoculation points (6 points/plant) and X = the highest value on the scale. For each treatment, 5 plants per treatment were evaluated.

### 2.6. Colonization Experiment

#### 2.6.1. Colonization in Leaves

Plants were prepared as described in Section 2.5.1 and transplanted to 500 mL pots with a mixture of peat and sterilized soil at a ratio of 1:1. After 5 days, plants were inoculated by immersing the leaves for 30 min in bacterial broth with a $10^7$ CFU/mL suspension of the BBC023 or BBC047 strain. The experimental design considered 5 plants for each strain and 10 control plants. Control leaves were immersed in MOLP medium. Colonization was evaluated at 7, 14 and 21 days post-inoculation (DPI).

#### 2.6.2. Colonization in Roots

Plants were prepared as described in Section 2.5.1 and transplanted into a rhizobox system consisting of 2 glass plates (40 cm high and 30 cm wide) with a 0.5 cm layer of substrate between the glasses. As substrate, a 1:1 mixture of peat and soil was used. The roots of the plants were submerged for 2 h in a bacterial suspension, and later, 2 plants were placed inside each rhizobox, which were then sealed. For each treatment, 12 rhizoboxes were assembled.

Colonization was evaluated at 7, 14 and 21 days post-inoculation (DPI). The evaluation was carried out at three root sections according to their growth: upper section (the third closest to the neck of the plant), middle section and lower section (last third of the roots) (Supplementary Figure S2). In each evaluation, three rhizoboxes per treatment ($n = 3$) were analyzed, of which one plant was used to obtain the population density and another was used to perform the observation with SEM. For the SEM analysis, the roots were separated according to the section. Finally, to identify the zone with the highest microbial density, each root of each section was divided into two observation points, the root tip and the middle zone, to identify the point with the highest concentration of microbial colonies.

*2.7. Evaluation of Population Dynamics in the Rhizoplane and Phylloplane*

2.7.1. Colonization Density

One gram of roots or leaves was ground in a mortar with 5 mL of phosphate-buffered saline (PBS) and placed in 15 mL tubes (Corning Inc., Corning, NY, USA) with 10 mL of $1\times$ PBS. The samples were heated for 30 min at 80 °C. From the obtained solution, a serial dilution was realized and, subsequently, 100 µL of each dilution was seeded in LB plates (triplicates) and incubated for 24 h at 37 °C. Subsequently, the CFU were counted with morphological characteristics corresponding to the strains BBC023 and BBC047. Each plant was considered as a replicate.

2.7.2. Scanning Electron Microscopy (SEM)

Structural analysis of each biofilm generated by the bacterial treatments was performed using SEM at 7, 14, and 21 DPI. The root or leaf samples were fixed in 0.1 mM of sodium cacodylate (pH 7.2) and 2.5 mM of glutaraldehyde for 12 to 24 h at room temperature; they were subsequently dehydrated in a series of 50% to 100% ethanol solutions and dried with a point dryer (Critical Bal-Tec CPD 030). The dried samples were coated with a thin layer of gold prior to observation using a Leica EM SCD050 coater. For visualization, a JEOL JSM-6490 LV SEM (JEOL Ltda, Tokyo, Japan) was used. The root analysis included primary and secondary roots as well as root hairs located in each of the previously described sections (upper, middle and lower), with 5 mm root pieces being analyzed.

*2.8. Statistical Analysis*

The obtained data were analyzed by one-way ANOVA, and the means were compared using the Tukey test. All analyses were performed using the agricolae package in R [31], considering $p < 0.05$ to indicate significance.

**3. Results**

*3.1. Attributes Associated with Colonization: Surfactin Production and Profiles, Biofilms, Motility and Adherence*

Surfactin production was quantified in strains BBC023, BBC047 and FZB42 (positive control) and in the mutant CH1 (negative control) (Figure 1A). Our results show that strain BBC047 produced the highest levels of surfactin ($633 \pm 14.1$ µg/mL), followed by strains FZB42 ($505 \pm 0.1$ µg/mL), BBC023 ($405.5 \pm 25$ µg/mL) and CH1 ($0 \pm 0$ µg/mL). BBC047 was also the strain with the highest capacity to assemble complex biofilms in the MSgg medium, while the CH1 strain, which does not produce surfactins, did not form secondary structures in the biofilm. BBC023 and FZB42 presented biofilms with evident secondary structures but with less complexity than that of BBC047.

Finally, strains BBC047, BBC023 and FZB42 showed no significant differences regarding their motility and adherence, while strain CH1 exhibited a significant reduction in these traits (Figure 1B,C).

The comparison of the surfactin profiles revealed that both strains produce a similar pattern of isoforms (Figure 2A), ranging from C12 to C17. The relative abundance was highest for surfactin isoforms C14, C15 and C16 (>88% of the total surfactin). Compared against the surfactin standard, BBC023 produced relatively less of each isoform (except C12) than BBC047 (Figure 2B). A particularly large difference was detected for surfactin isoforms C15 and C16 with 2.5% and 8.0% less in BBC023, respectively.

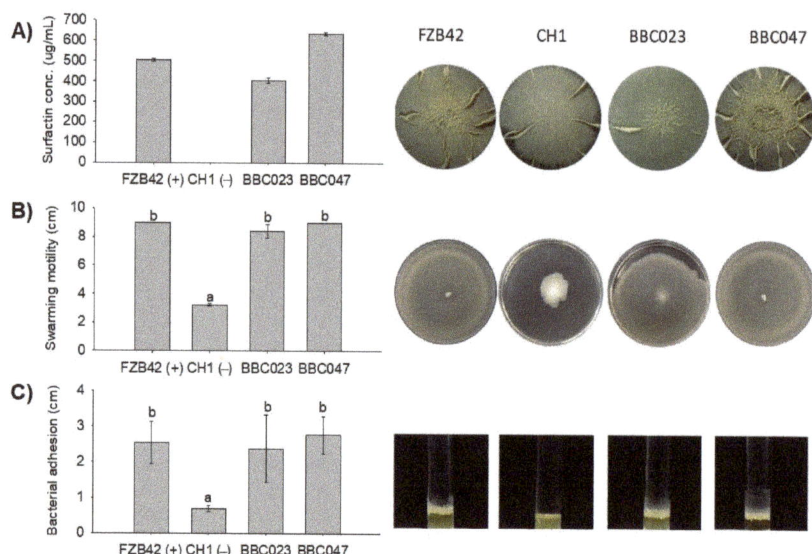

**Figure 1.** Surfactin production of *B. velezensis* strains BBC023 and BBC047, assemblage of biofilms in MSgg, motility and adherence. (**A**) Production of surfactins and biofilms in MSgg medium. (**B**) Swarming motility in a 9 cm plate with semisolid LB medium. (**C**) Surface adhesion in test tube (15 × 1 cm). Letters a and b indicate All evaluations were carried out at 48 h (images of biofilm structures were modified from Salvatierra-Martinez et al. [22]).

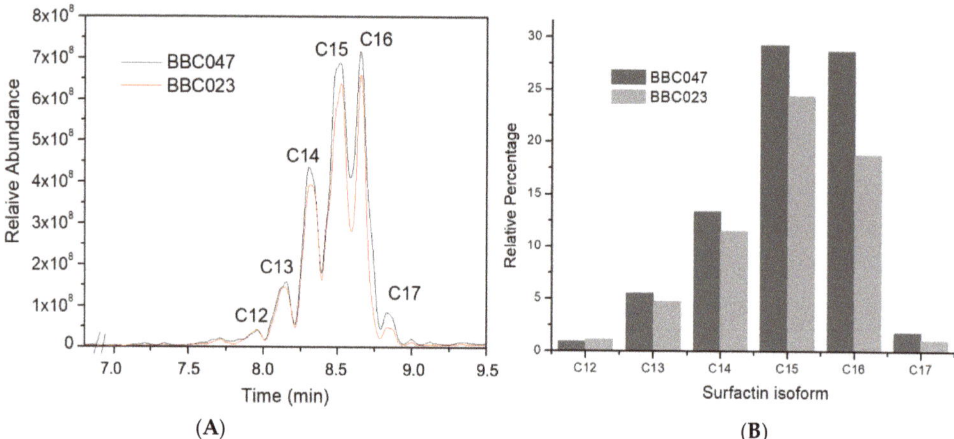

**Figure 2.** Surfactin profiles of *B. velezensis* strains BBC023 and BBC047 from UHPLC-MS. (**A**) Comparison of surfactin profiles from BBC023 and BBC047. (**B**) Relative proportion of surfactin isoforms produced by BBC023 and BBC047 compared against the surfactin standard. Detection was performed at 48 h liquid culture growth.

Interestingly, when looking at the relative distribution of the isoforms within the profile of each strain (Figure S3), BBC023 produced slightly more of all isoforms than BBC047, except C16 and C17. The isoform distribution for strain FZB42 (Supplementary Figure S3), used as reference, displayed a shift in the relative abundance: C13 to C15 isoforms represented 87% of total surfactin, whereas C17 was not detected.

*3.2. Protection (ISR) Experiment of Tomato by Inoculation of Plant Surfaces with Strains BBC023 and BBC047 against Disease Caused by B. cinerea*

The protection (ISR) responses of tomato plants, triggered by root and leaf treatments with strains BBC023 and BBC047, were evaluated with a disease severity index of the necrotrophic fungus *B. cinerea*. These results were compared with the direct effect of different concentrations of a surfactin standard.

All root treatments (both strains and both surfactin concentrations) reduced the disease severity of *B. cinerea* with respect to the control, where the most effective treatments were BBC047, surfactin at 25 µg/mL and BBC023 (Figure 3). In leaf applications, only strain BBC047 and surfactin at 25 µg/mL reduced DSI, whereas strain BBC023 and surfactin at 5 µg/mL presented no differences from the control. Particularly, the results for strain BBC023 are noticeable, as it was not capable of activating the induction of resistance via leaf application, but when applied to the roots, it managed to reduce the severity of *B. cinerea*.

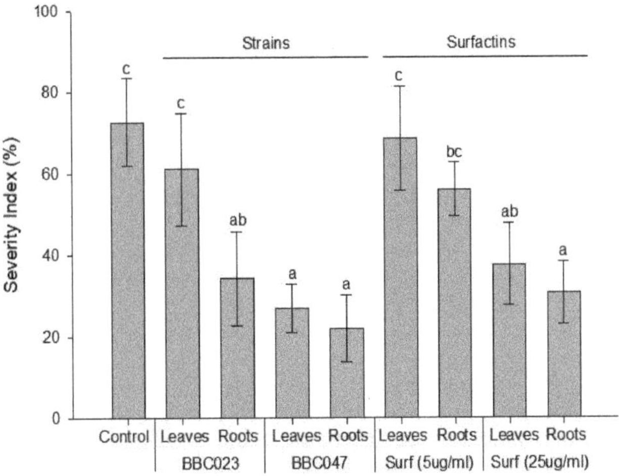


**Figure 3.** Disease severity index (DSI) (%) after *B. cinerea* infection in different bacterial and surfactin standard treatments on tomato leaves and roots via ISR (direct antagonism between treatments and pathogen was excluded). Each bar represents the mean ± standard deviation of the control index obtained from 30 observations in 5 plants. Different letters above the bars represent significant differences between the treatments according to the Tukey test ($p < 0.05$).

*3.3. Colonization and Population Dynamics in the Tomato Rhizoplane and Phylloplane*

The density and population dynamics of strains BBC023 and BBC047, on both the rhizoplane and phylloplane, were determined by SEM and quantified by CFU at 7, 14 and 21 DPI. We evaluated the effect of the application surface (root or leaf) on the density and population dynamics of each strain.

Root colonization by BBC023 and BBC047 resulted in similar population densities (Figure 4). Both strains showed a preference for the lower root section over the middle or upper section, whereas in the control treatment no colonization was observed (Supplementary Figure S4). The middle and upper sections presented only small bacterial groups up to 21 DPI (Supplementary Figures S5 and S6). On the other hand, from 7 DPI, strains BBC023 and BBC047 mainly colonized the secondary root tips of the lower section, while no marked colonization in the root hairs was registered. Finally, strain BBC047 stands out for developing marked colonization in the root maturation area.

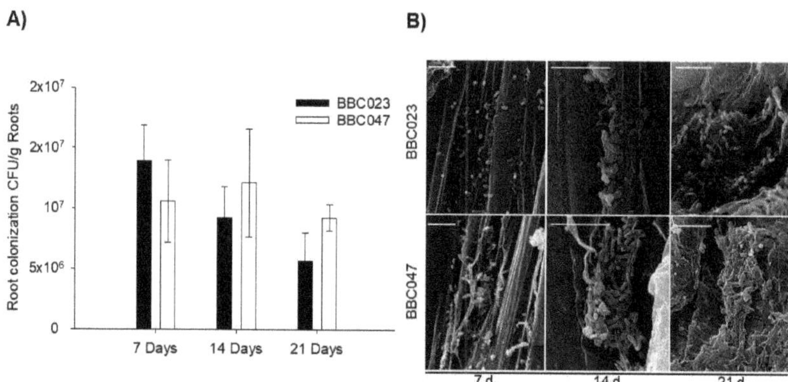



**Figure 4.** Population dynamics and colonization on the tomato rhizoplane by strains BBC023 and BBC047. (**A**) Quantification of CFU on tomato roots inoculated with BBC023 and BBC047. The bars represent the average of replicates with their standard deviation. (**B**) SEM images of root colonization. Photographs correspond to the meristematic zone of the secondary roots obtained from the lower third. Timeline represents the colony establishment (7 d), formation of extracellular matrix (14 d) and colony covered by this matrix (21 d). Scale bar represents 10 μm.

At 21 DPI, robust biofilms were observed at root tips, in the elongation zone and in the meristematic zone (Figure 5). In addition, a three-dimensional structure was observed at the meristematic and caliptral zones (Figure 5(Ac,Bc)), also highlighting the sporadic formation of microcolonies in concave zones of the primary roots. Specifically, the biofilm of BBC047 presented robust coverage, which ranged from the maturation zone to the meristematic zone of the roots. Additionally, for BBC023, numerous populations were observed in the meristematic zone, forming robust biofilms. In contrast, the close-up view of the maturation zone showed sporadic colonization but without the formation of robust films.

On the leaf, the strains showed opposite results. BBC023 concentration decreased after 7 days ($1 \times 10^7$ to $1.13 \times 10^6$), with a minimum quantification of $6.67 \times 10^5$ and $1.50 \times 10^5$ at 14 and 21 DPI, respectively. In contrast, BBC047 presented a stable quantification, close to 107 CFU, at 7 and 14 DPI ($8.38 \times 10^6$), with a slight decrease only at 21 DPI ($5.03 \times 10^6$) (Figure 6A).

Our SEM results showed that BBC023 did not establish or organize persistent communities on the leaf, being observed at 21 DPI in only a few dispersed populations (Figure 6B). BBC047 formed numerous populations, and from 7 DPI it was possible to observe the production of an extracellular matrix. Later, at 21 DPI, structured populations in biofilms were observed in the junctions between the walls of the epidermal cells of the leaves, supported by a network formed by a type of extracellular fibrillar matrix (Figure 6B).

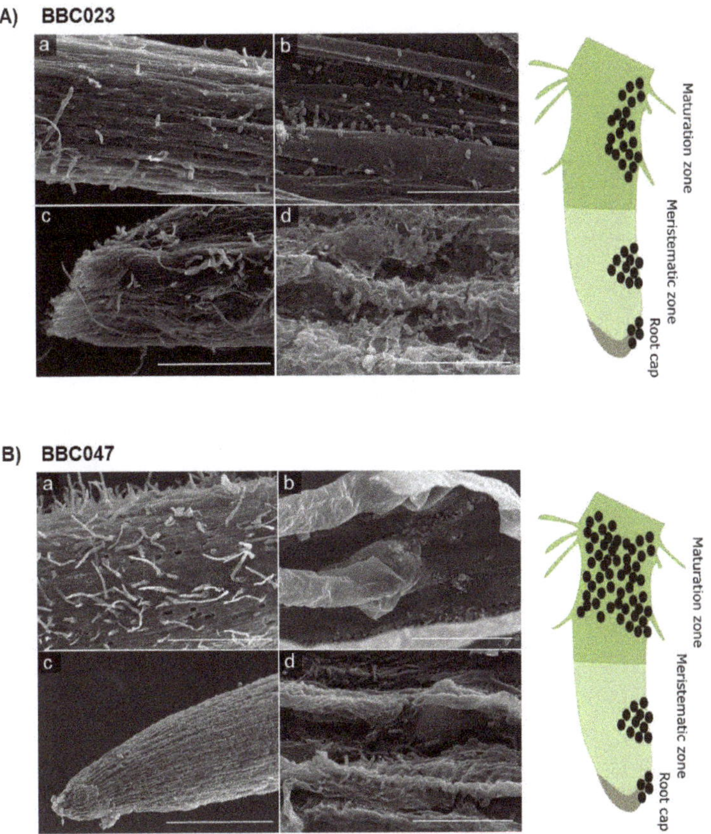



**Figure 5.** Colonization of tomato roots by *B. velezensis* strains BBC023 and BBC047 at 21 DPI. (**A**) BBC023: (**a**) secondary root maturation zone (bar 300 μm), (**b**) increased maturation zone (bar 30 μm), (**c**) zone of elongation and calyptra (bar 300 μm) and (**d**) close-up of elongation zone and calyptra (bar 10 μm). (**B**) BBC047: (**a**) secondary root maturation zone (bar 300 μm), (**b**) increased maturation zone (bar 30 μm), (**c**) elongation zone and calyptra (bar 300 μm) and (**d**) close-up of elongation zone and calyptra (bar 10 μm).

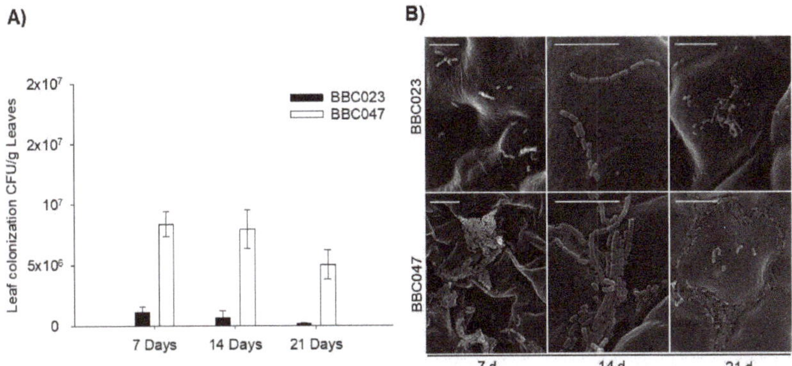



**Figure 6.** Population dynamics and colonization on the tomato rhizoplane by *B. velezensis* strains BBC023 and BBC047. (**A**) Quantification of CFU on tomato leaves inoculated with BBC023 and BBC047. The bars represent the average of replicates with their standard deviation. (**B**) SEM images of leave colonization. Scale bar represents 10 μm.

## 4. Discussion

Previously, our group published the first characterization of the lipopeptide production of strains BBC023 and BBC047, revealing their capability to generate all three families of LPs (surfactins, iturins and fengycins). In dual confrontation tests, synergistic antifungal activity of these LPs against *B. cinerea* was observed, revealing a direct control mechanism of this fungus [22]. Furthermore, both strains demonstrated the ability to inhibit *B. cinerea* disease via ISR when applied to tomato roots [22]. Both strains were isolated from the rhizosphere, so it is possible that they have strategies for efficient adaptation to this environment [32]. Julkowska et al. [33] reported that swarm movement (motility), adherence and biofilm production, three abilities important for efficient colonization, are linked to the production of surfactin. In this sense, the fact that BBC023 produces 1/3 less surfactin than BBC047 becomes relevant for the interpretation of our results. Despite the lower surfactin production, BBC023 and BBC047 demonstrated similar results for adherence and motility under in vitro conditions. This could be explained by the existence of other factors, which act in a coordinated manner to achieve swarming movement and adhesion, and which could compensate the lower surfactin production of BBC023 [29,34]. However, the biofilm of strain BBC047 presented a greater structural complexity than that of BBC023, which correlates with the respective surfactin production of these strains [18]. The surfactin profiles of both strains contain isoforms with carbon chains between C12 and C17, where longer carbon chains (C14, C15 and C16) represented more than 88% of the respective surfactin production. Compared to strain FZB42 (mainly producing C13 to C15 chains [24]), the surfactin profile of both strains used in this study is shifted towards longer C-chain isoforms. Particularly, these longer C-chains were reported to develop a stronger interaction with plant cell membranes and therefore are more likely involved in triggering ISR [20,35].

To elucidate the impact of the lower surfactin production of BBC023 with respect to its capacity to activate ISR, we performed a plant experiment, incorporating treatments with the application of exogenous surfactin (5 and 25 µg/mL), being the Surf 5 µg/mL treatment below the threshold reported to generate an effect on the plant [7].

Several authors mentioned a surfactin dose-dependent activation of ISR in the plant, which was confirmed by our results for the two treatments with different concentrations of exogenous surfactin [7,36]. All bacterial treatments on tomato roots activated ISR in defense against *B. cinera* disease on the leaves. In this context, Debois et al. [36] reported that the polysaccharides produced by tomato roots induce the bacterial production of surfactin. Such a root-exudate-mediated modulation of bacterial surfactin production could also explain the fact that strains BBC023 and BBC47 achieve a similar plant response, despite their differences in surfactin production under in vitro conditions.

Interestingly, for the bacterial treatments on tomato leaves the degree of disease inhibition correlated with the in vitro surfactin concentrations produced by each strain. In plants treated with BBC047, the disease severity of *B. cinerea* in root and leaf application of this strain is comparable—or in other words, the ISR response is similar. This contrasts with other studies, where surfactin showed low effectivity in activating ISR in dicot leaves [37,38]. Whereas BBC023 generated two different plant responses, on roots it triggered ISR similar to 25 µg/mL exogenous surfactin and BBC047; on leaves, disease severity is comparable to the control. These results suggest that in the interaction with the roots, other mechanisms could compensate for the lower production of surfactin and thus facilitate colonization, similar to, e.g., the production of exopolysaccharides [21,39]. Contrastingly, on leaves, the establishment of strain BBC023 might be insufficient to activate the formation of the extracellular matrix [40,41] as a consequence of its lower surfactin production [6].

At this point, it becomes important to consider the second known biological aspect of surfactin as a signaling molecule: its involvement in quorum sensing during biofilm formation [40]. Here, surfactin acts as a signaling molecule, activating the membrane receptor histidine kinase KinC and stimulating the formation of biofilms and, therefore, colonization [40,42,43]. We studied the colonization patterns of both strains in roots and leaves, correlating them with their surfactin production capacity.

After 21 DPI, in roots, BBC023 and BBC047 were able to maintain cell concentrations above $10^6$ CFU/mL, suggesting the existence of stable populations capable of inducing consistent effects [44,45]. Electron microscopy allowed us to visualize the structure and formation of the extracellular matrix of bacterial microcolonies in the different structures of the root [46]. In contrast to the findings of Fan et al. [47] for *B. velezensis* FZB42, strains BBC023 and BBC047 do not have a preference for junctions between primary and secondary roots, being found only sporadically in this area and not forming robust biofilms. However, they extensively and consistently colonized the lateral roots, forming biofilm structures.

The distribution of bacterial aggregates in different sections of the root was not uniform, similar to what was reported by Posada et al. [46]. The colonization pattern evidenced few cell aggregates (consisting of planktonic populations) in the upper and middle section, while in the lower section and specifically in the root tips robust biofilms were found.

In addition, differences in structure and robustness were detected between the biofilms formed in the maturation zone and those biofilms formed in the elongation or meristematic zones. Several authors have described that different root zones (maturation, meristematic or calyptra zones) can generate different types of compounds [48–50], with the highest exudation of compounds occurring in the meristematic zone [51]. Precisely in this area, a large colonization was found, although with less robust biofilm and a greater presence of planktonic cells [42]. Planktonic cells within subpopulations in biofilms are adapted for nutrient utilization rather than robustness and resistance [52]. Thus, the meristematic zone of the rhizosphere could participate in the attraction of bacteria through chemotaxis to later promote their multiplication and dispersal. On the other hand, the most robust biofilms were found in the calyptra [50]. These structures are rich in extracellular matrices, and their objective would be to protect the microbial population [53]. Interestingly, this area has the least availability of nutrients and presents the highest disturbance due to friction generated by root growth. Beyond the induction of biofilm formation, Chen et al. [40] relate surfactin to the relief of carbon catabolite repression (CCR) in *B. amyloliquefaciens* to enable the use of non-preferred carbon sources under nutrient limitation, as in the calyptra zone. These findings suggest a biological importance associated with surfactin, whose benefits for bacterial survival on plant surfaces require specific studies.

Leaf colonization patterns differed between the two strains. BBC047 presented enhanced extracellular matrix production, more structured populations and higher density. Biofilm formation occurred at the junctions of epidermal cells, similar to those reported by Zeriouh et al. [6] for the *B. amyloliquefaciens* UMAF6614 strain on melon leaves (*Cucumis sativus*). The opposite result was observed in strain BBC023. This strain showed a dramatic reduction in its population density at 7 DPI, registering only isolated and unstructured populations. This colonization pattern and behavior did not change during the experiment. In contrast to Luo et al. [19], the ability to produce other lipopeptides by BBC023 [20] did not compensate for the lower surfactin production, which resulted in an impediment to efficient and sustained leaf establishment. Similarly, the absence of a biofilm exposes microorganisms to an inhospitable environment subject to high temperatures, UV radiation and dehydration [6,54,55], reducing the population density. The leaf colonization pattern of BBC023 also could explain the results from the ISR experiment, where this strain was not able to stimulate resistance mechanisms via leaf application, as the low population density is probably insufficient to generate a physiologically relevant concentration of surfactin. Based on our results, we suggest that surfactin concentration contributes differentially not only to antifungal activity but also to biofilm formation and colonization.

## 5. Conclusions

Effective colonization of the plant surface is key for a biocontrol effect, both through antagonism and ISR. There is also a correlation between the surfactin production of a strain and its colonization, but not necessarily with its capacity to activate an ISR response. Our results suggest (1) the presence of environmental conditions that, depending on the plant surface (root or leaf), favor or limit colonization by a strain with less surfactin production; (2) the existence of other mechanisms that enable a strain with lower surfactin production to efficiently colonize, e.g., on roots; (3) that surfactin production is not a predictor for the colonization capacity of a bacterial strain; and (4) only efficient and sustained colonization enables the activation of ISR.

Finally, we conclude that the importance of surfactin lies in its impacts on biofilm formation and stable colonization, which finally enables its activity as an elicitor of ISR.

**Supplementary Materials:** The following are available online at https://www.mdpi.com/article/10.3390/microorganisms9112251/s1, Figure S1, Experimental design for ISR experiment with BBC023, BBC047 and fungal pathogen *B. cinerea*; Figure S2, Experimental design for colonization experiment in rhizobox; Figure S3, Relative proportion of surfactin isoforms produced by BBC023, BBC047 and FZB42 relative to the total amount of surfactin produced by each strain, respectively; Figure S4, Representative rhizobox and SEM photographs of control treatment at 21 DPI; Figure S5, Root photographs of the middle section at 14 DPI; Figure S6, Root photographs of the upper section at 14 DPI.

**Author Contributions:** A.S. and R.S.-M. conceptualized the experiments; R.S.-M. performed in vitro and plant experiments; M.A. performed HPLC- and UHPLC-MS analysis; M.G. performed statistical analysis; A.S., M.G. and R.S.-M. wrote the manuscript. All authors have read and agreed to the published version of the manuscript.

**Funding:** This research was funded by Projects BIP 30127532 and BIP 30403034 (Regional Fund of Innovation for Competitiveness Coquimbo Region FIC-R, Chile) (A.S.) and ANID Beca Doctorado Nacional Grant/Award N° 21140504 (R.S.-M.).

**Institutional Review Board Statement:** Not applicable.

**Data Availability Statement:** The data presented in this study are included this article and its supplements.

**Acknowledgments:** The authors are thankful to Stefanie Maldonado, Williams Arancibia and Juan Pablo Araya for technical support in the laboratory, UCN Coquimbo for granting access to the SEM platform as well as to Servicios y Almacigos SA for providing assistance with seedling growth.

**Conflicts of Interest:** The authors declare no conflict of interest.

## References

1. Borriss, R. Towards a new generation of commercial microbial disease control and plant growth promotion products. In *Principles of Plant-Microbe Interactions*; Lugtenberg, B., Ed.; Springer International Publishing: Cham, Switzerland, 2015; pp. 329–337.
2. Fan, H.; Zhang, Z.; Li, Y.; Zhang, X.; Duan, X.; Wang, Q. Biocontrol of bacterial fruit blotch by *Bacillus subtilis* 9407 via surfactin-mediated antibacterial activity and colonization. *Front. Microbiol.* **2017**, *8*, 1973. [CrossRef]
3. Pandin, C.; Le Coq, D.; Deschamps, J.; Védie, R.; Rousseau, T.; Aymerich, S.; Briandet, R. Complete genome sequence of *Bacillus velezensis* QST713: A biocontrol agent that protects *Agaricus bisporus* crops against the green mould disease. *J. Biotechnol.* **2018**, *278*, 10–19. [CrossRef] [PubMed]
4. Aloo, B.N.; Makumba, B.A.; Mbega, E.R. The potential of bacilli rhizobacteria for sustainable crop production and environmental sustainability. *Microbiol. Res.* **2019**, *219*, 26–39. [CrossRef]
5. Pal, K.; Gardener, B. Biological Control of Plant Pathogens. *Plant Health Instr.* **2006**. [CrossRef]
6. Zeriouh, H.; de Vicente, A.; Pérez-García, A.; Romero, D. Surfactin triggers biofilm formation of *Bacillus subtilis* in melon phylloplane and contributes to the biocontrol activity. *Environ. Microbiol.* **2014**, *16*, 2196–2211. [CrossRef]
7. Cawoy, H.; Mariutto, M.; Henry, G.; Fisher, C.; Vasilyeva, N.; Thonart, P.; Dommes, J.; Ongena, M. Plant defense stimulation by natural isolates of *Bacillus* depends on efficient surfactin production. *Mol. Plant-Microbe Interact.* **2014**, *27*, 87–100. [CrossRef]
8. Toral, L.; Rodríguez, M.; Béjar, V.; Sampedro, I. Antifungal activity of lipopeptides from *Bacillus* XT1 CECT 8661 against *Botrytis cinerea*. *Front. Microbiol.* **2018**, *9*, 1315. [CrossRef] [PubMed]
9. Farace, G.; Fernandez, O.; Jacquens, L.; Coutte, F.; Krier, F.; Jacques, P.; Clément, C.; Barka, E.; Jacquard, C.; Dorey, S. Cyclic lipopeptides from *Bacillus subtilis* activate distinct patterns of defence responses in grapevine. *Mol. Plant Pathol.* **2014**, *16*, 177–187. [CrossRef]
10. Rahman, A.; Uddin, W.; Wenner, N.G. Induced systemic resistance responses in perennial ryegrass against *Magnaporthe oryzae* elicited by semi-purified surfactin lipopeptides and live cells of *Bacillus amyloliquefaciens*. *Mol. Plant Pathol.* **2015**, *16*, 546–558. [CrossRef]

11. Timmusk, S.; Grantcharova, N.; Wagner, E.G.H. *Paenibacillus polymyxa* invades plant roots and forms biofilms. *Appl. Environ. Microbiol.* **2005**, *71*, 7292–7300. [CrossRef]
12. Pandin, C.; Le Coq, D.; Canette, A.; Aymerich, S.; Briande, R. Should the biofilm mode of life be taken into consideration for microbial biocontrol agents. *Microb. Biotechnol.* **2017**, *10*, 719–734. [CrossRef] [PubMed]
13. O'Toole, G.; Kaplan, H.B.; Kolter, R. Biofilm formation as microbial development. *Annu. Rev. Microbiol.* **2000**, *54*, 49–79. [CrossRef]
14. Kolter, R.; Greenberg, E.P. Microbial sciences—The superficial life of microbes. *Nature* **2006**, *441*, 300–302. [CrossRef]
15. Raaijmakers, J.M.; de Bruijn, I.; Nybroe, O.; Ongena, M. Natural functions of lipopeptides from *Bacillus* and *Pseudomonas*: More than surfactants and antibiotics. *FEMS Microbiol. Rev.* **2010**, *34*, 1037–1062. [CrossRef]
16. Beauregard, P.B.; Chai, Y.R.; Vlamakis, H.; Losick, R.; Kolter, R. *Bacillus subtilis* biofilm induction by plant polysaccharides. *Proc. Natl. Acad. Sci. USA* **2013**, *110*, E1621–E1630. [CrossRef]
17. Zhao, H.B.; Shao, D.Y.; Jiang, C.M.; Shi, J.L.; Li, Q.; Huang, Q.S.; Rajoka, M.; Yang, H.; Jin, M.L. Biological activity of lipopeptides from *Bacillus*. *Appl. Microbiol. Biotechnol.* **2017**, *101*, 5951–5960. [CrossRef] [PubMed]
18. Chen, Y.; Yan, F.; Chai, Y.; Liu, H.; Kolter, R.; Losick, R.; Guo, J.H. Biocontrol of tomato wilt disease by *Bacillus subtilis* isolates from natural environments depends on conserved genes mediating biofilm formation. *Environ. Microbiol.* **2013**, *15*, 848–864. [CrossRef]
19. Luo, H.; Zhou, H.; Zou, J.; Wang, X.; Zhang, R.; Xiang, Y.; Chen, Z. Bacillomycin L and surfactin contribute synergistically to the phenotypic features of *Bacillus subtilis* 916 and the biocontrol of rice sheath blight induced by *Rhizoctonia solani*. *Appl. Microbiol. Biotechnol.* **2015**, *99*, 1897–1910. [CrossRef] [PubMed]
20. Henry, G.; Deleu, M.; Jourdan, E.; Thonart, P.; Ongena, M. The bacterial lipopeptide surfactin targets the lipid fraction of the plant plasma membrane to trigger immune-related defence responses. *Cell Microbiol.* **2011**, *13*, 1824–1837. [CrossRef] [PubMed]
21. Therien, M.; Kiesewalter, H.T.; Auria, E.; Charron-Lamoureux, V.; Wibowo, M.; Maroti, G.; Kovacs, A.T.; Beauregard, P.B. Surfactin production is not essential for pellicle and root-associated biofilm development of *Bacillus subtilis*. *Biofilm* **2020**, *2*, 100021. [CrossRef]
22. Salvatierra-Martinez, R.; Arancibia, W.; Aguilera, S.; Olalde, V.; Bravo, J.; Stoll, A. Colonization ability as an indicator of enhanced biocontrol capacity—An example using two *Bacillus amyloliquefaciens* strains and *Botrytis cinerea* infection of tomatoes. *J. Phytotahol.* **2018**, *166*, 601–612. [CrossRef]
23. Aleti, G.; Lehner, S.; Bacher, M.; Compant, S.; Nikolic, B.; Plesko, M.; Schuhmacher, R.; Sessitsch, A.; Brader, G. Surfactin variants mediate species-specific biofilm formation and root colonization in *Bacillus*. *Environ. Microbiol.* **2016**, *18*, 2634–2645. [CrossRef]
24. Koumoutsi, A.; Chen, X.H.; Henne, A.; Liesegang, H.; Hitzeroth, G.; Franke, P.; Vater, J.; Borriss, R. Structural and functional characterization of gene clusters directing nonribosomal synthesis of bioactive cyclic lipopeptides in *Bacillus amyloliquefaciens* strain FZB42. *J. Bacterial.* **2004**, *186*, 1084–1096. [CrossRef]
25. Gu, X.B.; Zheng, Z.M.; Yu, H.Q.; Wang, J.; Liang, F.L.; Liu, R.L. Optimization of medium constituents for a novel lipopeptide production by Bacillus subtilis MO-01 by a response surface method. *Proc. Biochem.* **2005**, *40*, 3196–3201. [CrossRef]
26. Alvarez, F.; Castro, M.; Príncipe, A.; Borioli, G.; Fischer, S.; Mori, G.; Jofré, E. The plant-associated *Bacillus amyloliquefaciens* strains MEP2 18 and ARP2 3 capable of producing the cyclic lipopeptides iturin or surfactin and fengycin are effective in biocontrol of sclerotinia stem rot disease. *J. Appl. Microbiol.* **2011**, *112*, 159–174. [CrossRef] [PubMed]
27. Ali, S.; Hameed, S.; Imran, A.; Iqbal, M.; Lazarovits, G. Genetic, physiological and biochemical characterization of *Bacillus* sp. strain RMB7 exhibiting plant growth promoting and broad spectrum antifungal activities. *Microb. Cell Fact.* **2014**, *13*, 144. [CrossRef]
28. Branda, S.S.; Gonzalez-Pastor, J.E.; Ben-Yehuda, S.; Losick, R.; Kolter, R. Fruiting body formation by *Bacillus subtilis*. *Proc. Natl. Acad. Sci. USA* **2001**, *98*, 11621–11626. [CrossRef] [PubMed]
29. Ghelardi, E.; Salvetti, S.; Ceragioli, M.; Gueye, S.A.; Celandroni, F.; Senesi, S. Contribution of surfactin and SwrA to flagellin expression, swimming, and surface motility in *Bacillus subtilis*. *Appl. Environ. Microbiol.* **2012**, *78*, 6540–6544. [CrossRef] [PubMed]
30. Mariutto, M.; Duby, F.; Adam, A.; Bureau, C.; Fauconnier, M.L.; Ongena, M.; Thonart, P.; Dommes, J. The elicitation of a systemic resistance by *Pseudomonas putida* BTP1 in tomato involves the stimulation of two lipoxygenase isoforms. *BMC Plant Biol.* **2011**, *11*, 29. [CrossRef]
31. Mendiburu, F.; Simon, R. Agricolae—Ten years of an Open source Statistical tool for experiments in breeding, agriculture and biology. *PeerJ PrePr.* **2015**, *3*, e1404v1. [CrossRef]
32. Vlamakis, H.; Chai, Y.R.; Beauregard, P.; Losick, R.; Kolter, R. Sticking together: Building a biofilm the *Bacillus subtilis* way. *Nat. Rev. Microbiol.* **2013**, *11*, 157–168. [CrossRef] [PubMed]
33. Julkowska, D.; Obuchowski. M.; Holland, I.B.; Seror, S.J. Comparative analysis of the development of swarming communities of *Bacillus subtilis* 168 and a natural wild type: Critical effects of surfactin and the composition of the medium. *J. Bacteriol.* **2005**, *187*, 65–76. [CrossRef] [PubMed]
34. Falardeau, J.; Wise, C.; Novitsky, L.; Avis, T.J. Ecological and mechanistic insights into the direct and indirect antimicrobial properties of *Bacillus subtilis* lipopeptides on plant pathogens. *J. Chem. Ecol.* **2013**, *39*, 869–878. [CrossRef]
35. Schellenberger, R.; Touchard, M.; Clément, C.; Baillieul, F.; Cordelier, S.; Crouzet, J.; Dorey, S. Apoplastic invasion patterns triggering plant immunity: Plasma membrane sensing at the frontline. *Mol. Plant Pathol.* **2019**, *20*, 1602–1616. [CrossRef]
36. Debois, D.; Fernandez, O.; Franzil, L.; Jourdan, E.; de Brogniez, A.; Willems, L.; Clément, C.; Dorey, S.; De Pauw, E.; Ongena, M. Plant polysaccharides initiate underground crosstalk with bacilli by inducing synthesis of the immunogenic lipopeptide surfactin. *Environ. Microbiol. Rep.* **2015**, *7*, 570–582. [CrossRef]

37. Yamamoto, S.; Shiraishi, S.; Suzuki, S. Are cyclic lipopeptides produced by *Bacillus amyloliquefaciens* S13-3 responsible for the plant defence response in strawberry against *Colletotrichum gloeosporioides*? *Lett. Appl. Microbiol.* **2015**, *60*, 379–386. [CrossRef]
38. Pršić, J.; Ongena, M. Elicitors of Plant Immunity Triggered by Beneficial Bacteria. *Front. Plant Sci.* **2020**, *11*, 594530. [CrossRef]
39. Al-Ali, A.; Deravel, J.; Krier, F.; Béchet, M.; Ongena, M.; Jacques, P. Biofilm formation is determinant in tomato rhizosphere colonization by Bacillus velezensis FZB42. *Environ. Sci. Pollut. Res. Int.* **2018**, *25*, 29910–29920. [CrossRef] [PubMed]
40. Chen, B.; Wen, J.; Zhao, X.; Ding, J.; Qi, G. Surfactin: A quorum-sensing signal molecule to relieve CCR in *Bacillus amyloliquefaciens*. *Front. Microbiol.* **2020**, *11*, 631. [CrossRef]
41. Sharipova, M.R.; Mardanova, A.M.; Rudakova, N.L.; Pudova, D.S. Bistability and Formation of the Biofilm Matrix as Adaptive Mechanisms during the Stationary Phase of *Bacillus subtilis*. *Microbiology* **2021**, *90*, 20–36. [CrossRef]
42. Lopez-Bucio, J.; Campos-Cuevas, J.; Hernandez-Calderon, E.; Velasquez-Becerra, C.; Farias-Rodriguez, R.; Macias-Rodriguez, L.; Valencia-Cantero, E. *Bacillus megaterium* rhizobacteria promote growth and alter root-system architecture through an auxin- and ethylene-independent signaling mechanism in *Arabidopsis thaliana*. *Mol. Plant-Microbe Interact.* **2007**, *20*, 207–217. [CrossRef] [PubMed]
43. Ongena, M.; Jacques, P. *Bacillus* lipopeptides: Versatile weapons for plant disease biocontrol. *Trends Microbiol.* **2008**, *16*, 115–125. [CrossRef]
44. Bankhead, S.B.; Landa, B.B.; Lutton, E.; Weller, D.M.; McSpadden Gardener, B.B. Minimal changes in rhizobacterial population structure following root colonization by wild type and transgenic biocontrol strains. *FEMS Microbiol. Ecol.* **2004**, *49*, 307–318. [CrossRef] [PubMed]
45. Strigul, N.S.; Kravchenko, L.V. Mathematical modeling of PGPR inoculation into the rhizosphere. *Environ. Model. Softw.* **2006**, *21*, 1158–1171. [CrossRef]
46. Posada, L.F.; Álvarez, J.C.; Romero-Tabarez, M.; de-Bashan, L.; Villegas-Escobar, V. Enhanced molecular visualization of root colonization and growth promotion by *Bacillus subtilis* EA-CB0575 in different growth systems. *Microbiol. Res.* **2018**, *217*, 69–80. [CrossRef] [PubMed]
47. Fan, B.; Borriss, R.; Bleiss, W.; Wu, X. Gram-positive rhizobacterium *Bacillus amyloliquefaciens* FZB42 colonizes three types of plants in different patterns. *J. Microbiol.* **2012**, *50*, 38–44. [CrossRef] [PubMed]
48. Bengough, A.; McKenzie, B. Sloughing of root cap cells decreases the frictional resistance to maize (*Zea mays* L.) root growth. *J. Exp. Bot.* **1997**, *48*, 885–893. [CrossRef]
49. Kuzyakov, Y. Factors affecting rhizosphere priming effects. *J. Plant Nutr. Soil Sci.* **2002**, *165*, 382–396. [CrossRef]
50. Hassan, M.; McInroy, J.; Kloepper, J.W. The Interactions of Rhizodeposits with Plant Growth-Promoting Rhizobacteria in the Rhizosphere: A Review. *Agriculture* **2019**, *9*, 142. [CrossRef]
51. Marschner, P.; Crowley, D.; Rengel, Z. Rhizosphere interactions between microorganisms and plants govern iron and phosphorus acquisition along the root axis–model and research methods. *Soil Biol. Biochem.* **2011**, *43*, 883–894. [CrossRef]
52. Romero, D.; Aguilar, C.; Losick, R.; Kolter, R. Amyloid fibers provide structural integrity to *Bacillus subtilis* biofilms. *Proc. Natl. Acad. Sci. USA* **2010**, *107*, 2230–2234. [CrossRef] [PubMed]
53. Romero, D.; Sanabria-Valentín, E.; Vlamakis, H.; Kolter, R. Biofilm Inhibitors that Target Amyloid Proteins. *Chem. Biol.* **2013**, *20*, 102–110. [CrossRef] [PubMed]
54. Jacobs, J.; Sundin, G. Effect of solar UV-B radiation on a phyllosphere bacterial community. *Appl. Environ. Microbiol.* **2001**, *67*, 5488–5496. [CrossRef] [PubMed]
55. Carvalho, S.; Castillo, J. Influence of Light on Plant–Phyllosphere Interaction. *Front. Plant Sci.* **2018**, *9*, 1482. [CrossRef] [PubMed]

*microorganisms*

MDPI

*Article*

# Interactions of *Bacillus subtilis* Basement Spore Coat Layer Proteins

Daniela Krajčíková, Veronika Bugárová and Imrich Barák *

Department of Microbial Genetics, Institute of Molecular Biology, Slovak Academy of Sciences, Dúbravská cesta 21, 845 51 Bratislava, Slovakia; daniela.krajcikova@savba.sk (D.K.); veronika.bugarova@savba.sk (V.B.)
* Correspondence: imrich.barak@savba.sk; Tel.: +421-2-5930-7418

**Abstract:** *Bacillus subtilis* endospores are exceptionally resistant cells encircled by two protective layers: a petidoglycan layer, termed the cortex, and the spore coat, a proteinaceous layer. The formation of both structures depends upon the proper assembly of a basement coat layer, which is composed of two proteins, SpoIVA and SpoVM. The present work examines the interactions of SpoIVA and SpoVM with coat proteins recruited to the spore surface during the early stages of coat assembly. We showed that the alanine racemase YncD associates with two morphogenetic proteins, SpoIVA and CotE. Mutant spores lacking the *yncD* gene were less resistant against wet heat and germinated to a greater extent than wild-type spores in the presence of micromolar concentrations of L-alanine. In seeking a link between the coat and cortex formation, we investigated the interactions between SpoVM and SpoIVA and the proteins essential for cortex synthesis and found that SpoVM interacts with a penicillin-binding protein, SpoVD, and we also demonstrated that SpoVM is crucial for the proper localization of SpoVD. This study shows that direct contacts between coat morphogenetic proteins with a complex of cortex-synthesizing proteins could be one of the tools by which bacteria couple cortex and coat formation.

**Keywords:** spore coat; protein-protein interaction; cortex; PBP protein; alanine racemase; SpoVM; SpoVD

**Citation:** Krajčíková, D.; Bugárová, V.; Barák, I. Interactions of *Bacillus subtilis* Basement Spore Coat Layer Proteins. *Microorganisms* **2021**, *9*, 285. https://doi.org/10.3390/microorganisms9020285

Academic Editor: James Garnett
Received: 23 December 2020
Accepted: 25 January 2021
Published: 30 January 2021

**Publisher's Note:** MDPI stays neutral with regard to jurisdictional claims in published maps and institutional affiliations.

## 1. Introduction

*Bacillus subtilis* spores develop during a sporulation process that occurs when bacteria find themselves in an environment incompatible with their normal cell cycle. Spores are formed after an asymmetric division of rod-shaped cells from the smaller part of the cell, called the forespore, while the larger mother cell nourishes the spore until it becomes fully mature and can be released into the environment. Spores are known to be exceptionally resistant to a broad variety of conditions. Their resistance arises from the specific arrangement of several protective layers that surround the spore core, which harbors the chromosomal DNA. Two major structures are recognized within the spore: the cortex is a peptidoglycan layer embedded between the inner and outer forespore membranes onto which the spore coat, a thick lamellar protein layer, is formed [1]. More than 80 coat proteins are arrayed in four morphologically distinct coat layers: a basement layer localized directly on the outer forespore membrane, an inner coat, an outer coat and a crust, the outermost layer of the spore [2]. A few coat proteins, the morphogenetic proteins, SpoIVA, SpoVM, SpoVID, SafA, CotE and CotZ, greatly influence coat assembly by controlling the localization of individual proteins on the spore surface and the proper assembly of the four coat layers [3]. SpoIVA and SpoVM form the basement layer of the coat on top of which the inner coat is built under the control of SafA, followed by the outer coat governed by CotE. Assembly of the crust, the coat outermost layer, requires CotZ. The deletion or mutation of any of these proteins produces either coatless spores, or spores whose coat lacks the layer controlled by the missing morphogenetic protein. The basement layer proteins SpoVM and SpoIVA also affect the formation of the cortex [4,5].

These two morphogenetic proteins are synthesized under the control of the $\sigma^E$ transcription factor and are deposited onto the forespore surface very early after the asymmetric septum is formed. SpoVM is a small, 26-amino-acid-long peptide that is inserted into the outer forespore membrane via its amphipathic $\alpha$-helix, which has an affinity for membranes with a positive curvature [6]. SpoIVA, a structural protein with ATPase activity, directly interacts with SpoVM and undergoes conformational changes after ATP hydrolysis that enable it to polymerize and cover the forespore surface [7,8]. SpoVM and SpoIVA are co-dependent in their localization: through an interaction with SpoVM, SpoIVA reaches the local concentration required for its polymerization on the outer forespore membrane. Initially, SpoVM's association with the membrane is rather dynamic: it dips into and out of the membrane, but polymerized SpoIVA protein entraps SpoVM into its proper place, to the forespore membrane [9]. SpoIVA recruits SpoVID, a morphogenetic protein which is, together with SpoVM, essential for spore encasement [10]. During this process, individual coat proteins gradually cover the entire forespore surface in multiple waves, all depending on a cascade of transcriptional sigma factors which control the expression of coat components.

SafA, as noted above, directs inner coat formation. SafA has a peptidoglycan-binding LysM domain and the protein is localized in the coat fraction at the interface of the cortex and coat, and also in the cortex layer [11]. The ultrastructure of SafA mutant spores examined by thin sectioning transmission electron microscopy showed that the inner coat was impaired, losing its usual characteristics, and moreover, the cortex, inner coat and outer coat seemed to lack their regular tight adherence [11].

Outer coat formation is directed by CotE, a morphogenetic protein which localizes at the intersection of the outer and inner coat layers. In spores of *cotE* mutant, dark patches of mis-assembled proteins attached to the inner coat were observed [12]. CotE also controls the deposition of the crust layer proteins. Some inner coat proteins, like CotS, were also observed to depend on the presence of CotE [13,14].

Several studies demonstrated that morphogenetic proteins are in direct contact, indicating that they could form a basic, foundational structure, a scaffold to which the coat proteins are attached as the process of spore coat formation proceeds [15–19]. Other coat proteins are also produced at the same time as the key morphogenetic proteins, all of which are under the control of $\sigma^E$. Their role in spore development is often unknown, but the localization some of them have been shown to depend on the key morphogenetic proteins [20]. In the present study, we looked for contacts between the morphogenetic proteins and three coat components which appear to be recruited to the forespore surface during the first waves of coat assembly [20]. One of these is the spore alanine racemase YncD [21]. It is an early expressed protein whose deposition on the forespore surface is CotE-dependent. Alanine racemases catalyze the reversible racemization of L-alanine to D-alanine; they are ubiquitous in bacteria and are crucial for cell wall synthesis since D-alanine is a component of peptidoglycan. The role of YncD during the *B. subtilis* cell life cycle is presently unknown, but the *Bacillus anthracis* spore alanine racemase Alr, which is distributed in the exosporium and coat layer [22,23], protects developing spores against premature germination by converting L-alanine, a germinant, to D-alanine, a germination inhibitor. The function of other two proteins studied here, YhaX and YheD, is unknown.

There are a number of enzymes involved in cortex formation which are also under the control of $\sigma^E$, expressed early in sporulation and are localized to the outer forespore membrane, and they are not coat proteins. A link between coat and cortex formation was discovered by Ramamurthi's group, which found that a peptide called CmpA is an important factor participating in controlling coat formation [24]. CmpA monitors whether the basement layer consisting of SpoIVA and SpoVM is correctly assembled [25] and ensures that improperly developed spores are destroyed. However, there is no information on what processes take place in the absence of SpoIVA and the molecular mechanism and details of how SpoVM and SpoIVA affect cortex synthesis are still unknown. In this work, we sought to identify interactions between coat morphogenetic proteins and three proteins which are known to be essential for cortex production, SpoVB, SpoVD and SpoVE [26–28]. SpoVD, a class B, high-molecular-weight penicillin-binding protein (PBP) with transpeptidase activity, associates with SpoVE, which belongs to the SEDS (shape, elongation, division and sporulation) family of proteins [27]. SpoVE was thought to be a lipid II flippase, but later appeared to function as a glycosyltransferase [29]; the role of flippase is likely to be taken by SpoVB, which is a member of the multidrug/oligosaccharidyl-lipid/polysacharide (MOP) exporter superfamily [30].

In this work, we show that YncD associates with the morphogenetic proteins CotE and SpoIVA. The spores of *yncD* mutant are slightly less resistant to wet heat and can germinate to some level in a trace amount of alanine. We also find that the coat morphogenetic protein SpoVM directly binds SpoVD and that SpoVD localization to the spore surface is SpoVM dependent. Although control of the coat and cortex syntheses is certainly very complex, physical contacts between the key proteins involved could be one of the methods by which cells link the two processes. As a result, SpoIVA and SpoVM have a large impact on coat and cortex formation and spore maturation in general.

## 2. Materials and Methods

### 2.1. Bacterial Strains and Growth Conditions

The genomic DNA of *Bacillus subtilis* strain PY79 [30] was used for the amplification of the *yncD* gene and its upstream and downstream sequences. *Escherichia coli* strain MM294 (*endA1 hsdR17 supE44 thi-1 recA1*) was chosen as the host for the cloning and maintenance of all plasmids. The bacterial two hybrid assay was performed using *E. coli* strain BTH101 (F−, *cya-99, araD139, galE15, galK16, rpsL1 (Str r), hsdR2, mcrA1, mcrB1*) (Euromedex, Souffelweyersheim, France). Recombinant proteins were produced in *E. coli* strain BL21 (DE3) (F−, *ompT, hsdSB, (rB−mB−), gal, dcm*) (Novagen, (Novagen®, Merck KGaA, Darmstadt, Germany). Bacteria were grown in LB medium supplemented with appropriate antibiotics at 37 °C if not otherwise indicated.

### 2.2. Construction of Bacillus Subtilis yncD Strain

To delete the *yncD* gene from the *Bacillus subtilis* chromosome, we prepared the plasmid pUS_deltayncD. The plasmid was constructed by Gibson assembly [31] using four DNA fragments: a 349 bp DNA fragment of the *yncD* upstream sequence, including the first 42 bp of the gene, amplified with primers Gib upstream yncD5′ and Gib upstream yncD3′ from *B. subtilis* genomic DNA; a 1241 bp DNA fragment containing a gene for kanamycin resistance, amplified with primers Gib pUS19kan5′ and Gib pUS19kan3′ from plasmid pUK19; a 345 bp DNA fragment encompassing the downstream sequence of *yncD* and the last 36 bp of the gene, amplified with primers Gib downstream yncD5′ and Gib downstream yncD3′ from *B. subtilis* genomic DNA; and plasmid pUS19 [32], digested with the PstI and EcoRI restriction enzymes. The plasmid pUS_deltayncD was used to transform the *B. subtilis* competent cells. Null mutant clones generated after insertion of the kanamycin cassette by a double cross-over event were tested for spectinomycin sensitivity and verified by PCR; the PCR fragments were also sequenced.

### 2.3. Analysis of Protein–Protein Interactions Using a Bacterial Two Hybrid System (BACTH)

A bacterial two hybrid system based on adenylate cyclase was used [33]. All genes were cloned into four vectors—pUT18, pUT18C, pKT25 and pKNT25—which allowed coat proteins to be fused with the T18 or T25 domains of adenylate cyclase on their N- or C-termini. All genes were amplified by PCR using specific primers Table S1 and *B. subtilis* genomic DNA as the template. After cutting with appropriate restriction enzymes, the genes were cloned into the corresponding sites of the pKNT25, pKT25, pUT18 and pUT18C vectors. All plasmid constructs were verified by restriction analysis and by DNA sequencing. To examine the interactions between the individual proteins, the *E. coli* BTH1 strain was co-transformed with a pair of plasmids encoding the T25 or T18 fusions and transformants were plated onto LB indicator plates, supplemented with antibiotics and containing 40 µg/mL X-gal (5-bromo-4-chloro-3-indolyl-D-galactopyranoside) and 0.5 mM IPTG (isopropyl-β-D-1-thiogalactopyranoside). Bacteria were grown at 30 °C for 48 h. A blue color was observed in those bacterial colonies where direct contact between the expressed hybrid proteins occurred. The interactions were quantified by measuring their β-galactosidase activity using o-nitrophenol-D-galactopyranoside (ONPG) as a substrate. Bacteria grew on LB plates supplemented with antibiotics and 0.5 mM IPTG at 30 °C for 45 h. Bacterial colonies were scraped off, lysed with chloroform and SDS and the β-galactosidase activity was determined as described by Miller [34].

### 2.4. Protein Expression and Purification and Pull-Down Assay

To express His-tagged recombinant YncD, we used an expression plasmid based on pET28a or pETDuet vectors (Novagen). The *yncD* gene was amplified with specific primers Table S1 from *B. subtilis* genomic DNA and, after cleavage with restriction enzymes, cloned into the corresponding cloning site of a plasmid. To express un-tagged CotE and SpoIVA, the genes were amplified by PCR with the appropriate primers Table S1 using *B. subtilis* genomic DNA as a template, digested with the appropriate restriction enzymes and cloned into a pETDuet plasmid cut with corresponding restrictases; the proteins were then produced as before [35]. Briefly, *E. coli* BL21 (DE3) cells transformed with expression plasmids were grown in LB medium supplemented with antibiotics at 37 °C until the culture reached an OD600 of ≈0.6 and then induced by the addition of 1 mM IPTG. The cells were collected after 3 h of additional cultivation and stored at −80 °C until used. The YncD protein was isolated in a 50 mM Tris/HCl buffer, pH 8, containing 150 mM NaCl. Cells were lysed by sonication and the lysate was centrifuged at 70,000× g for 30 min and the protein was purified by metal affinity chromatography using a His Trap™ high performance (HP) column ( Merck KGaA, Darmstadt, Germany). For the pull-down assay, cells expressing the proteins were sonicated and the bacterial lysates were mixed and incubated on ice for approx. 1 h. After centrifugation, the proteins were co-purified using a Ni-agarose column. Interacting proteins were separated by sodium dodecyl sulfate-polyacrylamide gel electrophoresis (SDS-PAGE) and identified by Western blotting using polyclonal antibodies [35].

### 2.5. Spore Preparation

Spore crops were prepared by sporulation on Difco sporulation media (DSM) plates [36]. A single colony was used to inoculate LB medium and cultivated overnight at 37 °C. This culture was then used to inoculate liquid DSM and the bacteria were grown until OD600 = 0.8. 200 µL of this culture was spread on DSM plates and left at 37 °C for 3 days, followed by growth at room temperature for another 4–5 days. Spores were harvested, purified by the intensive washing with ice-cold water to remove cell debris and vegetative cells and stored at 4 °C until use.

### 2.6. Germination Assay

Before germination, spores were heat activated at 70 °C for 30 min followed by cooling on ice. Spores were resuspended in PBS buffer to reach an optical density OD600 = 1.0 and L-alanine was added to final concentrations of 100 mM, 1 mM or 0.01 mM. Germination was monitored by the loss of optical density at 600 nm at room temperature for up to 120 min.

### 2.7. Thermal Resistance Assay

The thermal resistance assay was performed in PCR Eppendorf tubes using a dry bath system. Spores (approx. $1 \times 10^8$ CFU/ mL) were suspended in PBS buffer and heated at 85 °C for 15 min to inactivate all vegetative or germinating spores. The wet heat resistance of at least three independent spore preparations was measured at 100 °C and 110 °C. 20 µL aliquots of spore suspension were used and at least 5 time points were performed for each spore prep and temperature. The Eppendorf tubes were immediately cooled on ice, the tube contents were transferred into 480 μL PBS and serial 10-fold dilutions were made to monitor the number of surviving spores by counting the bacterial colonies growing on LB plates. The *D*-value (the decimal reduction time or time required to kill 90% of the spores) was determined from a linear regression of survivors (log10 CFU/mL) versus exposure time.

### 2.8. Fluorescence Microscopy

The localization of mCherry-SpoVD in sporulating wild-type and *spoVM* mutant cells was visualized by fluorescence microscopy. Cells grown in DSM were examined three or five hours after reaching stationary phase. The samples were spotted onto a poly-L-lysine glass slide and covered with a poly-L-lysine coverslip. All images were obtained using an Olympus BX63 microscope equipped with an Andor Zyla 5.5 sCMOS camera (Olympus Europa SE & Co. KG., Hamburg, Germany). Olympus CellP imaging software was used for image acquisition and analysis. The final adjustment of the fluorescent images was done using Fiji ImageJ (open source software).

## 3. Results

### 3.1. Examination of Protein-Protein Interactions Using a Bacterial Two Hybrid System

To find potential direct contacts between different sets of proteins, we employed a bacterial two hybrid system. Previously, we performed a systematic study of potential interactions between the outer coat and crust proteins [31] and discovered a very complex protein interaction network within these coat layers. Here we investigated the interactions of morphogenetic coat proteins with three proteins, YncD, YhaX and YheD. We also examined three proteins essential for spore cortex synthesis: SpoVD, SpoVE and SpoVB.

Protein Interactions of YncD, YhaX and YheD:

- YncD contacts with morphogenetic coat proteins: YncD, an alanine racemase, is deposited onto a developing forespore under the control of CotE, as reported by McKenney [2]. Using BACTH, we observed that YncD physically contacts CotE. Screening all possible combinations of plasmids with the *cotE* and *yncD* genes fused to either the T18 or T25 domains of adenylate cyclase in plate assays showed that the highest expression of the β-galactosidase reporter gene arose from the T18-YncD/CotE-T25 combination (600 Mu). Some signals were also obtained for the combinations T18-YncD/T25-CotE, T25-YncD/CotE-T18, and T25-YncD/T18-CotE. As described previously [31], we compared these signals with the T18-CotE/CotE-T25 self-interaction used as the positive control (360 Mu in these experiments) [18,32]. The strength of the interaction of YncD and CotE observed by BACTH was almost twice that of CotE self-interaction and the observed blue color of the bacterial colonies harboring the YncD/CotE pair after 48 h of incubation was also more intense. A direct interaction between SpoVID or SpoVM and YncD was not detected, but a weak blue color could be observed in those bacterial colonies harboring combinations of plasmids containing the *spoIVA* and *yncD* genes. The highest β-galactosidase signal detected on X-gal plates for this pair was obtained from the YncD-T18/T25-SpoIVA and T18-SpoIVA/YncD-T25 combinations (160 Mu), which was approximately twice the negative control. A YncD/YncD self-interaction was also observed (220 Mu), indicating that the protein can form oligomers (Figure 1).
- Protein contacts of YheD and YhaX: Although the deposition of these two envelope proteins on the forespore surface has been shown to be dependent on SpoIVA [20], our bacterial two hybrid assays did not detect any direct binding to SpoIVA or to any other morphogenetic protein, only an YheD self-interaction could be observed. The positive signal was one of the strongest in this screening (1724 Mu) and arose from two plasmid combinations (T25–YheD/YheD–T18 and T25–YheD/T18–YheD). YhaX also formed homooligomers: β-galactosidase reporter gene activity was measured after complementation of the hybrid proteins T25–YhaX/YhaX–T18 and T25-YhaX/T18–YhaX (704 Mu) (Figure 1).

- Bacterial two hybrid assay of interactions of SpoVB, SpoVD and SpoVE with coat morphogenetic proteins: BACTH was previously used to successfully investigate interactions between protein involved in cell wall synthesis [33], many of which are integral membrane proteins or are associated with a membrane. Here we investigated the possible contacts between proteins essential for cortex peptidoglycan synthesis. An analysis of the topology of these proteins predicted that SpoVE and SpoVB have 10–15 transmembrane segments while SpoVD has one transmembrane segment at its N-terminus. Despite some difficulties with cloning, especially of SpoVE and SpoVD which may be toxic to *E. coli* cells, we managed to prepare the correct plasmid constructs. First, we tested their self-association. No homodimerization of SpoVB was observed, but positive signals were obtained for SpoVE and SpoVD; in both cases, β-galactosidase expression occurred in only one combination of plasmids. For SpoVE, we saw that bacterial colonies turned blue when the SpoVE–T18/T25–SpoVE hybrid proteins were co-expressed and we measured a β-galactosidase activity of up to 360 Mu (Figure 1). For SpoVD, it was important that, after fusion with the adenylate cyclase fragment, the protein be properly incorporated into the membrane such that the T18/T25 domain would be in the cytoplasm (on its N-terminus). Under these restrictions, we detected a positive signal from the combination T18–SpoVD/T25–SpoVD whose β-galactosidase activity reached 240 Mu (Figure 1). We next examined the interactions of these proteins with the coat morphogenetic proteins. Our experiments showed that SpoVD interacts with SpoVM when the hybrid protein T25–SpoVD was combined with T18–SpoVM; very faint blue bacterial colonies were also observed when T25–SpoVD was co-expressed with SpoVM–T18. Quantitative evaluation of the contact strength showed that the interaction of SpoVD with SpoVM was at roughly the same level as the CotE self-interaction (420 Mu vs. 360 Mu), which makes it an intermediate-strength interaction. Screening the interactions of SpoVE with the coat morphogenetic proteins indicated that there may also be an interaction of SpoVE and SpoVM, although the positive signals were very weak and could be identified only after prolonged incubation of the X-gal plate at 4 °C. The signals could be detected only for the T25–SpoVE and SpoVM–T18 and T18–SpoVM combinations (the β-galactosidase activity was approximately twice the negative control). The BACTH system did not detect any contact between these proteins and SpoIVA or between SpoVB and SpoVM or SpoIVA (Figure 1).

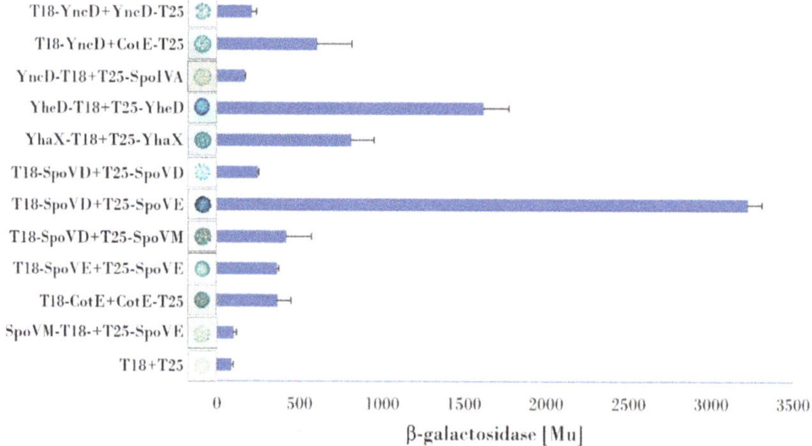


**Figure 1.** Bacterial two hybrid analysis of the interactions of spore proteins. The combinations of plasmids that provided the most intense signals on X-gal plates are shown. The β-galactosidase activity was determined as the average of at least three independent cell samples. All β-galactosidase values were subjected to the Dixon outlier test (*p* < 0.01).

### 3.2. Examination of the YncD–CotE and YncD–SpoIVA Interactions Using a Pull-Down Assay

To verify the physical contacts between YncD and CotE and SpoIVA, we performed a pull-down assay. We co-expressed His-tagged YncD and untagged CotE in a pETDuet plasmid. Co-purification of the tagged YncD and untagged CotE proteins was performed using nickel-affinity chromatography. The solubility of recombinant YncD when co-expressed together with CotE was unexpectedly greatly reduced, and the yield of soluble proteins after chromatography on the Ni-column was relatively low (Figure 2A). It seems likely that the complex formed by the two proteins during co-expression was less soluble than the proteins themselves. To circumvent this problem, both proteins were expressed separately, the bacterial lysates were mixed, applied to a Ni-column and the proteins were eluted with imidazole. In the elution fractions, a faint band of pulled-down CotE was visible on an SDS-PAGE gel after Coomasie blue staining (not shown). In addition, we identified the untagged CotE by Western blotting using a polyclonal anti-CotE antibody (Figure 2B). It is clear that untagged CotE was eluted from the column only in the presence of His-tagged YncD, whereas in the absence of YncD no CotE protein was detected in the elution fractions, thus confirming the YncD–CotE interaction observed in the bacterial two hybrid experiment.

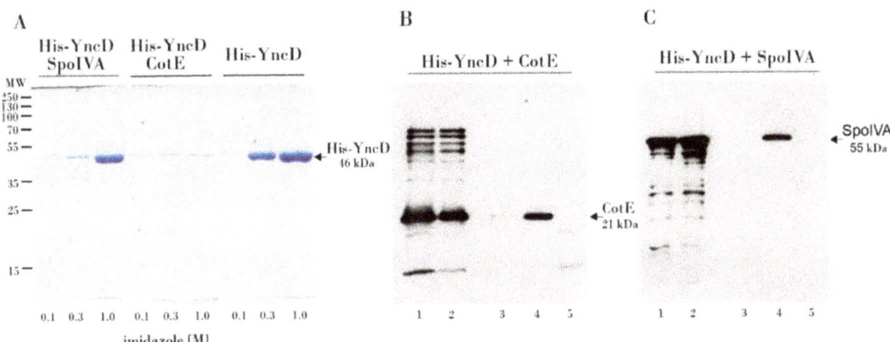


**Figure 2.** Pull-down analysis of the YncD interaction with morphogenetic proteins. (**A**) Sodium dodecyl sulfate-polyacrylamide gel electrophoresis (SDS-PAGE) gel of proteins co-expressed in *Escherichia coli*, loaded onto a HisTrap HP Ni-column and eluted with an imidazole step gradient. The imidazole concentration is shown in the figure. Lanes 1–3 contain fractions eluted after co-expression and co-purification of YncD+SpoIVA; lanes 4–6 contain fractions eluted after co-expression and co-purification of YncD+CotE; lanes 7–9 show fractions eluted after expression and purification of YncD alone. In all experiments we used the same volume of the same cell extract. (**B**) Western blot analysis of the YncD–CotE interaction. Proteins were probed with a polyclonal anti-CotE antibody. Lane 1–cell lysate with over-expressed, untagged CotE; lane 2–a mix of cell lysates with over-expressed CotE and His-tagged YncD; lane 3–negative control, elution from Ni-column, untagged CotE did not bind to the column; lane 4–CotE co-purified with His-tagged YncD; lane 5–His-tagged YncD does not provide a signal when using the anti-CotE antibody. (**C**) Western blot analysis of the YncD–SpoIVA interaction. Cell lysates with over-expressed proteins were mixed, loaded onto a HisTrap HP Ni-column and eluted with imidazole. Proteins were probed with a polyclonal anti-SpoIVA antibody. Lane 1–cell lysate with over-expressed SpoIVA; lane 2–mix of cell lysates with over-expressed SpoIVA and His-tagged YncD; lane 3–negative control, elution from Ni-column, untagged SpoIVA did not bind to the column; lane 4–SpoIVA co-purified with His-tagged YncD; lane 5–His-tagged YncD does not give a signal from an anti-SpoIVA antibody.

Similarly, we performed a pull-down assay of the YncD–SpoIVA interaction. Lysates containing untagged SpoIVA were mixed with His-tagged YncD and the proteins were purified on a Ni-column. We showed by Western blotting using a polyclonal anti-SpoIVA antibody that SpoIVA was eluted from the column in the presence of His-tagged YncD. Untagged SpoIVA did not bind to the Ni-resin on its own. The formation of a complex between SpoIVA and YncD is also indicated by a significant drop in the solubility of YncD in the presence of SpoIVA, although not as dramatic as in the case of CotE (Figure 2C).

### 3.3. Characterization of the ΔyncD Mutant Spores

BACTH experiments and pull-down assays showed that YncD can physically associate with two morphogenetic proteins, so we investigated whether deletion of its gene would cause any changes in spore properties. When growing in sporulation media, we did not see any significant difference between the ΔyncD and wild-type strain and the sporulation frequencies were essentially the same. Their resistance to chloroform was also identical (data not shown). In addition, when spores were treated with lysozyme and hydrogen peroxide, the percentage of survivals did not differ significantly Table 1.

**Table 1.** Resistance of wild-type and ΔyncD spores under various conditions. For resistance assays, spores were treated with lysozyme (1 mg/mL) in PBS buffer for 30 min at 37 °C and 15% $H_2O_2$ in PBS for 10 min. Resistance of spores was followed as the number of colonies forming units before and after treatment.

| Strain | Lysozyme | | Survival (%) | $H_2O_2$ | | Survival (%) |
|---|---|---|---|---|---|---|
| | Spore count (CFU/mL) | | | Spore Count (CFU/mL) | | |
| | − | + | | − | + | |
| PY79 | $1.1 \times 10^8$ | $2.6 \times 10^7$ | 24 | $4.2 \times 10^7$ | $1.5 \times 10^7$ | 36 |
| ΔyncD | $6.6 \times 10^7$ | $1.3 \times 10^7$ | 20 | $7.8 \times 10^7$ | $1.3 \times 10^7$ | 17 |

Additionally, the wet heat resistance of the purified spores was examined and the decimal reduction time or *D*-value (the time for 90% of the population to die) of the wild-type and mutant strains were determined. It was found that the heat resistance of the ΔyncD mutant was slightly lower than that of the PY79 strain measured at 100 °C (average *D*-value of three different *yncD* spore preparations, 3.2 min; wild-type spores, 8.0 min), at 110 °C the *D*-values for both strains practically did not differ (Figure 3, Table 2).

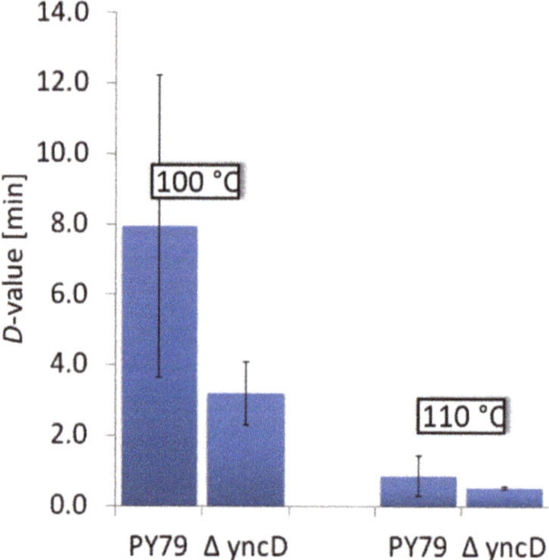


**Figure 3.** Heat resistance of wild-type and ΔyncD spores. The average D-values calculated for 100 °C and 110 °C are shown.

**Table 2.** Heat resistance of the wild-type and Δ*yncD* spores. *D*-values were calculated using a log-linear model for three independent spore preparations at 100 °C and 110 °C. $R^2$ is a regression statistic which reflects how well the regression predictions agree with the real data. The closer this value is to the value 1, the closer the measured points are to the calculated ones.

| Strain | Temperature | Prep 1 | | Prep 2 | | Prep 3 | |
|---|---|---|---|---|---|---|---|
| | | D-Value (min) | $R^2$ | D-Value (min) | $R^2$ | D-Value (min) | $R^2$ |
| PY79 | 100 °C | 5.4 | 0.96 | 5.6 | 0.90 | 12.9 | 0.87 |
| | 110 °C | 1.2 | 0.87 | 0.4 | 0.81 | 0.4 | 0.78 |
| Δ*yncD* | 100 °C | 2.9 | 0.94 | 2.5 | 0.90 | 4.2 | 0.94 |
| | 110 °C | 0.6 | 0.97 | 0.6 | 0.93 | 0.5 | 0.80 |

Another characteristic that may suggest a role for YncD in bacteria is germination. After heat activation at 70 °C, spores were induced by L-alanine in concentrations of 100 mM, 1 mM and 0.01 mM, and germination was monitored by the decrease in $OD_{600}$. Figure 4 shows a typical progression. The spores germinated similarly at the highest concentration of alanine; at 1 mM and 100 mM alanine concentration, only a subtle difference between the lag phases of the wild-type and mutant spores could be observed; at micromolar alanine concentrations, Δ*yncD* spores germinated faster than wild-type spores. We also noticed that mutant spores germinated slightly even at zero L-alanine concentration, while no germination at all could be observed in the wild-type strain. We also looked for a phenotype observed in studies of the spore-associated alanine racemase of *B. anthracis* where it was shown that spores began to germinate prematurely in the mother cytoplasm in the absence of the enzyme [23] (Chesnokova et al., 2009), although no similar effect was seen under the conditions we used.

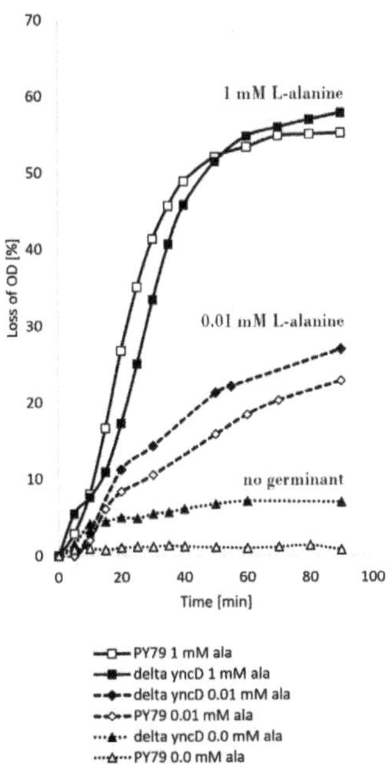


**Figure 4.** Germination assay. Spores were heat activated as described in Material and Methods, induced to germinate by L-alanine in PBS buffer and the $OD_{600}$ was monitored. The assays were repeated at least four times using independent spore preparations.

### 3.4. Localization of SpoVD-mCherry is Dependent on SpoVM

Fay et al. [27] proposed that SpoVD is recruited to the spore surface by SpoVE, which is subsequently stabilized by SpoVD. To investigate whether the interaction of SpoVD with SpoVM also affects the localization of SpoVD on the spore surface, we used *B. subtilis* strain IB1776, with a C-terminal fusion of SpoVD with mCherry [34] inserted into the *amyE* locus under the native *spoVD* promoter, and strain IB1790 with *spoVD-mCherry* and a disrupted *spoVM* gene. We found that after four hours of sporulation, in strain IB1776 SpoVD-mCherry was enriched in the forespore membrane, and later completely surrounded spore surface (Figure 5), whereas in strain IB1790 the fluorescence signal was observed mostly in the mother cell cytoplasm and localization in the forespore membrane occurred in only 20–30 % of the 216 cells observed. In addition, in about 30 % of cells SpoVD-mCherry was seen as a clump of protein close to the forespore (Figure 5).

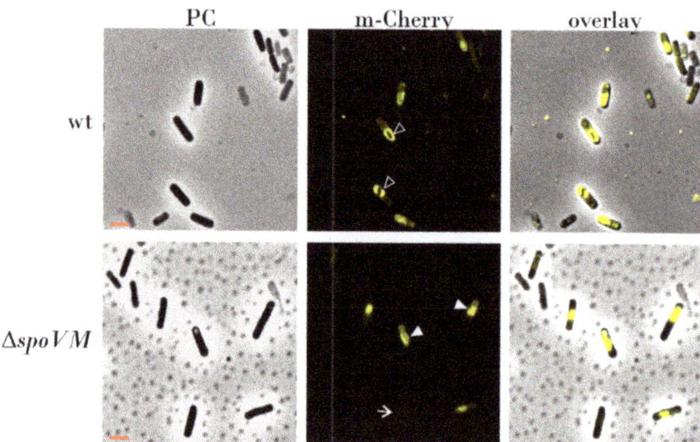


**Figure 5.** Localization of mCherry-SpoVD in sporulating cells. Cells grown in Difco sporulation media (DSM) were examined four hours after reaching stationary phase. The left panels show phase contrast images (PC); the middle panels show m-Cherry detection (false colored yellow); the right panels are merged images of PC and m-Cherry. The mCherry-SpoVD signal is around the forespore (open arrow heads) in wild type cells. In *spoVM* mutant cells, the fusion protein signals are mis-localized in the mother cell as clumps of proteins which are sometimes attached to the forespore (white arrowheads). The complete white arrow shows the proper localization of the mCherry-SpoVD protein. The red scale bar represents 2 μm.

## 4. Discussion

In this work we have investigated contacts between the basement layer morphogenetic proteins SpoVM and SpoIVA, the coat proteins which are synthesized at the onset of coat formation and some proteins essential for cortex synthesis. The initial stages of coat assembly require both these proteins and, as mentioned above, they interact with each other to form a platform for the deposition of the coat layers. The molecular mechanism of the contact between these two proteins has been described in detail and the last model of Peluso et al. [9] explains the process of their localization on the spore surface as the coordination of several events. According to this model, SpoVM localizes on the outer forespore membrane, followed by the recruitment of SpoIVA through its contact with SpoVM and the subsequent ATP-dependent polymerization of SpoIVA. The highly cross-linked SpoIVA then stitches SpoVM onto the membrane. In the present work, we asked if other coat proteins which are expressed early during spore coat formation and are not among the morphogenetic proteins, also interact with SpoIVA or SpoVM. Specifically, we were interested in three proteins, YhaX, YncD and YheD. YhaX is a protein of unknown function whose expression is induced from a $\sigma^E$ controlled promoter in response to phosphate starvation [35]. It was shown that this protein localizes to the nascent spore coat under the control of SpoIVA in the first wave of coat encasement. Using a BACTH system we did not detect any direct contacts to morphogenetic proteins, but YhaX was self-interacting, suggesting that it might form larger oligomers on the spore surface. Similarly, the function of YheD in *B. subtilis* cells has not yet been determined, although localization studies have clearly shown that it forms a shell around the spore [36]. The way in which YheD surrounds the spore surface is unusual: the protein initially forms two discrete circles with the one on the mother side of the forespore being more robust, and it later covers the entire surface of the spore. We observed a strong self-interaction for YheD and it is likely that the distribution of this protein throughout the spore surface occurs via polymerization of the protein itself. No interactions between YheD and the other tested proteins were observed. It may be that the association of YheD with proteins within the spore coat requires specific steric conditions which cannot be produced in the heterologous host used for the BACTH system. We showed that the third examined protein, the spore alanine racemase YncD, interacted with two morphogenetic proteins, CotE and SpoIVA (Figure 6). Although the localization of YncD to the spore surface was found to depend on CotE by McKenney et al. [20], their mutual interaction has not been previously reported. The association of CotE with YncD appears to be one of the strongest CotE interactions that we have discovered, including those in our earlier study of CotE contacts [31]. We have also found that the complex which these proteins form is almost insoluble, although both proteins are well soluble individually. This accords with our earlier study [16], in which we observed a tendency for coat proteins to change their solubility after co-expression in *E. coli*, indicating the possible formation of large polymeric structures.

More questions are raised by our discovery of the YncD–SpoIVA interaction. It is well known that SpoIVA directly or indirectly governs the localization of all coat proteins on the spore surface. However, except for the morphogenetic proteins SpoVM, SpoVID and SafA, no other physical contacts of SpoIVA have been found to date. Could YncD, part of the outer coat layer [20], be in contact with SpoIVA, which is a basement layer protein? Is it possible that the YncD interaction with SpoIVA is transient and exists just at the beginning of coat formation or could YncD be a component of more than one coat layer? Lastly, the exact role of YncD is not known. YncD seems to have no crucial structural function, since deletion of its gene does not significantly impair the coat's protective properties and such spores were still resistant to a hydrolyzing enzyme and oxidative stress. To understand why bacteria have a spore-specific alanine racemase, we also studied the germination process of mutant and wild-type spores. Butzin et al. [37] observed that with increasing alanine concentration, the germination of the strain in which both alanine racemase genes were deleted, the vegetative *dal* (*alrA*) and the spore protein *yncD* (or *alrB*), did not differ compared to wild-type spores. However, it is puzzling that in our experiments, spores of *yncD* mutant germinated at a slow rate even without the addition of alanine, whereas no germination was observed for wild-type spores. We hypothesize that germination of the mutant strain under this condition occurred as a result of the presence of a trace remnant of alanine. While YncD racemase in wild-type spores inhibits germination by converting L-alanine to D-alanine, residual alanine could initiate some germination of the mutant spores. Perhaps YncD is important in the non-laboratory, natural environment for monitoring whether the conditions for germination are optimal. At very low concentrations of alanine, YncD blocks this process. This function perhaps cannot be observed at the higher alanine concentrations normally used in germination assays.

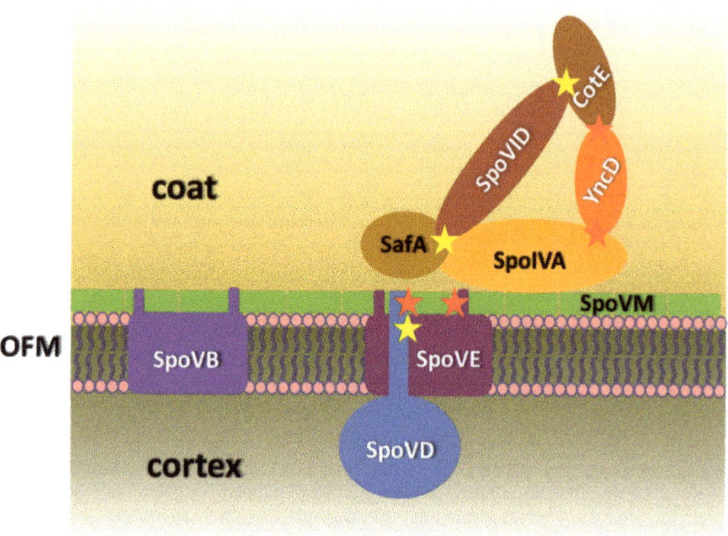



**Figure 6.** Model of the interactions between the coat proteins and the cortex synthesizing proteins. SpoIVA together with SpoVM, SpoVID, SafA and CotE build the basic structure of the spore coat through direct interactions. YncD is in contact with two of the morphogenetic proteins, SpoIVA and CotE. The localization of the SpoVD–SpoVE protein complex is dependent on SpoVM. The N-terminal domain (SpoVD) or N- and C-terminal domains (SpoVE) form physical contacts with SpoVM. The interactions discovered in this work are marked with red stars, previously known interactions are marked with yellow stars. The Bacterial Two Hybrid System (BACTH) assay did not reveal any physical contacts between SpoVB and these proteins. (OFM—outer forespore membrane.)

Given the contact between YncD and SpoIVA, we also considered whether the production of D-alanine, a peptidoglycan component, by YncD might be important for cortex synthesis. The cortex maintains spore dormancy by keeping the spore core dehydrated and thus resistant to heat and other physically damaging factors. We therefore tested the thermal resistance of *yncD* spores. We did not detect significant differences between the heat resistance of wild-type and *yncD* spores at lower temperatures, but *yncD* spores were less resistant against wet heat at 100 °C. Nevertheless, the deletion of the *yncD* gene did not have a dramatic effect on the heat resistance of the spores and we conclude that the activity of the vegetative *alrA* (dal) alanine racemase produces sufficient D-alanine for cortex peptidoglycan. To date, only a few genes have been identified that alter the cortex formation to such an extent that heat resistance changes radically. These include *spoVD*, *spoVE* and *spoVB*, whose deletion entirely stops or considerably diminishes peptidoglycan production. Here, we asked whether SpoIVA-SpoVM controls cortex synthesis simply by attaching these cortex-synthesizing proteins to the spore surface? It has long been known that the absence of SpoIVA or SpoVM impairs the assembly of both protective structures [4,5]. The mechanism behind this is not clear, although as mentioned above, *B. subtilis* blocks cortex formation using the small peptide CmpA if the spore coat is misassembled [38]. Therefore, we examined whether SpoVD, SpoVE and SpoVB interact with SpoIVA and SpoVM. We found that SpoVD physically contacts SpoVM (Figure 6) with a moderately strong interaction. After finding that they associate in *E. coli*, we tested whether they were in contact on the spore surface and we looked whether the localization of SpoVD-mCherry in wild-type and mutant cells differed. We found that the deposition of SpoVD-mCherry on the outer forespore membrane is dependent on SpoVM, confirming the observation obtained by BACTH analysis. The fluorescence signal in *spoVM* mutant cells was often mislocalized and formed a large clump in the mother cell cytoplasm that stuck to the forespore. In contrast, in wild-type cells SpoVD-mCherry nicely encircled the forespore, especially when spores entered the bright phase of spore development. These results indicate that the protein–protein interactions between SpoVM and SpoVD are important for anchoring SpoVD to the outer forespore membrane. Hence, it is reasonable to assume that the proper placement of the basement spore coat layer is necessary for the localization of key cortex synthesizing proteins. The mechanism by which cells control cortex and coat formation is certainly more sophisticated and interconnected than originally thought. Recent works by Fernandes et al. [25] and Pereira et al. [11] showed that SafA, an inner coat morphogenetic protein, is also recruited to the cortex layer and it affects the adherence of the cortex and coat layers. Moreover, the localization of SafA at the spore surface is influenced by SpoVE. However, the mechanism by which SafA is transferred through the outer forespore membrane and what the role of SafA is within the cortex remain to be determined.

### 5. Conclusions

In conclusion, our studies show that the spore alanine racemase YncD is associated with two morphogenetic proteins, CotE and SpoIVA. We propose that YncD affects spore germination at low concentrations of alanine. We also demonstrated a direct contact between SpoVM and SpoVD and suggested that the SpoVM/SpoIVA coat morphogenetic proteins could contribute to the proper localization of the proteins necessary for cortex synthesis, thereby linking the processes of cortex and coat formation. Unquestionably, further studies are needed to elucidate the details of the effect of coat morphogenetic proteins on cortex synthesis.

**Supplementary Materials:** The following are available online at https://www.mdpi.com/2076-260 7/9/2/285/s1, Table S1: Oligonucleotides, Table S2: Plasmids and strains.

**Author Contributions:** D.K.: Conceptualization; Investigation; Formal analysis; Methodology; Project administration, Supervision, Funding acquisition; Writing—original draft, Writing—review & editing. V.B.: Investigation. I.B.: Conceptualization, Investigation, Methodology; Supervision, Funding acquisition; Writing—review & editing. All authors have read and agreed to the published version of the manuscript.

**Funding:** This work was supported by VEGA–Grant No. 2/0133/18 from the Slovak Academy of Sciences to DK, and Grant from the Slovak Research and Development Agency under contract APVV-18-0104 to IB.

**Institutional Review Board Statement:** Not applicable.

**Informed Consent Statement:** Not applicable.

**Acknowledgments:** The authors thank Emília Chovancová for technical assistance and all members of I. Barák's laboratory for consultation and help; and Lars Hederstedt for kindly providing us with the LMD104 strain. We thank J. Bauer for critically reading the manuscript.

**Conflicts of Interest:** The authors declare no conflict of interest.

## References

1. Driks, A. Bacillus subtilis Spore Coat. *Microbiol. Mol. Biol. Rev.* **1999**, *63*, 1–20. [CrossRef]
2. McKenney, P.T.; Driks, A.; Eskandarian, H.A.; Grabowski, P.; Guberman, J.; Wang, K.H.; Gitai, Z.; Eichenberger, P. A Distance-Weighted Interaction Map Reveals a Previously Uncharacterized Layer of the Bacillus subtilis Spore Coat. *Curr. Biol.* **2010**, *20*, 934–938. [CrossRef]
3. Driks, A.; Eichenberger, P. The Spore Coat. *Microbiol Spectr.* **2016**. [CrossRef]
4. Levin, P.A.; Fan, N.; Ricca, E.; Driks, A.; Losick, R.; Cutting, S. An unusually small gene required for sporulation by *Bacillus subtilis*. *Mol. Microbiol.* **1993**, *9*, 761–771. [CrossRef]
5. Roels, S.; Driks, A.; Losick, R. Characterization of spoIVA, a Sporulation Gene Involved in Coat Morphogenesis in *Bacillus* subtilis. *J. Bacteriol.* **1992**, *174*, 575–585. [CrossRef]

6.  Ramamurthi, K.S.; Lecuyer, S.; Stone, H.A.; Losick, R. Geometric Cue for Protein Localization in a Bacterium. *Science* **2009**, *323*, 1354–1357. [CrossRef]
7.  Castaing, J.P.; Nagy, A.; Anantharaman, V.; Aravind, L.; Ramamurthi, K.S. ATP hydrolysis by a domain related to translation factor GTPases drives polymerization of a static bacterial morphogenetic protein. *Proc. Natl. Acad. Sci. USA* **2013**, *110*, E151–E160. [CrossRef] [PubMed]
8.  Ramamurthi, K.S.; Losick, R. ATP-Driven Self-Assembly of a Morphogenetic Protein in Bacillus subtilis. *Mol. Cell* **2008**, *31*, 406–414. [CrossRef] [PubMed]
9.  Peluso, E.A.; Updegrove, T.B.; Chen, J.; Shroff, H.; Ramamurthi, K.S. A 2-dimensional ratchet model describes assembly initiation of a specialized bacterial cell surface. *Proc. Natl. Acad. Sci. USA* **2019**, *116*, 21789–21799. [CrossRef] [PubMed]
10. Wang, K.H.; Isidro, A.L.; Domingues, L.; Eskandarian, H.A.; McKenney, P.T.; Drew, K.; Grabowski, P.; Chua, M.H.; Barry, S.N.; Guan, M.; et al. The coat morphogenetic protein SpoVID is necessary for spore encasement in Bacillus subtilis. *Mol. Microbiol.* **2009**, *74*, 634–649. [CrossRef] [PubMed]
11. Pereira, F.C.; Nunes, F.; Cruz, F.; Fernandes, C.; Isidro, A.L.; Lousa, D.; Soares, C.M.; Moran, C.P.; Henriques, A.O.; Serrano, M. A LysM Domain intervenes in sequential protein-protein and protein-peptidoglycan interactions important for spore coat assembly in bacillus subtilis. *J. Bacteriol.* **2019**, *201*, e00642-18. [CrossRef] [PubMed]
12. Zheng, L.B.; Donovan, W.P.; Fitz-James, P.C.; Losick, R. Gene encoding a morphogenic protein required in the assembly of the outer coat of the Bacillus subtilis endospore. *Genes Dev.* **1988**, *2*, 1047–1054. [CrossRef] [PubMed]
13. Takamatsu, H.; Chikahiro, Y.; Kodama, T.; Koide, H.; Kozuka, S.; Tochikubo, K.; Watabe, K. A Spore Coat Protein, CotS, of Bacillus subtilis Is Synthesized under the Regulation of K and GerE during Development and Is Located in the Inner Coat Layer of Spores. *J. Bacteriol.* **1998**, *180*, 2968–2974. [CrossRef] [PubMed]
14. Bauer, T.; Little, S.; Stöver, A.G.; Driks, A. Functional Regions of the Bacillus subtilis Spore Coat Morphogenetic Protein CotE. *J. Bacteriol.* **1999**, *181*, 7043–7051. [CrossRef] [PubMed]
15. Costa, T.; Isidro, A.L.; Moran, C.P.; Henriques, A.O. Interaction between coat morphogenetic proteins SafA and SpoVID. *J. Bacteriol.* **2006**, *188*, 7731–7734. [CrossRef]
16. Müllerová, D.; Krajčíková, D.; Barák, I. Interactions between Bacillus subtilis early spore coat morphogenetic proteins. *FEMS Microbiol. Lett.* **2009**, *299*, 74–85. [CrossRef]
17. De Francesco, M.D.; Jacobs, J.Z.; Nunes, F.; Serrano, M.; Mckenney, P.T.; Chua, M.H.; Henriques, A.O.; Eichenberger, P. Physical interaction between coat morphogenetic proteins spovid and cote is necessary for spore encasement in *Bacillus subtilis*. *J. Bacteriol.* **2012**, *194*, 4941–4950. [CrossRef]
18. Qiao, H.; Krajcikova, D.; Xing, C.; Lu, B.; Hao, J.; Ke, X.; Wang, H.; Barak, I.; Tang, J. Study of the interactions between the key spore coat morphogenetic proteins CotE and SpoVID. *J. Struct. Biol.* **2013**, *181*, 128–135. [CrossRef]
19. Liu, H.; Qiao, H.; Krajcikova, D.; Zhang, Z.; Wang, H.; Barak, I.; Tang, J. Physical interaction and assembly of *Bacillus subtilis* spore coat proteins CotE and CotZ studied by atomic force microscopy. *J. Struct. Biol.* **2016**, *195*, 245–251. [CrossRef]
20. McKenney, P.T.; Eichenberger, P. Dynamics of spore coat morphogenesis in *Bacillus subtilis*. *Mol. Microbiol.* **2012**, *83*, 245–260. [CrossRef]
21. Pierce, K.J.; Salifu, S.P.; Tangney, M. Gene cloning and characterization of a second alanine racemase from *Bacillus subtilis* encoded by yncD. *FEMS Microbiol. Lett.* **2008**, *283*, 69–74. [CrossRef] [PubMed]
22. Steichen, C.; Chen, P.; Kearney, J.F.; Turnbough, C.L. Identification of the immunodominant protein and other proteins of the *Bacillus anthracis* exosporium. *J. Bacteriol.* **2003**, *185*, 1903–1910. [CrossRef]
23. Chesnokova, O.N.; McPherson, S.A.; Steichen, C.T.; Turnbough, C.L. The spore-specific alanine racemase of Bacillus anthracis and its role in suppressing germination during spore development. *J. Bacteriol.* **2009**, *191*, 1303–1310. [CrossRef] [PubMed]
24. Ebmeier, S.E.; Tan, I.S.; Clapham, K.R.; Ramamurthi, K.S. Small proteins link coat and cortex assembly during sporulation in *Bacillus subtilis*. *Mol. Microbiol.* **2012**, *84*, 682–696. [CrossRef]
25. Tan, I.S.; Weiss, C.A.; Popham, D.L.; Ramamurthi, K.S. A Quality-Control Mechanism Removes Unfit Cells from a Population of Sporulating Bacteria. *Dev. Cell* **2015**, *34*, 682–693. [CrossRef]
26. Vasudevan, P.; Weaver, A.; Reichert, E.D.; Linnstaedt, S.D.; Popham, D.L. Spore cortex formation in *Bacillus subtilis* is regulated by accumulation of peptidoglycan precursors under the control of sigma K. *Mol. Microbiol.* **2007**, *65*, 1582–1594. [CrossRef] [PubMed]
27. Fay, A.; Meyer, P.; Dworkin, J. Interactions between late-acting proteins required for peptidoglycan synthesis during sporulation. *J. Mol. Biol.* **2010**, *399*, 547–561. [CrossRef] [PubMed]
28. Popham, D.L.; Stragier, P. Cloning, Characterization, and Expression of the spoVB Gene of *Bacillus subtilis*. *J. Bacteriol.* **1991**, *173*, 7942–7949. [CrossRef]
29. Meeske, A.J.; Riley, E.P.; Robins, W.P.; Uehara, T.; Mekalanos, J.J.; Kahne, D.; Walker, S.; Kruse, A.C.; Bernhardt, T.G.; Rudner, D.Z. SEDS proteins are a widespread family of bacterial cell wall polymerases. *Nature* **2016**, *537*, 634–638. [CrossRef]
30. Meeske, A.J.; Sham, L.T.; Kimsey, H.; Koo, B.M.; Gross, C.A.; Bernhardt, T.G.; Rudner, D.Z. MurJ and a novel lipid II flippase are required for cell wall biogenesis in *Bacillus subtilis*. *Proc. Natl. Acad. Sci. USA* **2015**, *112*, 6437–6442. [CrossRef]
31. Krajčíková, D.; Forgáč, V.; Szabo, A.; Barák, I. Exploring the interaction network of the *Bacillus subtilis* outer coat and crust proteins. *Microbiol. Res.* **2017**, *204*, 72–80. [CrossRef] [PubMed]
32. Krajčíková, D.; Lukáčová, M.; Müllerová, D.; Cutting, S.M.; Barák, I. Searching for protein-protein interactions within the *Bacillus subtilis* spore coat. *J. Bacteriol.* **2009**, *191*, 3212–3219. [CrossRef] [PubMed]

33. Jankute, M.; Byng, C.V.; Alderwick, L.J.; Besra, G.S. Elucidation of a protein-protein interaction network involved in Corynebacterium glutamicum cell wall biosynthesis as determined by bacterial two-hybrid analysis. *Glycoconj. J.* **2014**, *31*, 475–483. [CrossRef] [PubMed]
34. Bukowska-Faniband, E.; Hederstedt, L. Cortex synthesis during *Bacillus subtilis* sporulation depends on the transpeptidase activity of SpoVD. *FEMS Microbiol. Lett.* **2013**, *346*, 65–72. [CrossRef]
35. Prágai, Z.; Harwood, C.R. Regulatory interactions between the Pho and σB-dependent general stress regulons of *Bacillus subtilis*. *Microbiology* **2002**, *148*, 1593–1602. [CrossRef]
36. Van Ooij, C.; Eichenberger, P.; Losick, R. Dynamic patterns of subcellular protein localization during spore coat morphogenesis in *Bacillus subtilis*. *J. Bacteriol.* **2004**, *186*, 4441–4448. [CrossRef]
37. Butzin, X.Y.; Troiano, A.J.; Coleman, W.H.; Griffiths, K.K.; Doona, C.J.; Feeherry, F.E.; Wang, G.; Li, Y.Q.; Setlow, P. Analysis of the effects of a gerP mutation on the germination of spores of Bacillus subtilis. *J. Bacteriol.* **2012**, *194*, 5749–5758. [CrossRef]
38. Fernandes, C.G.; Moran, C.P.; Henriques, A.O. Autoregulation of SafA assembly through recruitment of a protein cross-linking enzyme. *J. Bacteriol.* **2018**, *200*, e00066-18. [CrossRef]

Article

# Molecular Physiological Characterization of a High Heat Resistant Spore Forming *Bacillus subtilis* Food Isolate

Zhiwei Tu [1,2], Peter Setlow [3], Stanley Brul [1,*,†] and Gertjan Kramer [2,†]

1   Laboratory for Molecular Biology and Microbial Food Safety, University of Amsterdam, 1098 XH Amsterdam, The Netherlands; Z.Tu@uva.nl
2   Laboratory for Mass Spectrometry of Biomolecules, University of Amsterdam, 1098 XH Amsterdam, The Netherlands; g.kramer@uva.nl
3   Department of Molecular Biology and Biophysics, UCONN Health, Farmington, CT 06030-3303, USA; setlow@nso2.uchc.edu
*   Correspondence: s.brul@uva.nl; Tel.: +31-20-525-7079 (ext. 6970)
†   These authors contributed equally to this work.

**Abstract:** Bacterial endospores (spores) are among the most resistant living forms on earth. Spores of *Bacillus subtilis* A163 show extremely high resistance to wet heat compared to spores of laboratory strains. In this study, we found that spores of *B. subtilis* A163 were indeed very wet heat resistant and released dipicolinic acid (DPA) very slowly during heat treatment. We also determined the proteome of vegetative cells and spores of *B. subtilis* A163 and the differences in these proteomes from those of the laboratory strain PY79, spores of which are much less heat resistant. This proteomic characterization identified 2011 proteins in spores and 1901 proteins in vegetative cells of *B. subtilis* A163. Surprisingly, spore morphogenic protein SpoVM had no homologs in *B. subtilis* A163. Comparing protein expression between these two strains uncovered 108 proteins that were differentially present in spores and 93 proteins differentially present in cells. In addition, five of the seven proteins on an operon in strain A163, which is thought to be primarily responsible for this strain's spores high heat resistance, were also identified. These findings reveal proteomic differences of the two strains exhibiting different resistance to heat and form a basis for further mechanistic analysis of the high heat resistance of *B. subtilis* A163 spores.

**Keywords:** *Bacillus subtilis* A163; proteome; high heat resistance

**Citation:** Tu, Z.; Setlow, P.; Brul, S.; Kramer, G. Molecular Physiological Characterization of a High Heat Resistant Spore Forming *Bacillus subtilis* Food Isolate. *Microorganisms* **2021**, *9*, 667. https://doi.org/10.3390/microorganisms9030667

Academic Editor: Imrich Barák

Received: 1 March 2021
Accepted: 20 March 2021
Published: 23 March 2021

**Publisher's Note:** MDPI stays neutral with regard to jurisdictional claims in published maps and institutional affiliations.

## 1. Introduction

Strains from *Bacillus subtilis* are causative agents of food spoilage, which can cause problems in the food industry [1–3]. This is largely due to the high stress resistance of the spores produced by this species. In response to nutritional and environmental stresses, vegetative cells of *B. subtilis* can form metabolically dormant spores which exhibit extreme resistance properties compared to their corresponding vegetative cells [4]. Surviving spores can, depending on the environmental conditions, quickly germinate, resume vegetative growth and subsequently cause food spoilage [5]. Compared to vegetative cells, spores are protected by multiple proteinaceous coat layers [6]. In particular, the water content within the spores' core is low, and DNA in the core is surrounded by dipicolinic acid (DPA) and saturated with protective α/β-type small acid-soluble proteins (SASPs). Many factors can affect spores' wet heat resistance. For example, spores prepared on solid media have higher wet heat resistance than those prepared in liquid [7]. Higher sporulation temperatures can also result in higher heat resistance of spores [8]. In addition, the protein composition and DPA levels of spores are affected by sporulation conditions and temperatures [8–10]. With respect to heat resistance of spores of different *Bacillus* strains, two distinct groups have been identified [11]. One group, which includes *B. subtilis* strains 168 and PY79, commonly used strains in laboratory research, has spores with low heat resistance, while the other group, which includes the foodborne isolate *B. subtilis* strain A163, has spores with high heat resistance. A mobile genetic element, Tn1546, is commonly present in the high resistance strains and this transposon contains the *spoVA*$^{2\,mob}$ operon which can profoundly heighten resistance of spores to heat and pressure, encoding four genes of unknown function and a three gene *spoVA* operon, which encodes proteins involved in DPA uptake into developing spores, and its release in spore germination [12–15]. Regrettably, the proteome of strain *B. subtilis* A163 has not yet been studied extensively using high resolution mass spectrometry-based proteomics so the protein expression profile has not been compared with that of low resistance spores.

First, we confirmed the higher thermal resistance of *B. subtilis* A163 spores compared to spores of *B. subtilis* PY79, and that rates of DPA release at elevated temperatures were much slower from A163 spores than from PY79 spores. We also determined the proteomes of spores and cells of *B. subtilis* A163. This work found 2011 spore and 1901 cell proteins, while 2045 cell and 2170 spore proteins were identified in *B. subtilis* strain PY79, a prototrophic derivative of strain 168 [16]. Among the proteome of spores of *B. subtilis* A163 were five of the seven proteins of the *spoVA*$^{2mob}$ operon found, and homologs of SpoVM, which are important in spore coat assembly in PY79 [17] were not found. Previous studies show that strain A163 and strain 168 are very similar at the genomic level [18]. By searching every protein sequence of *B. subtilis* PY79 against all protein sequences of *B. subtilis* A163, 4030 protein sequences of two strains showed more than 50% identity. Among them, 1312 and 1276 proteins were quantified between the two strains in spores and cells, respectively. Both high-abundance and low-abundance proteins were revealed in spores and cells. In spores of *B. subtilis* A163, the high-abundance proteins are enriched in the Uniprot categories glycosyltransferases and proteases, while low-abundance proteins are mainly enriched in membrane and sporulation proteins. In cells of *B. subtilis* A163, high-abundance proteins are enriched in the transport of proteins and peptides, as well as competence. Proteins involved in biosynthesis and metabolism of fatty acids and lipids, as well as oxidoreductase, are enriched in the low-abundant cellular proteins.

## 2. Materials and Methods

### 2.1. Strains and Sporulation

Low spore wet heat resistant *B. subtilis* strain PY79 and the high heat resistant spore forming food isolate *B. subtilis* A163 were used in this study [19]. Both strains were sporulated in shake flasks containing 3-(N-morpholino) propane sulfonic acid (MOPS) (Sigma-Aldrich, St. Louis, MI, USA) buffered defined liquid medium [20]. In brief, a single colony from a Lysogeny broth (LB) [21] agar plate was inoculated in 5 mL LB liquid medium and cultivated until its exponential phase at 37 °C and 200 rpm. The exponentially growing cells were then subjected to overnight growth in MOPS medium in a series of dilutions. The dilution with exponential growth was selected and 1 mL was transferred into 100 mL of MOPS medium and allowed to sporulate for 72 h. Spores were harvested and purified with Histodenz (Sigma-Aldrich, St. Louis, MI, USA) gradient centrifugation [22]. Vegetative cells were harvested from LB medium in the exponential phase. Three biological replicates of each strain were harvested for both spores and cells and stored at −80 °C for further experiments.

### 2.2. Heat Resistance and DPA Measurement in Spores

Heat resistance of spores was tested at 85 °C and 98 °C following an established protocol [9]. One ml of spores with an OD$_{600}$ of 2 (~2 × 10$^8$ spores/mL) was heat activated at 70 °C for 30 min in a water bath. After being placed on ice for 15 min, spores were injected into a metal screwcap tube with 9 mL sterile milli-Q water pre-warmed for 20 min in a glycerol bath (85 °C or 98 °C). The metal tube was then kept at 85 °C (or 98 °C) for another 10 min, after which, the tube was cooled on ice. The fraction of surviving spores after heat treatment was estimated by counting the number of colonies formed on LB plates. Three biological replicates were performed for each strain tested at both 85 °C and 98 °C.

Spore DPA content was calculated as µg of DPA per mg dry weight of spores. The protocol of DPA measurement was modified from [23]. One ml of spores with an $OD_{600}$ of 2 from each strain was suspended in a buffer containing 0.3 mM $(NH_4)_2SO_4$, 6.6 mM $KH_2PO_4$, 15 mM NaCl, 59.5 mM $NaHCO_3$ and 35.2 mM $Na_2HPO_4$. For total DPA measurement, the suspended spores were autoclaved at 121 °C for 15 min. After incubation, the sample was centrifuged at 15,000 rpm for 2 min and 10 µL supernatant was added to 115 µL of buffer 2 (1 mM Tris, 150 mM NaCl) with and without 0.8 mM terbium chloride (Sigma-Aldrich, St. Louis, MI, USA). After 15 min of incubation, the fluorescence of the samples was measured using a Synergy Mx microplate reader (BioTek; 270-nm excitation; 545-nm reading; gain, 100) (Bad Friedrichshall, Germany). The background fluorescence (without terbium, incubated at 37 °C) was subtracted from that of all the samples. A calibration curve of 0–125 µg/mL DPA (2,6-pyridinedicarboxylic acid) (Sigma-Aldrich, St. Louis, MI, USA) was used to calculate DPA concentrations of the sample. Samples incubated at 98 °C for 1–6 h were also used to measure the amount of DPA released at various heating times. Spore dry weights were determined by weighing overnight freeze-dried spores. Three biological replicates were measured for all conditions.

### 2.3. Proteome Databases and Comparison of Protein Sequences

Amino acid sequences encoded in the genome of *B. subtilis* A163 were acquired from [24] with accession no. JSXS00000000. The UniProt proteome database UP000001570 was used for *B. subtilis* PY79 [25]. Every protein sequence within UP000001570 was searched against the database of *B. subtilis* A163 to find the best match(es) using NCBI BLAST+ BLASTP (Galaxy version 0.3.3) embedded in the web-based platform Galaxy Europe (https://usegalaxy.eu/, accessed at 5 January 2021) with the E-value set at 0.00001 [26–28]. The protein sequences encoded in the $spoVA^{2mob}$ operon in *B. subtilis* B4417 were acquired from NCBI with the reference sequence NZ_LJSM01000045.1.

### 2.4. Data Acquisition for Proteomic Analysis

Processing of samples and fractionation of every trypsin digested sample into 10 fractions was done following the protocol described by Tu et al. [29]. Every fraction was reconstituted in 0.1% formic acid in water and 200 ng equivalent (set by measuring absorbance at a wavelength of 205 nm [30]) was injected by a Ultimate 3000 RSLCnano UHPLC system (Thermo Scientific, Germeringen, Germany) onto a 75 µm × 250 mm analytical column (C18, 1.6 µm particle size, Aurora, Ionopticks, Australia) kept at 50 °C at 400 nL/min for 15 min in 3% solvent B before being separated by a multi-step gradient (Solvent A: 0.1% formic acid in water, Solvent B: 0.1% formic acid in acetonitrile) to 5% B at 16 min, 17% B at 38 min, 25% B at 43 min, 34% B at 46 min, 99% B at 47 min held until 54 min returning to initial conditions at 55 min equilibrating until 80 min.

Eluting peptides were sprayed by the emitter coupled to the column into a captive spray source (Bruker, Bremen, Germany) with a capillary voltage of 1.5 kV, a source gas flow of 3 L/min of pure nitrogen and a dry temperature setting of 180 °C, attached to a timsTOF pro (Bruker, Bremen, Germany) trapped ion mobility, quadrupole, time of flight mass spectrometer. The timsTOF was operated in PASEF mode of acquisition. The TOF scan range was 100–1700 *m/z* with a tims range of 0.6–1.6 V·s/cm$^2$. In PASEF mode a filter was applied to the *m/z* and ion mobility plane to select features most likely representing peptide precursors, the quad isolation width was 2 Th at 700 *m/z* and 3 Th at 800 *m/z*, and the collision energy was ramped from 20–59 eV over the tims scan range to generate fragmentation spectra. A total number of 10 PASEF MS/MS scans scheduled with a total cycle time of 1.16 s, scheduling target intensity $2 \times 10^4$ and intensity threshold of $2.5 \times 10^3$ and a charge state range of 0–5 were used. Active exclusion was on (release after 0.4 min), reconsidering precursors if ratio current/previous intensity >4.

*2.5. Data Processing*

Generated data for spores (of two strains) and cells (of two strains) were processed with MaxQuant (Version 1.6.14, Martinsried, Germany) in two separate analyses [31]. 10 fractions from the same sample were set as one experiment. Proteome databases for *B. subtilis* A163 and *B. subtilis* PY79 were included in the analysis. The proteolytic enzyme used was trypsin/p, and the maximum missed cleavages was set to 2. Carbamidomethyl (C) was set as fixed modification with variable modifications of Oxidation (M) and Acetyl (Protein N-term). The type of Group specific parameters was set as TIMS-DDA. The rest of the parameters were set using the default. Since two databases were used in the analysis, the quantified values of proteins from two strains with high percentage of identity were sometimes reported as two separate proteins in the output proteinGroup.txt (vide infra). To quantitatively compare the homologous protein from two strains, we re-assembled the identified peptides in the evidence.txt file and quantified the protein amounts using the R software package *iq* [32]; the R-script and evidence.txt files used can be found in Supplementary File S1. The minimum number of peptides for the quantification was 2. The differentially presented proteins in cells and spores were determined by using R/Bioconductor software package *limma* [33]. DAVID Bioinformatics Resources tool (version 6.8) was used to retrieve the UniProt keyword enrichment of the differentially presented proteins [34,35]. The protein list of coat proteins was retrieved from *Subti*Wiki (http://subtiwiki.uni-goettingen.de/, accessed at 5 January 2021) [36].

## 3. Results and Discussion

*3.1. Heat Resistance and DPA Measurement of B. subtilis Spores*

Spores of *B. subtilis* A163 are reported to show extreme wet heat resistance [11,18,19]. To confirm this, spores of *B. subtilis* A163 and PY79 were heat treated at 85 °C and 98 °C, and as expected there was a significant decrease in surviving PY79 spores treated at 98 °C, but no such difference for A163 spores (Figure 1A). Our measurements also showed that while *B. subtilis* A163 spores gave slightly higher values for DPA than PY79 spores, the difference was not statistically significant (Figure 1B). This latter finding is consistent with the essentially identical DPA levels found recently in *B. subtilis* with and without transposon Tn1546 [15]. However, the rate of release of DPA during heat treatment of A163 spores was much slower than from PY79 spores (Figure 1C). Treatment of spores of PY79 spores at 98 °C resulted in release of ~80% DPA at 1 h and almost all DPA at 5 h. However, only trace amounts of DPA were released from A163 spores during the first hour of heat treatment and only ~20% of the DPA was released at 6 h. This finding confirms a previous report [19], that A163 spores need much higher temperatures to release their DPA than low wet heat resistance spores. Killing of low wet heat resistance spores by wet heat appears due to damage to key spore proteins, and this takes place before loss of DPA presumably due to spore inner membrane (IM) damage but prevents the outgrowth of spores [37,38]. However, the high wet heat resistant A163 spores have features to maintain the native state of key spore proteins as well is the IM impermeability during heat treatment at higher temperatures.

*3.2. Identification and Quantification of Proteomes of Spores and Cells*

One important field in the study of spore forming bacteria is identification and quantification of the spore and growing cell proteome. In this study, we investigated the proteome of both spores and cells of *B. subtilis* A163. 2011 and 1901 proteins were separately identified in at least two of three biological replicates of spores and cells of *B. subtilis* A163, while with *B. subtilis* PY79, 2170 spore proteins and 2045 cellular proteins were identified. Lists of identified proteins in spores and cells can be found in Supplementary Tables S1 and S2. In terms of identification of proteins from the *spoVA*$^{2mob}$ operon in *B. subtilis* A163 spores, five proteins were identified (Table 1), excluding the protein with both a predicted DUF 421 domain and a DUF 1657 domain (2Duf protein), which is thought to be the most important one in the *spoVA*$^{2mob}$ operon [12], as well as the SpoVAEb protein. In addition, homologs for the two DUF 1657 domain-containing proteins and SpoVAC$^{2mob}$ were also identified.

This could be because multiple copies of *spoVA*$^{2mob}$ were present in *B. subtilis* A163 [12]. To compare proteomes of spores or cells of two strains, we first checked how much similarity there was between the protein sequences of the two strains. Among 4800 protein coding genes of *B. subtilis* A163, amino acid sequences of 4141 genes show a minimum 22% of identity with the lab strain, and 4030 genes show more than 50% identity (Figure 2). Quite a number of proteins from the two strains show a high percentage of identity, indicating that these proteins could be considered homologs. Coat proteins identified in *B. subtilis* spores are shown in Table 2. SpoVM, CotU, CotR and YjdH have no homologous proteins found in *B. subtilis* A163. SpoVM is a key protein for the proper assembly of the spore coat [17]. Two homologous genes were found for *oxdD*, *yjqC* and *cotF* in *B. subtilis* A163, but only one homolog of CotF and two homologs of YjqC were identified. For the proteins involved in germination and the endogenous SpoVA channel [5], some germinant receptor proteins, most notably GerB proteins, were only identified in *B. subtilis* PY79, but not in *B. subtilis* A163 (Table 3). Moreover, strain specific proteins were identified in both strains. Among the *B. subtilis* A163 specific spore proteins, none of them show predicted functions except an alpha-glucosidase (protein id = KIL30593.1).

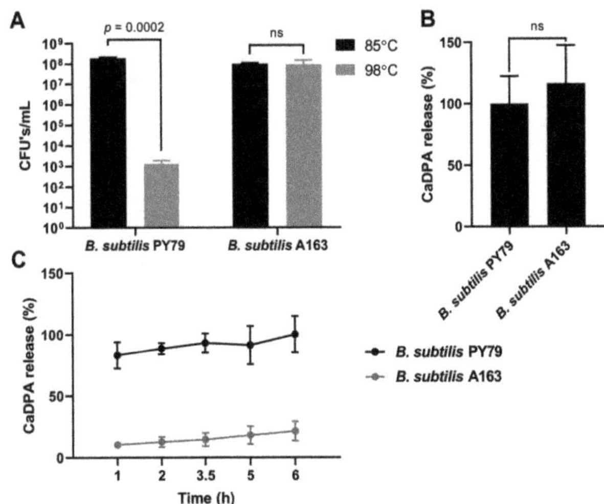


**Figure 1.** Heat resistance test and DPA content of *B. subtilis* spores. The standard deviation is shown in the graphs. Statistical significance was determined using Student's *t*-test. ns, not significant. (**A**) Numbers of colonies formed on LB agar plates of *B. subtilis* spores wet heat treated at 85 °C and 98 °C for 10 min. (**B**) Amount of CaDPA released by *B. subtilis* spores that were autoclaved at 121 °C. The amount of CaDPA released by spores was calculated as % of CaDPA released by *B. subtilis* PY79. (**C**) CaDPA released by *B. subtilis* spores heat treated at 98 °C for 1–6 h. The amount of CaDPA released by spores was calculated as % of the total CaDPA.

**Table 1.** *spoVA*$^{2mob}$ proteins identified in *B. subtilis* A163 spores.

| Genes | Identifier in *B. subtilis* B4417 [a] | Identifier in *B. subtilis* A163 [b] | Percentage of Identity (%) [c] | Number of Peptides |
|---|---|---|---|---|
| *DUF1657* | prot_WP_009336483.1_679 | prot_KIL30783.1_1 <br> prot_KIL30530.1_3 | 100 <br> 80.882 | 27 <br> 17 |
| *spoVAD*$^{2mob}$ | prot_WP_013352386.1_681 | prot_KIL30785.1_3 | 100 | 8 |
| *spoVAC*$^{2mob}$ | prot_WP_013352385.1_682 | prot_KIL30782.1_3 <br> prot_KIL32093.1_7 | 100 <br> 97.183 | 5 <br> 11 |
| *Yhcn/YlaJ* | prot_WP_017697692.1_683 | prot_KIL30781.1_2 | 100 | 9 |
| *DUF1657* | prot_WP_009336488.1_684 | prot_KIL30780.1_1 <br> prot_KIL32090.1_4 | 100 <br> 91.176 | 17 <br> 16 |

[a] *B. subtilis* 168 carrying the *spoVA*$^{2mob}$ operon [12]. [b] identified by searching protein sequences of *spoVA*$^{2mob}$ operon of *B. subtilis* B4417 against the protein sequences of *B. subtilis* A163. [c], percentage of identity of the proteins between *B. subtilis* B4417 and *B. subtilis* A163.

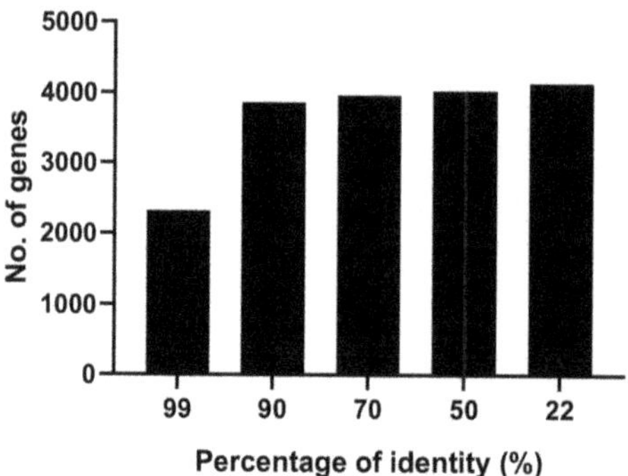



**Figure 2.** Amino acid sequence comparisons between *B. subtilis* strains. Every protein sequence in the genome of *B. subtilis* PY79 was searched against the database containing all the protein sequences of *B. subtilis* A163. The match with the highest percentage of identity was included in the figure.

**Table 2.** Identified coat proteins in spores of *B. subtilis* PY79 and A163.

| Proteins | UniProt IDs | Homologous Proteins in *B. subtilis* A163 | Percentage of Identity (%) | Number of Peptides | |
|---|---|---|---|---|---|
| | | | | *B. subtilis* A163 | *B. subtilis* PY79 |
| SpoIVA | P35149 | prot_KIL33240.1_118 | 99.797 | 131 | 90 |
| SpoVID | P37963 | prot_KIL31245.1_33 | 95.549 | 22 | 9 |
| SpoVM | P37817 | NA | NA | NA | 2 |
| YaaH | P37531 | prot_KIL32101.1_8 | 96.721 | 117 | 132 |
| YuzC | O32089 | prot_KIL29514.1_5 | 96.721 | 13 | 14 |
| CotE | P14016 | prot_KIL29844.1_246 | 100 | 58 | 53 |
| CotM | Q45058 | prot_KIL31177.1_55 | 98.387 | 5 | 5 |
| CotO | O31622 | prot_KIL30325.1_87 | 99.111 | 9 | 8 |
| YhjR | O07572 | prot_KIL33619.1_38 | 100 | 21 | 24 |
| YknT | O31700 | prot_KIL30843.1_42 | 98.754 | 2 | 4 |
| YncD | P94494 | prot_KIL31326.1_4 | 99.746 | 40 | 38 |
| CotZ | Q08312 | prot_KIL30327.1_89 | 100 | 46 | 35 |
| CwlJ | P42249 | prot_KIL33394.1_41 | 95.775 | 36 | 26 |
| YisY | O06734 | prot_KIL33591.1_10 | 98.134 | 81 | 73 |
| YsxE | P37964 | prot_KIL31244.1_32 | 98.534 | 10 | 4 |
| YutH | O32123 | prot_KIL29568.1_59 | 99.11 | 16 | 2 |
| CotT | P11863 | prot_KIL30718.1_123 | 98.78 | 48 | 31 |
| YybI | P37495 | prot_KIL31595.1_71 | 93.893 | 47 | 51 |
| CotA | P07788 | prot_KIL30080.1_5 | 99.61 | 170 | 166 |
| CotB | P07789 | prot_KIL31029.1_37 | 68.116 | 4 | 61 |
| CotG | P39801 | prot_KIL31027.1_35 | 92.308 | 86 | 45 |
| CotP | P96698 | prot_KIL30034.1_53 | 97.203 | 24 | 14 |
| CotQ | O06997 | prot_KIL33096.1_104 (a, b) | 25 | 0 | 111 |
| CotS | P46914 | prot_KIL33875.1_18 | 99.145 | 133 | 123 |
| CotW | Q08310 | prot_KIL30330.1_92 | 98.095 | 22 | 25 |

Table 2. Cont.

| Proteins | UniProt IDs | Homologous Proteins in *B. subtilis* A163 | Percentage of Identity (%) | Number of Peptides | |
|---|---|---|---|---|---|
| | | | | *B. subtilis* A163 | *B. subtilis* PY79 |
| LipC | P42969 | prot_KIL29460.1_9 | 100 | 72 | 55 |
| OxdD | O34767 | prot_KIL33540.1_11 (a)<br>prot_KIL33541.1_12 (a) | 94.231<br>99.642 | 0<br>0 | 70 |
| Tgl | P40746 | prot_KIL32205.1_19 | 99.184 | 56 | 40 |
| YjqC | O34423 | prot_KIL32111.1_3 (b)<br>prot_KIL30071.1_90 | 34.426<br>51.163 | 151<br>1 | 46 |
| YjzB | O34891 | prot_KIL30279.1_41 | 96.104 | 1 | 4 |
| YmaG | O31793 | prot_KIL29859.1_261 | 96.703 | 7 | 19 |
| YppG | P50835 | prot_KIL33183.1_61 | 97.6 | 4 | 6 |
| YtxO | P46916 | prot_KIL33874.1_17 | 95.804 | 52 | 47 |
| YxeE | P54944 | prot_KIL32257.1_32 | 100 | 17 | 15 |
| CotU | O31802 | NA | NA | NA | 27 |
| CgeA | P42089 | prot_KIL31959.1_13 | 96.241 | 12 | 8 |
| CgeB | P42090 | prot_KIL31960.1_14 | 94.386 | 11 | 3 |
| CgeC | P42091 | prot_KIL31958.1_12 | 98.02 | 2 | 3 |
| CgeE | P42093 | prot_KIL31956.1_10 | 99.228 | 14 | 21 |
| CmpA | P14204 | prot_KIL33555.1_2 | 100 | 2 | 10 |
| CotC | P07790 | prot_KIL31334.1_12 | 100 | 48 | 30 |
| CotF | P23261 | prot_KIL31586.1_62 (a)<br>prot_KIL31587.1_63 | 100<br>94.643 | 0<br>3 | 51 |
| CotH | Q45535 | prot_KIL31028.1_36 | 97.238 | 101 | 77 |
| CotI | O34656 | prot_KIL33877.1_20 | 98.3 | 115 | 107 |
| CotJA | Q45536 | prot_KIL30149.1_74 | 98.78 | 67 | 53 |
| CotJB | Q45537 | prot_KIL30150.1_75 | 100 | 10 | 12 |
| CotJC | Q45538 | prot_KIL30151.1_76 | 100 | 56 | 41 |
| CotR | O06996 | NA | NA | NA | 70 |
| CotSA | P46915 | prot_KIL33876.1_19 | 99.469 | 185 | 116 |
| CotX | Q08313 | prot_KIL30329.1_91 | 100 | 72 | 48 |
| CotY | Q08311 | prot_KIL30328.1_90 | 100 | 50 | 39 |
| GerQ | P39620 | prot_KIL31830.1_12 | 98.895 | 21 | 17 |
| GerT | Q7WY67 | prot_KIL32533.1_42 | 95.541 | 33 | 7 |
| SafA | O32062 | prot_KIL32613.1_30 | 98.45 | 77 | 74 |
| SpsB | P39622 | prot_KIL31833.1_15 | 98.911 | 27 | 23 |
| YabG | P37548 | prot_KIL32152.1_41 | 98.276 | 67 | 54 |
| YdhD | O05495 | prot_KIL30052.1_71 | 99.048 | 65 | 53 |
| YgaK | Q796Y5 | prot_KIL30263.1_25 (a, b) | 24.691 | 0 | 49 |
| YhbB | O31589 | prot_KIL31183.1_5 | 99.016 | 39 | 43 |
| YheC | O07544 | prot_KIL33710.1_129 | 100 | 48 | 2 |
| YjdH | O31649 | NA | NA | NA | 19 |
| YkvP | O31681 | prot_KIL33950.1_72 | 99.248 | 59 | 49 |

**Table 2.** *Cont.*

| Proteins | UniProt IDs | Homologous Proteins in *B. subtilis* A163 | Percentage of Identity (%) | Number of Peptides | |
|---|---|---|---|---|---|
| | | | | *B. subtilis* A163 | *B. subtilis* PY79 |
| YobN | O34363 | prot_KIL30953.1_14 | 98.536 | 26 | 14 |
| YodI | O34654 | prot_KIL32546.1_55 | 95.181 | 10 | 11 |
| YpeP | P54164 | prot_KIL33154.1_32 | 99.558 | 13 | 17 |
| YqfT | P54477 | prot_KIL32864.1_5 | 100 | 14 | 18 |

NA, not available, no homologous proteins were found in *B. subtilis* A163 through searching using BLASTP; (a), not identified in the spore proteome, but showing a high percentage of identity with the coat protein; (b), low percentage of identity, more research is necessary.

**Table 3.** Identified *B. subtilis* PY79 and A163 proteins involved in germination.

| Proteins | UniProt IDs | Homologous proteins in *B. subtilis* A163 | Percentage of Identity (%) | Number of Peptides | |
|---|---|---|---|---|---|
| | | | | *B. subtilis* A163 | *B. subtilis* PY79 |
| GerAA | P07868 | prot_KIL32731.1_49 | 97.303 | 14 | 24 |
| GerAB | P07869 | prot_KIL32730.1_48 (a) | 98.082 | 0 | 3 |
| GerAC | P07870 | prot_KIL32729.1_47 | 94.906 | 19 | 30 |
| GerBA | P39569 | prot_KIL31057.1_65 (a) | 98.324 | 0 | 15 |
| GerBC | P39571 | prot_KIL31055.1_63 (a) | 97.861 | 0 | 27 |
| GerKA | P49939 | prot_KIL29364.1_41 | 98.162 | 9 | 15 |
| GerKB | P49940 | prot_KIL29362.1_39 | 96.783 | 1 | 3 |
| GerKC | P49941 | prot_KIL29363.1_40 (a, b) | 24.378 | 0 | 6 |
| GerD | P16450 | prot_KIL31648.1_3 | 100 | 31 | 52 |
| GerPA | O06721 | prot_KIL33608.1_27 | 98.63 | 6 | 9 |
| GerPB | O06720 | prot_KIL33609.1_28 | 100 | 6 | 9 |
| GerPC | O06719 | prot_KIL33610.1_29 | 99.024 | 9 | 3 |
| GerPD | O06718 | prot_KIL33611.1_30 (a) | 100 | 0 | 1 |
| GerPE | O06717 | prot_KIL33612.1_31 | 99.115 | 5 | 1 |
| GerPF | O06716 | prot_KIL33613.1_32 | 100 | 3 | 4 |
| CwlJ | P42249 | prot_KIL33394.1_41 | 95.775 | 36 | 26 |
| SleB | P50739 | prot_KIL33255.1_133 | 90.12 | 46 | 49 |
| SpoVAA | P40866 | prot_KIL33311.1_189 | 99.515 | 1 | 1 |
| SpoVAC | P40868 | prot_KIL33309.1_187 | 99.333 | 13 | 10 |
| SpoVAD | P40869 | prot_KIL33308.1_186 | 99.408 | 94 | 89 |
| SpoVAEa | P40870 | prot_KIL33306.1_184 | 100 | 13 | 11 |
| SpoVAF | P31845 | prot_KIL33305.1_183 | 99.189 | 31 | 27 |

(a), not identified in the spore proteome, but showing a high percentage of identity with the protein; (b), low percentage of identity, more research is necessary.

Since two databases were used in this study, homologous proteins from the two strains were often reported as two separate results, for example RsmE (ribosomal RNA small subunit methyltransferase E, Uniprot ID P54461) identified in the cellular proteome. In a comparison of amino acid sequences, RsmE from the two strains had more than 99% identity with one amino acid difference, T18A (Figure 3A). However, two proteins were in the output as two items with their own quantitative values (Figure 3B). By checking their peptide composition, we found they both contain shared peptides which can be identified from either of the two proteins, and specific peptides caused by the T18A change. Quantification of RsmE using either of the two outputs may result in an incorrect conclusion. To overcome this issue, we re-assembled the identified peptides to include both the shared peptides and specific peptides and calculated the protein abundances accordingly [32]. In the new output, proteins from two strains having shared peptides were treated as homologous proteins for the moment and their protein identifiers were both shown in the column of Protein IDs (such as RsmE in Figure 3C). In total, 1312 and 1276 proteins were quantified between two strains in spores and cells, respectively, with at least two quantified values in each strain (Tables S1 and S2).

Among the proteins quantitated, proteins between two strains with an identity higher than 50% were subjected to further analysis. Some proteins have multiple homologs identified in *B. subtilis* A163, such as YjqC, but only one homolog in our data is quantitively compared with the protein in *B. subtilis* PY79. That is because in a comparison of one protein between two samples, the quantification algorithm requires that some peptides identified in one sample must be also identified in the other sample [39]. In our data, since not enough peptides were identified for the other homologs, this makes them impossible to quantitively analyze. In the quantified spore proteins, 39 proteins were found to be highly abundant in *B. subtilis* A163, while 69 were low abundance (Figure 4A). For the analysis of cellular proteins, 32 and 61 proteins in *B. subtilis* A163 were present at high and low abundance, respectively (Figure 4A). In the known spore coat proteins retrieved from SubtiWiki [36], YmaG and CgeA were low abundance and CotJC, CotH, CotSA, SpoVID and GerT were high abundance in spores of *B. subtilis* A163. CgeA is a protein located in the spore crust, the outermost layer, and is considered to play a role in crust glycosylation [40]. SpoVID and CotH are essential for spore coat morphogenesis, and a *spoVID* mutant fails to encase the spore inner and outer coat layers [41,42]. *cotH* mutant spores have normal heat resistance but are deficient in several coat proteins [43]. CotJC upregulation was observed in spores from a sporulation that was *kinA*-induced, and these spores had higher wet heat resistance than when sporulation was induced by nutrient depletion [29]. However, how increased levels of SpoVID, CotH and CotJC affect spore resistance is not known. GerT is also a component of the spore coat and Δ*gerT* spores respond poorly to multiple germinants [44]. For the small acid soluble proteins and the proteins involved in spore germination (germinant receptors, SpoVA channel proteins, SleB and CwlJ) [5], none were quantified to be more or less abundant in the two strains analyzed (Table S2). For proteins encoded in the *spoVA*$^{2mob}$ operon, none of them are present in PY79 strain and thus were not quantitatively compared with any proteins identified in PY79 strain.

**A**

| | | |
|---|---|---|
| P54461 | 1 | MQRYFIELTKQQIEEAPTFSITGEEVHHIVNVMRMNEGDQIICCSQDGFEAKCELQSVSK |
| | | MQRYFIELTKQQIEEAP FSITGEEVHHIVNVMRMNEGDQIICCSQDGFEAKCELQSVSK |
| prot_KIL34095.1_38 | 1 | MQRYFIELTKQQIEEAPAFSITGEEVHHIVNVMRMNEGDQIICCSQDGFEAKCELQSVSK |
| P54461 | 61 | DKVSCLVIEWTNENRELPIKVYIASGLPKGDKLEWIIQKGTELGAHAFIPFQAARSVVKL |
| | | DKVSCLVIEWTNENRELPIKVYIASGLPKGDKLEWIIQKGTELGAHAFIPFQAARSVVKL |
| prot_KIL34095.1_38 | 61 | DKVSCLVIEWTNENRELPIKVYIASGLPKGDKLEWIIQKGTELGAHAFIPFQAARSVVKL |
| P54461 | 121 | DDKKAKKKRERWTKIAKEAAEQSYRNEVPRVMDVHSFQQLLQRMQDFDKCVVAYEESSKQ |
| | | DDKKAKKKRERWTKIAKEAAEQSYRNEVPRVMDVHSFQQLLQRMQDFDKCVVAYEESSKQ |
| prot_KIL34095.1_38 | 121 | DDKKAKKKRERWTKIAKEAAEQSYRNEVPRVMDVHSFQQLLQRMQDFDKCVVAYEESSKQ |
| P54461 | 181 | GEISAFSAIVSSLPKGSSLLIVFGPEGGLTEAEVERLTEQDGVTCGLGPRILRTETAPLY |
| | | GEISAFSAIVSSLPKGSSLLIVFGPEGGLTEAEVERLTEQDGVTCGLGPRILRTETAPLY |
| prot_KIL34095.1_38 | 181 | GEISAFSAIVSSLPKGSSLLIVFGPEGGLTEAEVERLTEQDGVTCGLGPRILRTETAPLY |
| P54461 | 241 | ALSAISYQTELLRGDQ 256 |
| | | ALSAISYQTELLRGDQ |
| prot_KIL34095.1_38 | 241 | ALSAISYQTELLRGDQ 256 |

**Figure 3.** *Cont.*

## B

Protein intensities

| Protein IDs | Log$_2$(LFQ intensity) | | | | | |
| | *B. subtilis* A163 | | | *B. subtilis* PY79 | | |
| | Replicate 1 | Replicate 2 | Replicate 3 | Replicate 1 | Replicate 2 | Replicate 3 |
| --- | --- | --- | --- | --- | --- | --- |
| KIL34095.1_38 | 15.37818 | 15.4571 | 15.72669 | 16.37598 | 16.52611 | 16.37271 |
| P54461 | 15.27847 | 15.35567 | 15.58086 | 16.30919 | 16.40818 | 16.21047 |

Peptide intensities

| Peptides | Log$_2$(intensity) | | | | | | Protein IDs |
| | *B. subtilis* A163 | | | *B. subtilis* PY79 | | | |
| | Replicate 1 | Replicate 2 | Replicate 3 | Replicate 1 | Replicate 2 | Replicate 3 | |
| --- | --- | --- | --- | --- | --- | --- | --- |
| _GDKLEWIIQK__2 | 15.38555 | | | | 16.72864 | | prot_KIL34095.1_38 P54461 |
| _GTELGAHAFIPFQAAR__2 | 14.9411 | 14.90776 | 15.43042 | 16.10634 | 15.92261 | 16.56553 | prot_KIL34095.1_38 P54461 |
| _LTEQDGVTCGLGPR__2 | 18.41169 | 13.99105 | | 18.30617 | 17.98735 | 18.14961 | prot_KIL34095.1_38 P54461 |
| _LTEQDGVTCGLGPR__3 | 13.83015 | | | | 14.43017 | 15.14749 | prot_KIL34095.1_38 P54461 |
| _M(Oxidation (M))NEGDQIICCSQDGFEAK__2 | | 15.72834 | 15.95324 | | 13.46957 | | prot_KIL34095.1_38 P54461 |
| _MNEGDQIICCSQDGFEAK__2 | 14.56736 | 17.08857 | 17.16921 | 16.80275 | 16.94359 | 16.35128 | prot_KIL34095.1_38 P54461 |
| _MNEGDQIICCSQDGFEAK__3 | | 13.81956 | | 15.83651 | | | prot_KIL34095.1_38 P54461 |
| _QGEISAFSAIVSSLPK__2 | 13.79521 | | | 15.1003 | 16.51937 | 16.82737 | prot_KIL34095.1_38 P54461 |
| _TETAPLYALSAISYQTELLR__2 | 17.73187 | 18.35464 | | 17.41283 | 18.79549 | | prot_KIL34095.1_38 P54461 |
| _TETAPLYALSAISYQTELLR__3 | 15.53881 | 15.43475 | 14.42246 | 17.07862 | 16.29354 | 16.09578 | prot_KIL34095.1_38 P54461 |
| _TETAPLYALSAISYQTELLRGDQ__2 | 16.14173 | | | | | | prot_KIL34095.1_38 P54461 |
| _VM(Oxidation (M))DVHSFQQLLQR__2 | | | | | | 14.11502 | prot_KIL34095.1_38 P54461 |
| _VM(Oxidation (M))DVHSFQQLLQR__3 | | | 15.33555 | 15.92746 | | 16.07843 | prot_KIL34095.1_38 P54461 |
| _VMDVHSFQQLLQR__2 | | | | 15.73163 | 16.17888 | 14.17939 | prot_KIL34095.1_38 P54461 |
| _VMDVHSFQQLLQR__3 | 14.66901 | 14.57133 | 14.96693 | 15.30314 | 17.48346 | 14.40023 | prot_KIL34095.1_38 P54461 |
| _YFIELTK__2 | 15.87567 | | 14.71861 | 14.93808 | 16.17358 | | prot_KIL34095.1_38 P54461 |
| _QQIEEAPTFSITGEEVHHIVNVM(Oxidation (M))R__3 | | | | | 14.0555 | | P54461 |
| _QQIEEAPTFSITGEEVHHIVNVM(Oxidation (M))R__4 | | | | | 15.98962 | | P54461 |
| _QQIEEAPTFSITGEEVHHIVNVMR__4 | | | | 16.4606 | 16.17382 | 16.01658 | P54461 |

Peptide intensities

| Peptides | Log$_2$(intensity) | | | | | | Protein IDs |
| | *B. subtilis* A163 | | | *B. subtilis* PY79 | | | |
| | Replicate 1 | Replicate 2 | Replicate 3 | Replicate 1 | Replicate 2 | Replicate 3 | |
| --- | --- | --- | --- | --- | --- | --- | --- |
| _GDKLEWIIQK__2 | 15.38555 | | | | 16.72864 | | prot_KIL34095.1_38 P54461 |
| _GTELGAHAFIPFQAAR__2 | 14.9411 | 14.90776 | 15.43042 | 16.10634 | 15.92261 | 16.56553 | prot_KIL34095.1_38 P54461 |
| _LTEQDGVTCGLGPR__2 | 18.41169 | 13.99105 | | 18.30617 | 17.98735 | 18.14961 | prot_KIL34095.1_38 P54461 |
| _LTEQDGVTCGLGPR__3 | 13.83015 | | | | 14.43017 | 15.14749 | prot_KIL34095.1_38 P54461 |
| _M(Oxidation (M))NEGDQIICCSQDGFEAK__2 | | 15.72834 | 15.95324 | | 13.46957 | | prot_KIL34095.1_38 P54461 |
| _MNEGDQIICCSQDGFEAK__2 | 14.56736 | 17.08857 | 17.16921 | 16.80275 | 16.94359 | 16.35128 | prot_KIL34095.1_38 P54461 |
| _MNEGDQIICCSQDGFEAK__3 | | 13.81956 | | 15.83651 | | | prot_KIL34095.1_38 P54461 |
| _QGEISAFSAIVSSLPK__2 | 13.79521 | | | 15.1003 | 16.51937 | 16.82737 | prot_KIL34095.1_38 P54461 |
| _TETAPLYALSAISYQTELLR__2 | 17.73187 | 18.35464 | | 17.41283 | 18.79549 | | prot_KIL34095.1_38 P54461 |
| _TETAPLYALSAISYQTELLR__3 | 15.53881 | 15.43475 | 14.42246 | 17.07862 | 16.29354 | 16.09578 | prot_KIL34095.1_38 P54461 |
| _TETAPLYALSAISYQTELLRGDQ__2 | 16.14173 | | | | | | prot_KIL34095.1_38 P54461 |
| _VM(Oxidation (M))DVHSFQQLLQR__2 | | | | | | 14.11502 | prot_KIL34095.1_38 P54461 |
| _VM(Oxidation (M))DVHSFQQLLQR__3 | | | 15.33555 | 15.92746 | | 16.07843 | prot_KIL34095.1_38 P54461 |
| _VMDVHSFQQLLQR__2 | | | | 15.73163 | 16.17888 | 14.17939 | prot_KIL34095.1_38 P54461 |
| _VMDVHSFQQLLQR__3 | 14.66901 | 14.57133 | 14.96693 | 15.30314 | 17.48346 | 14.40023 | prot_KIL34095.1_38 P54461 |
| _YFIELTK__2 | 15.87567 | | 14.71861 | 14.93808 | 16.17358 | | prot_KIL34095.1_38 P54461 |
| _QQIEEAPAFSITGEEVHHIVNVM(Oxidation (M))R__3 | | | 17.31028 | | | | prot_KIL34095.1_38 |
| _QQIEEAPAFSITGEEVHHIVNVM(Oxidation (M))R__4 | 16.33674 | 15.6472 | 17.00137 | | | | prot_KIL34095.1_38 |
| _QQIEEAPAFSITGEEVHHIVNVMR__3 | | 16.38696 | | | | | prot_KIL34095.1_38 |
| _QQIEEAPAFSITGEEVHHIVNVMR__4 | 17.61909 | 17.74196 | | | | | prot_KIL34095.1_38 |

**Figure 3.** *Cont.*

**C**

Protein intensities

| Protein IDs | Log$_2$(LFQ intensity) | | | | | |
| --- | --- | --- | --- | --- | --- | --- |
| | *B. subtilis* A163 | | | *B. subtilis* PY79 | | |
| | Replicate 1 | Replicate 2 | Replicate 3 | Replicate 1 | Replicate 2 | Replicate 3 |
| P54461;prot_KIL34095.1_38 | 15.32688486 | 15.40759781 | 15.5423869 | 16.5316273 | 16.5996403 | 16.3665443 |

Peptide intensities

| Peptides | Log$_2$(intensity) | | | | | | Protein IDs |
| --- | --- | --- | --- | --- | --- | --- | --- |
| | *B. subtilis* A163 | | | *B. subtilis* PY79 | | | |
| | Replicate 1 | Replicate 2 | Replicate 3 | Replicate 1 | Replicate 2 | Replicate 3 | |
| _GDKLEWIIQK__2 | 15.38555 | | | | 16.72864 | | prot_KIL34095.1_38 P54461 |
| _GTELGAHAFIPFQAAR__2 | 14.9411 | 14.90776 | 15.43042 | 16.10634 | 15.92261 | 16.56553 | prot_KIL34095.1_38 P54461 |
| _LTEQDGVTCGLGPR__2 | 18.41169 | 13.99105 | | 18.30617 | 17.98735 | 18.14961 | prot_KIL34095.1_38 P54461 |
| _LTEQDGVTCGLGPR__3 | 13.83015 | | | | 14.43017 | 15.14749 | prot_KIL34095.1_38 P54461 |
| _M(Oxidation (M))NEGDQIICCSQDGFEAK__2 | | 15.72834 | 15.95324 | | 13.46957 | | prot_KIL34095.1_38 P54461 |
| _MNEGDQIICCSQDGFEAK__2 | 14.56736 | 17.08857 | 17.16921 | 16.80275 | 16.94359 | 16.35128 | prot_KIL34095.1_38 P54461 |
| _MNEGDQIICCSQDGFEAK__3 | | 13.81956 | | 15.83651 | | | prot_KIL34095.1_38 P54461 |
| _QGEISAFSAIVSSLPK__2 | 13.79521 | | | 15.1003 | 16.51937 | 16.82737 | prot_KIL34095.1_38 P54461 |
| _TETAPLYALSAISYQTELLR__2 | 17.73187 | 18.35464 | | 17.41283 | 18.79549 | | prot_KIL34095.1_38 P54461 |
| _TETAPLYALSAISYQTELLR__3 | 15.53881 | 15.43475 | 14.42246 | 17.07862 | 16.29354 | 16.09578 | prot_KIL34095.1_38 P54461 |
| _TETAPLYALSAISYQTELLRGDQ__2 | 16.14173 | | | | | | prot_KIL34095.1_38 P54461 |
| _VM(Oxidation (M))DVHSFQQLLQR__2 | | | | | | 14.11502 | prot_KIL34095.1_38 P54461 |
| _VM(Oxidation (M))DVHSFQQLLQR__3 | | | 15.33555 | 15.92746 | | 16.07843 | prot_KIL34095.1_38 P54461 |
| _VMDVHSFQQLLQR__2 | | | | 15.73163 | 16.17888 | 14.17939 | prot_KIL34095.1_38 P54461 |
| _VMDVHSFQQLLQR__3 | 14.66901 | 14.57133 | 14.96693 | 15.30314 | 17.48346 | 14.40023 | prot_KIL34095.1_38 P54461 |
| _YFIELTK__2 | 15.87567 | | 14.71861 | 14.93808 | 16.17358 | | prot_KIL34095.1_38 P54461 |
| _QQIEEAPTFSITGEEVHHIVNVM(Oxidation (M))R__3 | | | | | 14.0555 | | P54461 |
| _QQIEEAPTFSITGEEVHHIVNVM(Oxidation (M))R__4 | | | | | 15.98962 | | P54461 |
| _QQIEEAPTFSITGEEVHHIVNVMR__4 | | | | 16.4606 | 16.17382 | 16.01658 | P54461 |
| _QQIEEAPAFSITGEEVHHIVNVM(Oxidation (M))R__3 | | | 17.31028 | | | | prot_KIL34095.1_38 |
| _QQIEEAPAFSITGEEVHHIVNVM(Oxidation (M))R__4 | 16.33674 | 15.6472 | 17.00137 | | | | prot_KIL34095.1_38 |
| _QQIEEAPAFSITGEEVHHIVNVMR__3 | | 16.38696 | | | | | prot_KIL34095.1_38 |
| _QQIEEAPAFSITGEEVHHIVNVMR__4 | 17.61909 | 17.74196 | | | | | prot_KIL34095.1_38 |

**Figure 3.** Quantitative comparison of RsmE peptide levels in growing *B. subtilis* PY79 and A163 cells. prot_KIL34095.1_38, identifier of the homologous protein of RsmE in *B. subtilis* A163; LFQ, Relative label-free quantification [39]. (**A**) Alignment of amino acids of RsmE between two strains. (**B**) Protein intensities of RsmE and their identified peptides in the default output. (**C**) Protein intensities of RsmE and their peptide components in the new output.

The Uniprot terms enriched from the most differentially presented spore and cellular proteins are shown in Figure 4B. Glycosyltransferases and proteases are enriched in the high abundance spore proteins of *B. subtilis* A163. Of the glycosyltransferases, YtcC is a product of the *ytcABC* operon, which could be involved in the extensive glycosylation of the spore surface [45]. YdhE plays roles in the resistance to bacterial toxins [46]. The pyrimidine biosynthetic (*pyr*) gene cluster includes the gene for PyrE [47], one of the high abundance glycosyltransferases. The last high abundance glycosyltransferase is the coat protein CotSA [48]. Among the proteases, IspA is an intracellular serine protease, and an *ispA* null mutant showed a decreased sporulation in at least one medium [49]. While AprE is one of the major extracellular alkaline proteases [50], serine protease YtrC has been reported to be in the spore IM fraction and to play a pivotal role in spore germination [51,52]. Proteins enriched in Uniprot terms sporulation and (cell) membrane are the major group among the low abundance spore proteins of *B. subtilis* A163. The proteins enriched in sporulation are listed in Table 4. Their contribution to spore resistance is unknown. Among them, SpoIIIAG, YabP, OppA, OppB, OppC, OppF, DppE and PbpE are also membrane proteins. High abundance proteins in cells of *B. subtilis* A163 are enriched in the transport of proteins and peptides, as well as competence. Proteins involved in biosynthesis and metabolism of fatty acids and lipids, as well as oxidoreductase, are enriched in the low-abundance cellular proteins of *B. subtilis* A163.

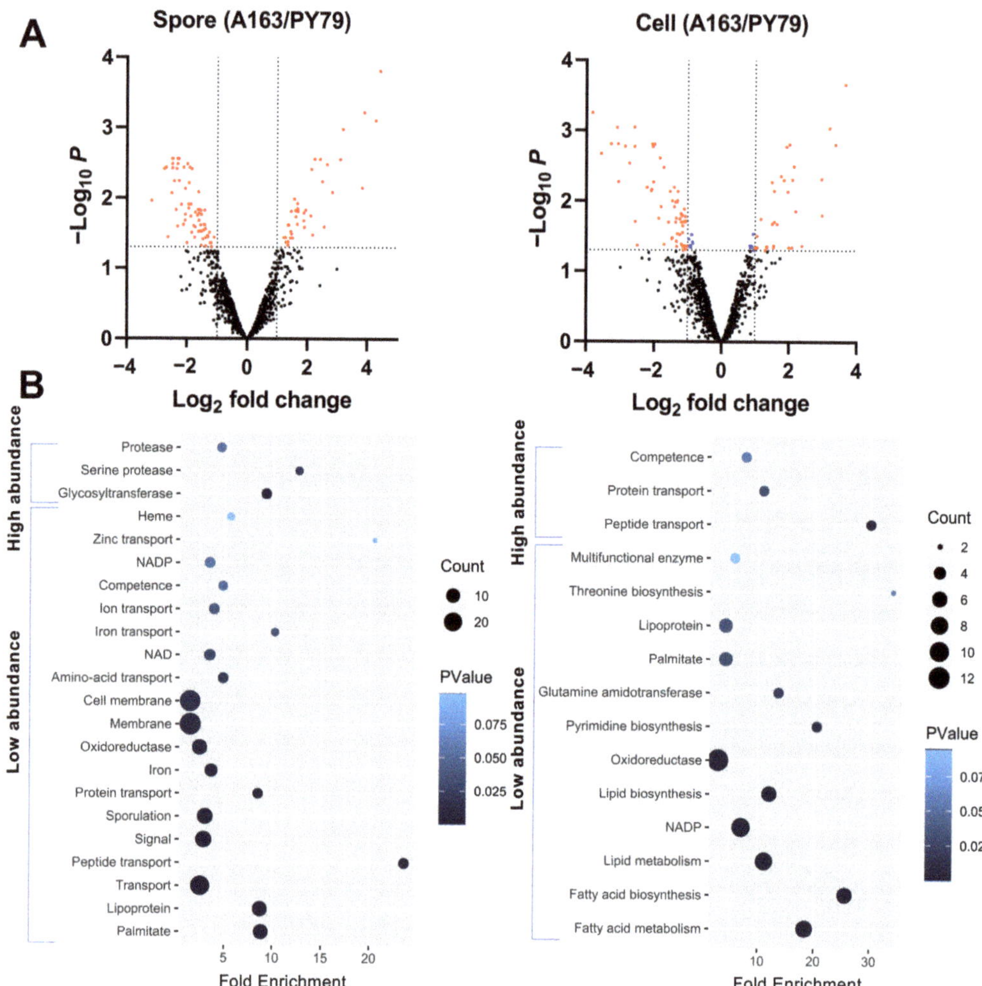



**Figure 4.** Quantitative comparison of proteomes in spores and cells of *B. subtilis* PY79 and A163. (**A**) Volcano plots of the quantified proteins in spores of the two strains and the proteome comparison of their correspondent cells. Log2 fold changes smaller than 0 (or larger than 0) indicate proteins with low (or high) abundance in *B. subtilis* A163. Dots in red indicate proteins in *B. subtilis* A163 that were differentially present more than twofold with $p < 0.05$. Dots in blue indicate proteins that were present with Scheme 0. but less than twofold. (**B**) Uniprot categories enrichment of the differentially presented proteins in spores and cells of *B. subtilis* A163. The fold enrichment is defined as the ratio of two proportions. The first proportion is the quantified proteins belonging to a UniProt category divided by all high- (or low-) abundant proteins. The second proportion is all proteins belonging to the UniProt category in the genome divided by the total proteins in the genome. The size of the dots is indicative of the number of quantified proteins (Count) belonging to a particular term, as shown in the legend. The color of the dots is corresponding to the Fisher exact p-value (PValue), again as shown in the legend.

**Table 4.** Proteins enriched in the Uniprot term of sporulation.

| Proteins | Descriptions | References |
|---|---|---|
| SpoIIIAG | a key component of a feeding tube apparatus creating a direct conduit between the developing forespore and the mother cell | [53,54] |
| Spo0M | regulating progress of sporulation and expression of Spo0A, but the mechanisms is still unknow | [55,56] |
| SpoVIF | involved in assembly of spore coat proteins that have roles in lysozyme resistance | [57] |
| YabP | a coat-associated protein | [58] |
| SplB | UV resistance of spores, DNA repair in spore germination | [59] |
| SinR | the master regulator of biofilm formation | [60] |
| OppA, OppB, OppC, OppF, DppE | the ATP binding cassette (ABC) transporter systems | [61] |
| PbpE | penicillin-binding protein PBP 4 | [62,63] |
| YraD | forespore-specific sporulation protein, similar to spore coat protein | [64] |

Multiple factors can contribute to the wet heat resistance of spores [65]. Among them, a *spoVA*$^{2mob}$ operon is considered to play roles in elevation of their resistance to heat and pressure. Measurement of the DPA content of spores of *B. subtilis* A163, the parental strain containing the *spoVA*$^{2mob}$ operon, indicated that these spores do not contain statistically significantly higher levels of DPA than those of a low resistance strain, and recent work has also found that the core water content of spores with the *spoVA*$^{2mob}$ operon is identical in a strain lacking this operon [15]. Of the seven proteins encoded in the *spoVA*$^{2mob}$ operon, we identified five in A163 spores, but none were the 2Duf protein thought to be of most importance in these spores high heat resistance or SpoVAEb. Furthermore, a number of germinant receptors were not identified in spores of *B. subtilis* A163. A practical approach might be trying to focus on identification of the proteome of the spore IM, as this analysis has been done on *B. subtilis* strain 1A700 [51]. On the other hand, the rate of DPA release from spores of *B. subtilis* A163 during heat treatment is low, although the mechanism preventing faster DPA release during heat treatment is unknown. Presumably this is due to the integrity and impermeability of the IM and protection by coat layers. In addition, the A163 spore proteome contains proteins specific to this strain, but with unknown function. The location of these proteins in spores and their contribution to spore resistance are also unknown. In addition, some proteins have multiple homologs identified in *B. subtilis* A163, for example coat protein YiqC. However, no homologs were found in *B. subtilis* A163 for proteins important for spore morphogenesis, such as SpoVM. What proteins would supplement the function of SpoVM or how the spore completes the coat encasement without SpoVM homologs would certainly be worth investigating. In addition, high and low abundant A163 spore proteins were revealed for the coat layers and a number of the Uniprot categories for both cellular and spore proteins. Those could also play some role in the observed high thermal resistance of A163 spores.

### 4. Conclusions

*B. subtilis* A163 arouses the interest of researchers due to its ability to produce high heat resistant spores. In this study, we found the release of spore DPA from *B. subtilis* A163 at 98 °C was much slower than that from *B. subtilis* PY79 spores. How the spore prevented the DPA from rapidly being released during heat treatment and if this is related with the high heat resistance of A163 spores is unknown. Through the extensive study of the proteomes of *B. subtilis* A163 and PY79 spores and cells, the proteomic differences of the two strains are revealed. This provides novel insights on the putative molecular basis of spore high wet heat resistance. Open questions include whether 2Duf is really not expressed during the *B. subtilis* A163 life-cycle, and more generally, how the proteins encoded by the *spoVA*$^{2mob}$ operon contribute to the observed high wet heat resistance of *B. subtilis* A163 spores.

**Supplementary Materials:** The following are available online at https://www.mdpi.com/2076-260 7/9/3/667/s1, Table S1: proteins identified and quantified in *Bacillus subtilis* cells; Table S2: proteins identified and quantified in *Bacillus subtilis* spores; File S1: evidence.txt files and R-scripts. The raw proteomic data of this study are available in MassIVE with access number MSV000087080.

**Author Contributions:** Conceptualization, S.B. and G.K.; formal analysis, Z.T.; writing—original draft preparation, Z.T.; writing—review and editing, S.B., G.K. and P.S.; visualization, Z.T.; supervision, S.B. and G.K.; All authors have read and agreed to the published version of the manuscript.

**Funding:** This research was funded by the China Scholarship Council, grant number 201606170125. The APC was funded by Laboratory for Molecular Biology and Microbial Food Safety, University of Amsterdam.

**Institutional Review Board Statement:** Not applicable.

**Informed Consent Statement:** Not applicable.

**Data Availability Statement:** No new data were created or analyzed in this study. Data sharing is not applicable to this article.

**Acknowledgments:** We thank Winfried Roseboom and Henk L. Dekker from the laboratory for mass spectrometry of biomolecules, University of Amsterdam, for expert technical assistance.

**Conflicts of Interest:** The authors declare no conflict of interest.

# References

1.    Oomes, S.; Van Zuijlen, A.C.M.; Hehenkamp, J.O.; Witsenboer, H.; Van der Vossen, J.; Brul, S. The characterisation of *Bacillus* spores occurring in the manufacturing of (low acid) canned products. *Int. J. Food Microbiol.* **2007**, *120*, 85–94. [CrossRef]

2.    Rosenkvist, H.; Hansen, Å. Contamination profiles and characterisation of *Bacillus* species in wheat bread and raw materials for bread production. *Int. J. Food Microbiol.* **1995**, *26*, 353–363. [CrossRef]

3.    Scheldeman, P.; Pil, A.; Herman, L.; De Vos, P.; Heyndrickx, M. Incidence and diversity of potentially highly heat-resistant spores isolated at dairy farms. *Appl. Environ. Microbiol.* **2005**, *71*, 1480–1494. [CrossRef] [PubMed]

4.    Setlow, P. Spore resistance properties. *Microbiol. Spectr.* **2014**, *2*. [CrossRef] [PubMed]

5.    Setlow, P. Summer meeting 2013–when the sleepers wake: The germination of spores of *Bacillus* species. *J. Appl. Microbiol.* **2013**, *115*, 1251–1268. [CrossRef] [PubMed]

6.    Henriques, A.O.; Moran, C.P. Structure, assembly, and function of the spore surface layers. *Annu. Rev. Microbiol.* **2007**, *61*, 555–588. [CrossRef] [PubMed]

7.    Rose, R.; Setlow, B.; Monroe, A.; Mallozzi, M.; Driks, A.; Setlow, P. Comparison of the properties of *Bacillus subtilis* spores made in liquid or on agar plates. *J. Appl. Microbiol.* **2007**, *103*, 691–699. [CrossRef] [PubMed]

8.    Melly, E.; Genest, P.C.; Gilmore, M.E.; Little, S.; Popham, D.L.; Driks, A.; Setlow, P. Analysis of the properties of spores of *Bacillus subtilis* prepared at different temperatures. *J. Appl. Microbiol.* **2002**, *92*, 1105–1115. [CrossRef] [PubMed]

9.    Abhyankar, W.R.; Kamphorst, K.; Swarge, B.N.; van Veen, H.; van der Wel, N.N.; Brul, S.; de Koster, C.G.; de Koning, L.J. The influence of sporulation conditions on the spore coat protein composition of *Bacillus subtilis* spores. *Front. Microbiol.* **2016**, *7*, 1636. [CrossRef]

10.   Isticato, R.; Lanzilli, M.; Petrillo, C.; Donadio, G.; Baccigalupi, L.; Ricca, E. *Bacillus subtilis* builds structurally and functionally different spores in response to the temperature of growth. *Environ. Microbiol.* **2020**, *22*, 170–182. [CrossRef] [PubMed]

11.   Berendsen, E.M.; Zwietering, M.H.; Kuipers, O.P.; Wells-Bennik, M.H. Two distinct groups within the *Bacillus subtilis* group display significantly different spore heat resistance properties. *Food Microbiol.* **2015**, *45*, 18–25. [CrossRef]

12.   Berendsen, E.M.; Boekhorst, J.; Kuipers, O.P.; Wells-Bennik, M.H. A mobile genetic element profoundly increases heat resistance of bacterial spores. *ISME J.* **2016**, *10*, 2633–2642. [CrossRef] [PubMed]

13.   Li, Z.; Schottroff, F.; Simpson, D.J.; Gänzle, M.G. The copy number of the *spoVA*$^{2mob}$ operon determines pressure resistance of *Bacillus* endospores. *Appl. Environ. Microbiol.* **2019**, *85*, e01596-19. [CrossRef]

14.   Berendsen, E.M.; Koning, R.A.; Boekhorst, J.; de Jong, A.; Kuipers, O.P.; Wells-Bennik, M.H. High-level heat resistance of spores of *Bacillus amyloliquefaciens* and *Bacillus licheniformis* results from the presence of a *spoVA* operon in a Tn1546 transposon. *Front. Microbiol.* **2016**, *7*, 1912. [CrossRef]

15.   Luo, Y.; Korza, G.; DeMarco, A.M.; Kuipers, O.P.; Li, Y.; Setlow, P. Properties of spores of Bacillus subtilis with or without a transposon that decreases spore germination and increases spore wet heat resistance. Unpublished.

16. Youngman, P.; Perkins, J.B.; Losick, R. Construction of a cloning site near one end of Tn917 into which foreign DNA may be inserted without affecting transposition in *Bacillus subtilis* or expression of the transposon-borne *erm* gene. *Plasmid* **1984**, *12*, 1–9. [CrossRef]
17. Ramamurthi, K.S.; Clapham, K.R.; Losick, R. Peptide anchoring spore coat assembly to the outer forespore membrane in *Bacillus subtilis*. *Mol. Microbiol.* **2006**, *62*, 1547–1557. [CrossRef] [PubMed]
18. Brul, S.; van Beilen, J.; Caspers, M.; O'Brien, A.; de Koster, C.; Oomes, S.; Smelt, J.; Kort, R.; Ter Beek, A. Challenges and advances in systems biology analysis of *Bacillus* spore physiology; molecular differences between an extreme heat resistant spore forming *Bacillus subtilis* food isolate and a laboratory strain. *Food Microbiol.* **2011**, *28*, 221–227. [CrossRef]
19. Kort, R.; O'brien, A.C.; Van Stokkum, I.H.; Oomes, S.J.; Crielaard, W.; Hellingwerf, K.J.; Brul, S. Assessment of heat resistance of bacterial spores from food product isolates by fluorescence monitoring of dipicolinic acid release. *Appl. Environ. Microbiol.* **2005**, *71*, 3556–3564. [CrossRef] [PubMed]
20. Abhyankar, W.; Beek, A.T.; Dekker, H.; Kort, R.; Brul, S.; de Koster, C.G. Gel-free proteomic identification of the *Bacillus subtilis* insoluble spore coat protein fraction. *Proteomics* **2011**, *11*, 4541–4550. [CrossRef]
21. Bertani, G. Studies on Lysogenesis I. *J. Bacteriol.* **1951**, *62*, 293–300. [CrossRef] [PubMed]
22. Ghosh, S.; Korza, G.; Maciejewski, M.; Setlow, P. Analysis of metabolism in dormant spores of *Bacillus species* by $^{31}$P nuclear magnetic resonance analysis of low-molecular-weight compounds. *J. Bacteriol.* **2015**, *197*, 992–1001. [CrossRef]
23. Donnelly, M.L.; Fimlaid, K.A.; Shen, A. Characterization of *Clostridium difficile* spores lacking either SpoVAC or dipicolinic acid synthetase. *J. Bacteriol.* **2016**, *198*, 1694–1707. [CrossRef] [PubMed]
24. Berendsen, E.M.; Wells-Bennik, M.H.; Krawczyk, A.O.; de Jong, A.; van Heel, A.; Eijlander, R.T.; Kuipers, O.P. Draft genome sequences of 10 *Bacillus subtilis* strains that form spores with high or low heat resistance. *Genome Announc.* **2016**, *4*. [CrossRef]
25. Kunst, F.; Ogasawara, N.; Moszer, I.; Albertini, A.M.; Alloni, G.O.; Azevedo, V.; Bertero, M.G.; Bessières, P.; Bolotin, A.; Borchert, S. The complete genome sequence of the gram-positive bacterium *Bacillus subtilis*. *Nature* **1997**, *390*, 249–256. [CrossRef] [PubMed]
26. Jalili, V.; Afgan, E.; Gu, Q.; Clements, D.; Blankenberg, D.; Goecks, J.; Taylor, J.; Nekrutenko, A. The Galaxy platform for accessible, reproducible and collaborative biomedical analyses: 2020 update. *Nucleic Acids Res.* **2020**, *48*, W395–W402. [CrossRef]
27. Camacho, C.; Coulouris, G.; Avagyan, V.; Ma, N.; Papadopoulos, J.; Bealer, K.; Madden, T.L. BLAST+: Architecture and applications. *BMC Bioinform.* **2009**, *10*, 421. [CrossRef] [PubMed]
28. Cock, P.J.; Chilton, J.M.; Grüning, B.; Johnson, J.E.; Soranzo, N. NCBI BLAST+ integrated into Galaxy. *Gigascience* **2015**, *4*, s13742-015. [CrossRef] [PubMed]
29. Tu, Z.; RAbhyankar, W.; NSwarge, B.; van der Wel, N.; Kramer, G.; Brul, S.; de Koning, L.J. Artificial sporulation induction (ASI) by *kinA* overexpression affects the proteomes and properties of *Bacillus subtilis* spores. *Int. J. Mol. Sci.* **2020**, *21*, 4315. [CrossRef]
30. Scopes, R.K. Measurement of protein by spectrophotometry at 205 nm. *Anal. Biochem.* **1974**, *59*, 277–282. [CrossRef]
31. Tyanova, S.; Temu, T.; Cox, J. The MaxQuant computational platform for mass spectrometry-based shotgun proteomics. *Nat. Protoc.* **2016**, *11*, 2301. [CrossRef]
32. Pham, T.V.; Henneman, A.A.; Jimenez, C.R. *Iq*: An R package to estimate relative protein abundances from ion quantification in DIA-MS-based proteomics. *Bioinformatics* **2020**, *36*, 2611–2613. [CrossRef]
33. Ritchie, M.E.; Phipson, B.; Wu, D.I.; Hu, Y.; Law, C.W.; Shi, W.; Smyth, G.K. *Limma* powers differential expression analyses for RNA-sequencing and microarray studies. *Nucleic Acids Res.* **2015**, *43*, e47. [CrossRef]
34. Huang, D.W.; Sherman, B.T.; Lempicki, R.A. Bioinformatics enrichment tools: Paths toward the comprehensive functional analysis of large gene lists. *Nucleic Acids Res.* **2009**, *37*, 1–13. [CrossRef] [PubMed]
35. Sherman, B.T.; Lempicki, R.A. Systematic and integrative analysis of large gene lists using DAVID bioinformatics resources. *Nat. Protoc.* **2009**, *4*, 44.
36. Zhu, B.; Stülke, J. SubtiWiki in 2018: From genes and proteins to functional network annotation of the model organism *Bacillus subtilis*. *Nucleic Acids Res.* **2018**, *46*, D743–D748. [CrossRef] [PubMed]
37. Coleman, W.H.; Chen, D.; Li, Y.; Cowan, A.E.; Setlow, P. How moist heat kills spores of *Bacillus subtilis*. *J. Bacteriol.* **2007**, *189*, 8458–8466. [CrossRef] [PubMed]
38. Coleman, W.H.; Zhang, P.; Li, Y.-Q.; Setlow, P. Mechanism of killing of spores of *Bacillus cereus* and *Bacillus megaterium* by wet heat. *Lett. Appl. Microbiol.* **2010**, *50*, 507–514. [CrossRef] [PubMed]
39. Cox, J.; Hein, M.Y.; Luber, C.A.; Paron, I.; Nagaraj, N.; Mann, M. Accurate proteome-wide label-free quantification by delayed normalization and maximal peptide ratio extraction, termed MaxLFQ. *Mol. Cell. Proteom.* **2014**, *13*, 2513–2526. [CrossRef] [PubMed]
40. Bartels, J.; Blüher, A.; López Castellanos, S.; Richter, M.; Günther, M.; Mascher, T. The *Bacillus subtilis* endospore crust: Protein interaction network, architecture and glycosylation state of a potential glycoprotein layer. *Mol. Microbiol.* **2019**, *112*, 1576–1592. [CrossRef] [PubMed]
41. Beall, B.; Driks, A.; Losick, R.; Moran, C.P. Cloning and characterization of a gene required for assembly of the *Bacillus subtilis* spore coat. *J. Bacteriol.* **1993**, *175*, 1705–1716. [CrossRef] [PubMed]
42. Nunes, F.; Fernandes, C.; Freitas, C.; Marini, E.; Serrano, M.; Moran Jr, C.P.; Eichenberger, P.; Henriques, A.O. SpoVID functions as a non-competitive hub that connects the modules for assembly of the inner and outer spore coat layers in *Bacillus subtilis*. *Mol. Microbiol.* **2018**, *110*, 576–595. [CrossRef]

43. Naclerio, G.; Baccigalupi, L.; Zilhao, R.; De Felice, M.; Ricca, E. *Bacillus subtilis* spore coat assembly requires *cotH* gene expression. *J. Bacteriol.* **1996**, *178*, 4375–4380. [CrossRef] [PubMed]

44. Ferguson, C.C.; Camp, A.H.; Losick, R. *GerT*, a newly discovered germination gene under the control of the sporulation transcription factor $\sigma^K$ in *Bacillus subtilis*. *J. Bacteriol.* **2007**, *189*, 7681–7689. [CrossRef] [PubMed]

45. Steil, L.; Serrano, M.; Henriques, A.O.; Völker, U. Genome-wide analysis of temporally regulated and compartment-specific gene expression in sporulating cells of *Bacillus subtilis*. *Microbiology* **2005**, *151*, 399–420. [CrossRef] [PubMed]

46. Thierbach, S.; Sartor, P.; Yücel, O.; Fetzner, S. Efficient modification of the *Pseudomonas aeruginosa* toxin 2-heptyl-1-hydroxyquinolin-4-one by three *Bacillus* glycosyltransferases with broad substrate ranges. *J. Biotechnol.* **2020**, *308*, 74–81. [CrossRef]

47. Turner, R.J.; Lu, Y.; Switzer, R.L. Regulation of the *Bacillus subtilis* pyrimidine biosynthetic (*pyr*) gene cluster by an autogenous transcriptional attenuation mechanism. *J. Bacteriol.* **1994**, *176*, 3708–3722. [CrossRef] [PubMed]

48. Takamatsu, H.; Kodama, T.; Watabe, K. Assembly of the CotSA coat protein into spores requires CotS in *Bacillus subtilis*. *FEMS Microbiol. Lett.* **1999**, *174*, 201–206. [CrossRef] [PubMed]

49. Koide, Y.; Nakamura, A.; Uozumi, T.; Beppu, T. Cloning and sequencing of the major intracellular serine protease gene of *Bacillus subtilis*. *J. Bacteriol.* **1986**, *167*, 110–116. [CrossRef]

50. Han, X.; Shiwa, Y.; Itoh, M.; Suzuki, T.; Yoshikawa, H.; Nakagawa, T.; Nagano, H. Molecular cloning and sequence analysis of an extracellular protease from four *Bacillus subtilis* strains. *Biosci. Biotechnol. Biochem.* **2013**, *77*, 870–873. [CrossRef] [PubMed]

51. Zheng, L.; Abhyankar, W.; Ouwerling, N.; Dekker, H.L.; van Veen, H.; van der Wel, N.N.; Roseboom, W.; de Koning, L.J.; Brul, S.; de Koster, C.G. *Bacillus subtilis* spore inner membrane proteome. *J. Proteome Res.* **2016**, *15*, 585–594. [CrossRef] [PubMed]

52. Bernhards, C.B.; Chen, Y.; Toutkoushian, H.; Popham, D.L. HtrC is involved in proteolysis of YpeB during germination of *Bacillus anthracis* and *Bacillus subtilis* spores. *J. Bacteriol.* **2015**, *197*, 326–336. [CrossRef]

53. Doan, T.; Morlot, C.; Meisner, J.; Serrano, M.; Henriques, A.O.; Moran, C.P., Jr.; Rudner, D.Z. Novel secretion apparatus maintains spore integrity and developmental gene expression in *Bacillus subtilis*. *PLoS Genet.* **2009**, *5*, e1000566. [CrossRef]

54. Rodrigues, C.D.; Henry, X.; Neumann, E.; Kurauskas, V.; Bellard, L.; Fichou, Y.; Schanda, P.; Schoehn, G.; Rudner, D.Z.; Morlot, C. A ring-shaped conduit connects the mother cell and forespore during sporulation in *Bacillus subtilis*. *Proc. Natl. Acad. Sci. USA* **2016**, *113*, 11585–11590. [CrossRef] [PubMed]

55. Vega-Cabrera, L.A.; Guerrero, A.; Rodríguez-Mejía, J.L.; Tabche, M.L.; Wood, C.D.; Gutierrez-Rios, R.-M.; Merino, E.; Pardo-Lopez, L. Analysis of Spo0M function in *Bacillus subtilis*. *PLoS ONE* **2017**, *12*, e0172737. [CrossRef] [PubMed]

56. Han, W.-D.; Kawamoto, S.; Hosoya, Y.; Fujita, M.; Sadaie, Y.; Suzuki, K.; Ohashi, Y.; Kawamura, F.; Ochi, K. A novel sporulation-control gene (*spo0M*) of *Bacillus subtilis* with a $\sigma^H$-regulated promoter. *Gene* **1998**, *217*, 31–40. [CrossRef]

57. Kuwana, R.; Yamamura, S.; Ikejiri, H.; Kobayashi, K.; Ogasawara, N.; Asai, K.; Sadaie, Y.; Takamatsu, H.; Watabe, K. *Bacillus subtilis spoVIF (yjcC)* gene, involved in coat assembly and spore resistance. *Microbiology* **2003**, *149*, 3011–3021. [CrossRef]

58. Van Ooij, C.; Eichenberger, P.; Losick, R. Dynamic patterns of subcellular protein localization during spore coat morphogenesis in *Bacillus subtilis*. *J. Bacteriol.* **2004**, *186*, 4441–4448. [CrossRef] [PubMed]

59. Rebeil, R.; Sun, Y.; Chooback, L.; Pedraza-Reyes, M.; Kinsland, C.; Begley, T.P.; Nicholson, W.L. Spore photoproduct lyase from *Bacillus subtilis* spores is a novel iron-sulfur DNA repair enzyme which shares features with proteins such as class III anaerobic ribonucleotide reductases and pyruvate-formate lyases. *J. Bacteriol.* **1998**, *180*, 4879–4885. [CrossRef]

60. Newman, J.A.; Rodrigues, C.; Lewis, R.J. Molecular basis of the activity of SinR protein, the master regulator of biofilm formation in *Bacillus subtilis*. *J. Biol. Chem.* **2013**, *288*, 10766–10778. [CrossRef] [PubMed]

61. Quentin, Y.; Fichant, G.; Denizot, F. Inventory, assembly and analysis of *Bacillus subtilis* ABC transport systems. *J. Mol. Biol.* **1999**, *287*, 467–484. [CrossRef] [PubMed]

62. Popham, D.L.; Setlow, P. Cloning, nucleotide sequence, and regulation of the *Bacillus subtilis pbpE* operon, which codes for penicillin-binding protein 4* and an apparent amino acid racemase. *J. Bacteriol.* **1993**, *175*, 2917–2925. [CrossRef] [PubMed]

63. Scheffers, D.-J. Dynamic localization of penicillin-binding proteins during spore development in *Bacillus subtilis*. *Microbiology* **2005**, *151*, 999–1012. [CrossRef] [PubMed]

64. Wang, S.T.; Setlow, B.; Conlon, E.M.; Lyon, J.L.; Imamura, D.; Sato, T.; Setlow, P.; Losick, R.; Eichenberger, P. The forespore line of gene expression in *Bacillus subtilis*. *J. Mol. Biol.* **2006**, *358*, 16–37. [CrossRef] [PubMed]

65. Bressuire-Isoard, C.; Broussolle, V.; Carlin, F. Sporulation environment influences spore properties in *Bacillus*: Evidence and insights on underlying molecular and physiological mechanisms. *FEMS Microbiol. Rev.* **2018**, *42*, 614–626. [CrossRef] [PubMed]

*microorganisms*

*Review*

# Not Just Transporters: Alternative Functions of ABC Transporters in *Bacillus subtilis* and *Listeria monocytogenes*

Jeanine Rismondo *[ID] and Lisa Maria Schulz

Department of General Microbiology, GZMB, Georg-August-University Göttingen, Grisebachstr. 8, D-37077 Göttingen, Germany; lisamaria.schulz@uni-goettingen.de
* Correspondence: jrismon@gwdg.de; Tel.: +49-551-39-33796

**Abstract:** ATP-binding cassette (ABC) transporters are usually involved in the translocation of their cognate substrates, which is driven by ATP hydrolysis. Typically, these transporters are required for the import or export of a wide range of substrates such as sugars, ions and complex organic molecules. ABC exporters can also be involved in the export of toxic compounds such as antibiotics. However, recent studies revealed alternative detoxification mechanisms of ABC transporters. For instance, the ABC transporter BceAB of *Bacillus subtilis* seems to confer resistance to bacitracin via target protection. In addition, several transporters with functions other than substrate export or import have been identified in the past. Here, we provide an overview of recent findings on ABC transporters of the Gram-positive organisms *B. subtilis* and *Listeria monocytogenes* with transport or regulatory functions affecting antibiotic resistance, cell wall biosynthesis, cell division and sporulation.

**Keywords:** ABC transporter; antibiotic resistance; Gram-positive bacteria; cell wall

**Citation:** Rismondo, J.; Schulz, L.M. Not Just Transporters: Alternative Functions of ABC Transporters in *Bacillus subtilis* and *Listeria monocytogenes*. *Microorganisms* **2021**, *9*, 163. https://doi.org/10.3390/microorganisms9010163

Received: 17 December 2020
Accepted: 10 January 2021
Published: 13 January 2021

**Publisher's Note:** MDPI stays neutral with regard to jurisdictional claims in published maps and institutional affiliations.

## 1. Introduction

ATP-binding cassette (ABC) transporters can be found in all kingdoms of life and can be subdivided in three main groups: eukaryotic transporters, bacterial importers and bacterial exporters. In these transporters, the translocation of cognate substrates is driven by ATP hydrolysis [1]. In bacteria, ABC transporters can be involved in the import or export of a wide range of substrates, such as amino acids, ions, sugars or complex organic molecules [2]. For instance, the ABC transporter ZnuABC of *Bacillus subtilis* is required for the uptake of zinc [3], and the transporter TagGH is responsible for the export of wall teichoic acid, a secondary cell wall polymer of Gram-positive bacteria [4–6]. ABC exporters can also be involved in the export of toxic compounds, such as antibiotics—for instance, the multidrug transporter Sav1866 of *Staphylococcus aureus* or BmrA of *B. subtilis* [7–9].

ABC transporters are typically composed of four core domains: two nucleotide binding domains (NBDs), also referred to as ATP-binding proteins, which hydrolyze ATP, and two transmembrane domains (TMDs), which allow the transport of the substrate across the cell membrane. NBDs and TMDs can be formed by either homodimers or heterodimers [8]. Additionally to NBDs and TMDs, ABC importers possess an extracellular substrate binding protein (SBP), which is required for the capturing and delivery of substrates to the transporter [10]. Energy-coupling factor (ECF) transporters use a membrane-integrated S-component for substrate binding [11].

In past years, it became apparent that ABC transporters are not only involved in the import or export of their cognate substrates. For some ABC transporters it was shown that they use alternative drug detoxification mechanisms or possess regulatory functions, such as FtsEX of *B. subtilis*, which actives the D,L-endopeptidase CwlO via direct protein–protein interaction [12]. In this review, we want to highlight some ABC transporters with alternative detoxification mechanisms or regulatory functions of *B. subtilis* and the human pathogen *Listeria monocytogenes*, which affect antibiotic resistance, cell wall biosynthesis, cell division and sporulation.

## 2. ABC Transporters Involved in Drug Export, Drug Sensing and Detoxification

Antibiotic resistance is a rising problem in modern society, which is associated with enormous costs for the public health sector. Bacteria found different strategies to combat antibiotics: inactivating or degrading the antibiotic itself, changing the antibiotic target, preventing the penetration of the drug into the cell and exporting the drug via efflux pumps. ABC exporters can also be involved in the export of antibiotics; however, different detoxification mechanisms have been proposed in recent years. Here we summarize the findings for the multidrug transporters BmrA and BmrCD of *B. subtilis* and the proposed mechanisms of action for BceAB-BceRS systems of *B. subtilis* and *L. monocytogenes*.

### 2.1. The Multidrug Transporters BmrA and BmrCD

The ABC exporter BmrA of *B. subtilis* is composed of one NBD and one TMD, which are linked by long intracellular domains, and functions as a homodimer [13,14]. Different techniques, such as cryo-electron microscopy, solid state nuclear magnetic resonance (NMR), electron paramagnetic resonance (EPR) spectroscopy and biochemical studies, revealed an inward-facing conformation of apo-BmrA, whereas ATP-bound BmrA is outward-facing [14–18]. Interestingly, the transition from inward to outward-facing conformation only seems to require ATP binding and does not depend on ATP hydrolysis. Whether the release of the substrate of BmrA during translocation is facilitated solely by the switch to the outward-facing conformation or whether this process requires the hydrolysis of ATP is currently unknown [15]. In vitro and in vivo studies showed that BmrA is able to transport several substrates, such as Hoechst 33342, ethidium bromide, doxorubicin and cervimycin C [9,19]. Cervimycin C resistant mutants of *B. subtilis* were isolated, which harbored two mutations in the intergenic region proceeding *bmrA* leading to enhanced *bmrA* transcription and BmrA production. In addition, deletion of *bmrA* leads to a reduction of cervimycin C resistance of *B. subtilis*. Based on these observations, it was suggested that BmrA is involved in cervimycin C resistance in *B. subtilis* [19].

The half transporters BmrC and BmrD consist of one NBD fused to one TMD. Co-purification studies revealed that both proteins form a heterodimer [20]. Alterations in the sequence of the consensus motif of the NBDs result in the presence of one degenerate nucleotide binding site (NBS) and one consensus NBS in the BmrCD complex. In the apo-form, the BmrCD transporter is proposed to have an inward-facing conformation with one ATP bound to the degenerate NBS. Substrate binding to the TMD either occurs before or after the first ATP is bound to the NBS. The binding of a second ATP molecule to the consensus NBS results in a conformational change of the NBDs, which is not transferred to the TMDs. The inward to outward-facing transition of the TMDs is solely driven by the hydrolysis of the ATP molecule bound to the consensus NBS and leads to the release of the substrate [21].

Using inside-out membrane vesicles, Torres et al. could show that BmrCD is able to transport the fluorescent substances Hoechst 33342, doxorubicin and mitoxantrone [20]. BmrCD was also able to translocate ethidium bromide into reconstituted giant unilamellar vesicles, and the transport activity could be inhibited by orthovanadate, an ATPase inhibitor [22]. The genes encoding the BmrCD transporter are induced in the presence of a number of antibiotics, most of which target the ribosomes (e.g., chloramphenicol, erythromycin, gentamycin) [20,23,24]. Surprisingly, deletion of *bmrCD* had no effect on growth in the presence of the tested antibiotics; however, other transporters could also be involved in the detoxification of these drugs and could mask the phenotype [20]. In-depth analysis of the regulation of *bmrCD* expression revealed that it is controlled at two levels: the induction of *bmrCD* is only possible during the transition and stationary growth phase, and the antibiotic-induced expression is regulated by a ribosome-mediated transcriptional attenuation mechanism, which is controlled by the leader peptide BmrB [25]. *bmrB* encodes a small protein of 54 amino acids, which is co-transcribed with *bmrCD* [25]. Upstream of *bmrB* is a binding site for the main transition stage regulator AbrB [26], which is responsible for the growth phase dependent expression of the *bmrBCD* operon [25]. An intrinsic terminator-like structure and alternative anti-terminator and anti-anti-terminator structures are located within the coding region of *bmrB*, which are required for the antibiotic-mediated expression of *bmrCD*. In the absence of antibiotics, the *bmrB* terminator forms, thereby blocking the expression of *bmrCD*. The binding of ribosome-targeting antibiotics such as lincomycin leads to the formation of an anti-termination structure, thereby resulting in the expression of *bmrCD* [25].

The biological role of the growth phase-dependent expression of *bmrCD* is still unknown; however, it was speculated that BmrCD might have a role in sporulation [25]. Indeed, transcriptome data show that *bmrCD* transcript levels are elevated during sporulation [27]. In *B. subtilis*, initiation of sporulation depends on the phosphorylation of the transcription factor Spo0A, which is accomplished by a phosphorelay including the phosphotransferases Spo0F and Spo0B. Spo0F is phosphorylated by several histidine kinases, the major ones of which are KinA and KinB [28,29]. BmrD was identified as an interaction partner of the histidine kinase KinA, and it was speculated that the transporter BmrCD might transfer signals to KinA to initiate sporulation. Furthermore, it was shown that the overexpression of BmrCD reduced sporulation efficiency of a *kinB* deletion strain in a Spo0A-dependent manner, and deletion of *bmrCD* had no effect on sporulation [30]. Further studies are required to identify the exact biological role of BmrCD for sporulation initiation in *B. subtilis*. Homologs of BmrCD can also be found in the non-spore forming Gram-positive bacterium *L. monocytogenes*; however, the functions of Lmo1651 and Lmo1652 have not been investigated so far.

### 2.2. BceAB-BceRS Systems in B. subtilis and L. monocytogenes

In the past few years, several ABC transporters have been identified, which are not only required to inactivate or export drugs, but which are essential for drug sensing. The best-characterized system is the BceAB-BceRS system of *B. subtilis*, which is required for the resistance against bacitracin, actagardine, mersacidin and plectasin [31–34]. Bacitracin blocks the recycling of lipid carriers during the lipid II cycle of cell wall biosynthesis by binding to the diphosphate lipid carrier undecaprenyl pyrophosphate (UPP), and actagardine, mersacidin and plectasin directly bind lipid II. A phylogenetic analysis revealed that BceAB-like transporters are nearly exclusively found in bacteria of the phylum Firmicutes and confer resistance to a subset of antimicrobials [35–37]. These transporters belong to the peptide 7 exporter family, usually consist of two ATP-binding proteins (BceA), a large permease (BceB) and are mostly associated with a BceRS-like two-component system. It has been shown that both components, the transporter BceAB and the two-component system BceRS, are required for sensing of and resistance to bacitracin in *B. subtilis* [35,38,39]. The histidine kinase BceS belongs to the family of intramembrane-sensing histidine kinases whose members lack the extracellular sensing domain [36,40]. *B. subtilis* BceB and other BceB-like permeases possess a large extracellular loop of around 200 to 250 amino acids, which is thought to contain the ligand-binding site [32,34,39]. In the absence of bacitracin, *B. subtilis* produces basal levels of BceAB transporters, which control the conformation of the histidine kinase BceS, thereby keeping BceS in its inactive state [41]. Upon binding of bacitracin, BceAB and BceS form a sensory complex, which detects the activity of BceAB and activates the promoter $P_{bceA}$ via the response regulator BceR in a dose-dependent manner [33,42,43]. The mechanism by which the BceAB transporter confers resistance to antibiotics, however, has been investigated for nearly two decades and is still highly discussed. In the past, BceAB was thought to function as an efflux transporter [31], but other studies suggested that bacitracin is imported and degraded by BceAB [39]; experimental evidence for both hypotheses is missing. Kingston et al. proposed that BceAB transports UPP across the membrane to the cytoplasmic leaflet, thereby rendering it inaccessible to bacitracin [44]. However, this mechanism would not explain how BceAB confers resistance to the lipid II-targeting antibiotics. A recent study revealed that BceAB specifically recognizes complexes of bacitracin and UPP. The authors further suggest that BceAB breaks this interaction using the energy derived from ATP hydrolysis by BceA, thereby releasing

UPP from the inhibitory grip of bacitracin [45]. Thus, BceAB seems to provide resistance to antimicrobials through target protection rather than degradation.

The closest homolog of *B. subtilis* BceB in *L. monocytogenes* is AnrB (27% sequence identity), encoded by *lmo2115*. AnrB is part of the ABC transporter AnrAB and forms a multicomponent resistance module together with a second ABC transporter, VirAB, and the two-component system VirRS (Figure 1) [46,47]. This system confers resistance towards nisin, bacitracin, several β-lactam antibiotics, benzalkonium chloride and ethidium bromide [46–49]. Activation of the histidine kinase VirS depends on the activity of the ABC transporter VirAB; however, it has not been determined yet whether VirAB and VirS form a sensory complex as described for the BceAB-BceRS system of *B. subtilis*, or whether VirAB transports its substrate to a place where VirS can sense it [46,47]. In *L. monocytogenes* strain EGD-e, VirAB is required for sensing of both nisin and bacitracin, whereas sensing of bacitracin seems to be VirAB-independent in *L. monocytogenes* strain 10403S [46,47]. Upon activation, VirS phosphorylates the response regulator VirR, which subsequently induces the expression of the *anrAB* operon. The ABC transporter AnrAB then detoxifies antimicrobials such as nisin and bacitracin [46,48]; however, the exact mechanism by which AnrAB confers resistance is unknown. It is tempting to speculate that AnrAB might also provide resistance via target protection rather than by inactivation or export of the drug. In addition to its role in drug sensing, VirAB is involved in the detoxification of kanamycin and tetracycline in a VirRS-independent manner [46]. Furthermore, it was suggested that Lm.G_1771, the VirB homolog in *L. monocytogenes* serotype 4b strain G, acts as a negative regulator of biofilm formation, potentially by the export of a signal that prevents biofilm formation [50].

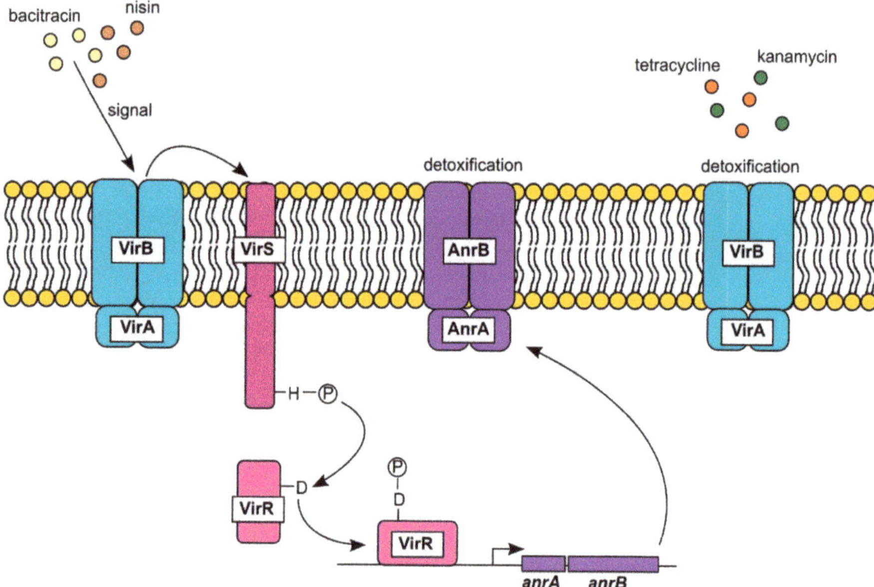


**Figure 1.** The VirAB-VirRS-AnrAB resistance module of *L. monocytogenes* EGD-e. The ABC transporter VirAB senses the presence of bacitracin, nisin and a number of β-lactam antibiotics. This leads to activation of the intramembrane-sensing histidine kinase VirS, which subsequently phosphorylates the response regulator VirR [46,47]. VirR binds to the VirR consensus sequence present in the promoter region of *anrAB* and induces the expression of both genes [46,51]. The ABC transporter AnrAB detoxifies bacitracin and nisin by a so far unknown mechanism. The detoxification of kanamycin and tetracycline by VirAB occurs in a VirRS-independent manner [46].

In addition to AnrAB, VirR regulates the expression of DltABCD required for D-alanylation of the cell wall polymers teichoic acids and the lysinyl-transferase MprF, which attaches lysine residues onto phosphatidyl-glycerol in the membrane [51–54]. D-alanylation of teichoic acids and lysinylation of phosphatidyl-glycerol reduce the negative charge of the cell surface, thereby leading to a diminished binding of cationic peptides and human defensins [49,55–57]. In addition to its role in conferring resistance towards antimicrobials, the VirAB-VirRS-AnrAB resistance module is also required for virulence of *L. monocytogenes*. Transcriptome studies revealed enhanced expression of *anrAB* during in vitro and in vivo infections [58,59]. Mutants lacking either the histidine kinase VirS or the response regulator VirR produce shorter actin tails during intracellular infection, thereby leading to a reduced ability to spread from cell-to-cell via actin-based motility [47]. Absence of either VirR or VirS also leads to a reduced adherence and entry into human epithelial cells [51]. Taken together, BceAB-BceRS-like systems are involved in drug sensing and detoxification and can have crucial functions for the virulence of pathogenic bacteria.

## 3. ABC Transporters Affecting Cell Wall Biosynthesis and Remodeling

The most prominent target of clinically used antibiotics is the bacterial cell wall. In Gram-positive bacteria, the cell wall consists of a thick layer of peptidoglycan (PG) and secondary cell wall polymers, the so-called teichoic acids. The synthesis of PG starts in the cytoplasm with the production of lipid II, a disaccharide-pentapeptide, which is linked to an undecaprenyl-carrier. In *B. subtilis*, lipid II is then transported across the membrane by the lipid II flippases MurJ and Amj and incorporated into the growing glycan strand by the glycosyltransferase activity of bifunctional penicillin binding proteins or FtsW and RodA. The glycan strands are subsequently crosslinked by the transpeptidase activity of penicillin binding proteins (reviewed in: [60,61]). Several steps of the PG biosynthesis process can be inhibited by antibiotics, for instance, moenomycin blocks the glycosyltransferase activity and β-lactam antibiotics inhibit the transpeptidase activity of penicillin binding proteins [39,62]. During bacterial growth and cell division, PG is constantly remodeled, degraded and recycled [63,64]. These processes depend on the activity of a diverse set of autolytic enzymes, whose activity needs to be tightly controlled and adjusted to the growth stage and environmental growth condition to prevent cell lysis [60,64].

### 3.1. The YtrBCDEF Transporter of B. subtilis

Recent studies suggest that ABC transporters can also have a direct or indirect effect on PG biosynthesis and remodeling. For instance, it has been shown that the overexpression of the YtrBCDEF ABC transporter of *B. subtilis* leads to the production of a thicker PG layer [65]. The transporter YtrBCDEF is encoded in the *ytrGABCDEF* operon and the GntR family repressor YtrA controls its expression [66,67]. YtrA also represses the expression of the *ywoBCD* operon, which codes for a membrane protein of unknown function, a hydrolase and a major facilitator superfamily transporter, respectively [67]. Both operons, *ytrGABCDEF* and *ywoBCD*, are induced in the presence of several cell wall-acting antibiotics, such as ramoplanin, bacitracin and vancomycin [40,67,68]. YtrE was also identified as a marker protein for the inhibition of membrane-bound steps of PG biosynthesis [68]. In addition, it was shown that the expression of the *ytrGABCDEF* operon is also induced after *B. subtilis* was subjected to cold shock [69].

The YtrBCDEF ABC transporter is composed of two nucleotide binding proteins; YtrB and YtrE; two transmembrane domain proteins, YtrC and YtrD and the solute binding protein YtrF. However, it is still unknown whether these components form one or two separate ABC transporters [66,70]. Yoshida et al. suggested that YtrBCDEF might be involved in the import of acetoin, and thus, in the acetoin utilization of *B. subtilis* [66]. *B. subtilis* produces and excretes acetoin during growth in the presence of excess carbohydrates, which is reused during stationary phase and sporulation [71]. The degradation of acetoin depends on the activity of the multicomponent acetoin dehydrogenase enzyme system (AoDH ES) encoded in the *acoABCL* operon [72]. In addition, it has been shown that absence of the acetyltransferase AcuA, which is encoded in the *acuABC* operon and is involved in the inactivation of the acetyl coenzyme A synthetase AscA [73,74], leads to reduced growth and sporulation on acetoin [75]. Utilization of acetoin is also diminished in a *B. subtilis* and *Bacillus licheniformis acuA* mutant [76,77]. How AcuA affects the acetoin catabolism in *B. subtilis* and *B. licheniformis* is still unknown. The utilization of acetoin is also reduced in a *B. subtilis* strain lacking the putative acetoin importer encoded in the *ytrGABCDEF* operon; however, the acetoin catabolism was not completely abolished in this strain, suggesting that *B. subtilis* encodes at least one additional acetoin import system. Interestingly, the expression of the *ytrGABCDEF* operon was not induced in presence of acetoin [66], suggesting that the YtrBCDEF transporter might only have an indirect effect on acetoin utilization. The absence of the transcriptional regulator YtrA further led to a complete loss of genetic competence [65,78]. The loss of competence of the *ytrA* mutant could potentially be explained by an inability of the DNA to reach the competence pilus ComG due to the thicker PG layer produced by this strain [65]; however, further experiments are required to investigate this. In addition, it remains to be elucidated how the function of the ABC transporter YtrBCDEF is linked to PG biosynthesis and acetoin utilization in *B. subtilis*.

### 3.2. The Putative ABC Transporter EslABC of L. monocytogenes

The ATP binding protein EslA and the transmembrane domain protein EslB of *L. monocytogenes* are weak homologs of YtrB and YtrC/YtrD of *B. subtilis* with a sequence identity of 35% and 22%, respectively. *L. monocytogenes* EslA and EslB are encoded in the *eslABCR* operon together with the transmembrane protein EslC and the RpiR regulator EslR [79]. EslA and EslB are predicted to form an ABC transporter; however, it is still unknown whether the transmembrane protein EslC is also part of the ABC transporter or whether it has an independent role of EslAB. Bacterial two-hybrid experiments indicated that EslB and EslC interact with each other [80], and we will thus refer to the transporter as the EslABC transporter. Phenotypic analysis of strains lacking either EslA or EslB revealed a strong increase in lysozyme sensitivity, and deletion of *eslC* does not affect lysozyme resistance of *L. monocytogenes* [80–82]. This observation does not rule out that EslC might be part of the ABC transporter; however, it suggests that EslA and EslB have an independent role of EslC. In *L. monocytogenes*, lysozyme resistance mainly depends on the activity of two PG modifying enzymes: the *N*-deacetylase PgdA and the *O*-acetyltransferase OatA. PgdA is required for the deacetylation of GlcNAc residues of the PG backbone, and OatA modifies MurNAc residues with *O*-acetyl groups [83,84]. The PG produced by an *eslB* mutant is more deacetylated and less *O*-acetylated as compared to the PG of the *L. monocytogenes* wildtype strain. Furthermore, deletion of *eslB* results in the production of a thinner PG layer, and thus, the lysozyme sensitivity of the *eslB* mutant is likely caused by both, the decrease in *O*-acetylation and PG layer thickness [80]. The reduction in cell wall thickness in the *eslB* mutant also leads to an increased cell lysis and EslB is thus required for the maintenance of cell wall integrity in *L. monocytogenes* [80]. In addition to its importance for PG biosynthesis, EslB is required for proper cell division in *L. monocytogenes* as cells lacking EslB form elongated cells. Interestingly, localization studies with the early cell division protein ZapA, which is required for FtsZ filament stabilization [85], suggest that these elongated cells form multiple Z-rings [80]. Thus, a process downstream of the recruitment of early cell division proteins, but prior to septum formation seems to be disturbed in absence of EslB. It also remains to be elucidated what the cellular function of the EslABC transporter is. The genes coding for EslABC are co-transcribed in an operon with *eslR* encoding an RpiR transcriptional regulator [79]. RpiR transcriptional regulator are usually involved in the regulation of sugar phosphate metabolic pathways [86–92], which indicates that EslABC could acts as a sugar importer. However, ABC importers usually require a substrate binding protein for the recognition of their cognate substrates. None of the *esl* genes or genes adjacent to the *eslABCR* operon code for a putative substrate binding protein; thus, it is unlikely that EslABC is involved in the import of sugars. Another possibility is that EslABC is required for the export of certain PG components, which would explain the production of a thinner PG layer by the *eslB* mutant. It is also possible that EslABC has an alternative function, for instance EslABC could be required for the proper localization of other proteins or the regulation of protein activity. Further studies need to be performed to evaluate these hypotheses.

### 3.3. FtsEX of B. subtilis Regulates the D,L-endopeptidase CwlO

The type VII ABC transporter FtsEX is a transporter involved in PG remodelling. It consists of the transmembrane protein FtsX and the ATP binding protein FtsE and was first identified in the Gram-negative model organism *Escherichia coli*, in which it is essential for cell division as conditional mutants formed filamentous cells (*fts* for filamentous temperature-sensitive) [93]. In contrast to the observation in *E. coli*, initial screens in the Gram-positive bacterium *B subtilis* associated the FtsEX protein complex to polar division and was hence thought to be involved in spore formation in *Bacillus* [94]. The absence of the putative transporter FtsEX in *B. subtilis* led to delayed entry into sporulation and medial septum formation. At the time, it was hypothesized that FtsEX imports a sporulation signal which in turn leads to activation of the master regulator of sporulation, Spo0A. However, later analysis showed that FtsEX stimulates entry into sporulation through its regulation of the autolysin CwlO [12]. CwlO is the major autolysin for cell wall elongation in *B. subtilis*, which hydrolyses the peptide bond between γ-D-glutamate and meso-diaminopimelic acid linkages [95]. Deletion of *cwlO* and *ftsEX* leads to wider and shorter cells with many cells exhibiting a curved or twisted shape, indicating that CwlO and FtsEX share the same pathway [12,96]. Interestingly, the phenotypes associated with *ftsEX* and *cwlO* deletion could be rescued by addition of magnesium to the medium, which is often observed in correlation with disrupted PG biosynthesis. This is in agreement with results of bacterial two hybrid experiments that revealed an interaction between FtsEX and enzymes involved in PG or teichoic acid synthesis and proteins associated with cell elongation [12,96]. An independent study also revealed that absence of CwlO or FtsEX results in decreased competence due to reduced expression of the major regulator of competence, ComK [97]. To further analyse the impact of FtsEX on the hydrolase activity of CwlO, deletions in *ftsEX* or *cwlO* were combined with the deletion of *lytE*, encoding the second major autolysin involved in cell elongation in *B. subtilis*. *lytE* and *cwlO* are synthetic lethal as lack of both autolysins resulted in a disruption of cell elongation and subsequent cell lysis [98]. Synthetic lethality was also observed for the *lytE ftsEX* double mutant. In contrast, a strain lacking FtsEX and CwlO phenocopied the *ftsEX* and *cwlO* single mutants, which again suggests that FtsEX and CwlO are part of the same pathway in *B. subtilis* [12,96]. Deletion of *ftsEX* does not lead to altered CwlO levels and it was thus hypothesized that FtsEX might be involved in the regulation of CwlO hydrolase activity [12]. An in-depth analysis suggests that both, the transmembrane component of the transporter FtsX and the ATP-binding protein FtsE, are required for CwlO functionality. Studies with FtsE variants carrying mutations in residues involved in ATP binding (K41) and hydrolysis (D162) showed that both processes are essential for CwlO activity. In addition, it was shown that CwlO interacts with FtsEX and that FtsX is required for proper localization of CwlO at the cell wall [12]. In addition, SweC and SweD were identified as essential co-factors and binding partners of FtsX, which are required for FtsEX-dependent regulation of CwlO (Figure 2) [99]. Comprehensive analysis of the type VII ABC transporter MacB of *Aggregatibacter actinomycetemcomitans* and *E. coli*, the structural homologue of FtsEX, suggested that it does not seem to transport molecules across the membrane, but rather uses a mechanotransmission mechanism and it was proposed that FtsEX works in a similar manner [100]. Precisely, it was hypothesized, that cytoplasmic ATP hydrolysis via FtsE results in a conformational change of the extra-cytoplasmic part of the transmembrane protein FtsX, which in turn leads to recruitment and/or activation of CwlO.

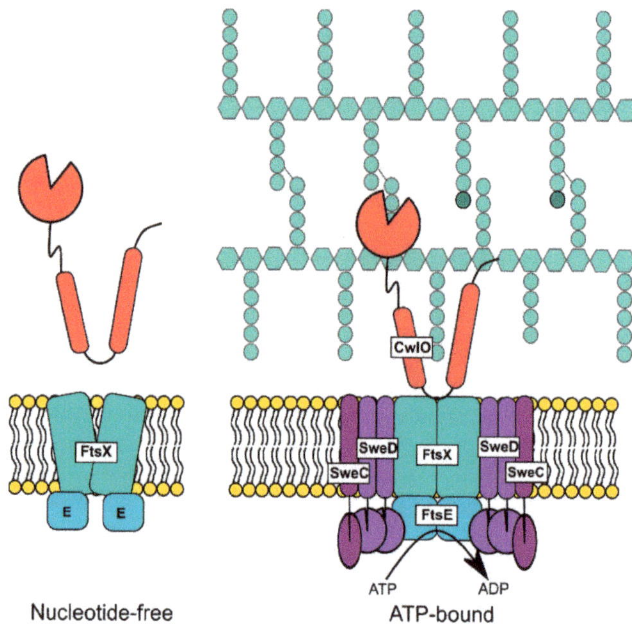


**Figure 2.** The FtsEX system of *B. subtilis*. The type VII ABC transporter FtsEX is an essential activator of the major cell wall hydrolase CwlO in *B. subtilis*. ATP hydrolysis by FtsE in the cytoplasm is proposed to result in conformational changes of FtsX which allows direct interaction with CwlO and subsequent regulation of its D-L-endopeptidase activity, and the complex is relaxed when in the nucleotide free state. The complex resides in the membrane together with SweC and SweD, that act as essential cofactors (adapted from [99]).

## 4. Conclusions

Genomes of Gram-positive and Gram-negative bacteria encode a variety of transport proteins required for the import of molecules such as sugars, amino acids, ions, peptides and for the export of toxic compounds. The classification of transporters into different categories such as ABC transporters or multidrug resistance transporters does not necessarily indicate anything about their function in vivo. Bacteria possess "classical" multidrug resistance ABC exporters, which translocate drugs across the bacterial membrane such as BmrA of *B. subtilis*. ABC transporters can also be involved in the detoxification of drugs by alternative mechanisms—e.g., by target production—as has been described for *B. subtilis* BceAB, or can have an additional impact on the physiology of the organism, as seen for the ABC transporter BmrCD. Some transporters are not even required for the import or export of any substrate; instead, they are involved in the regulation of proteins. The ABC transporter FtsEX is the best-characterized example for such transporters, which directly affects the activity of a PG hydrolase in *B. subtilis* and other Gram-positive and Gram-negative bacteria.

Molecular genetics are helpful to get first insights into the relevance of transporters for the physiology of bacteria; however, an in-depth biochemical analysis is required to fully understand the biological functions of ABC transporters and other transport systems, and to identify their mechanisms of action. These analyses are challenging, as they are often hampered by difficulties in protein preparations required for in vitro assays or for the determination of the 3D structure of the transporters. The genomes of pathogenic and non-pathogenic bacteria harbor numerous transporters, which have not been characterized yet and which might be crucial for the growth and/or lifestyle of these organisms and should thus not be overlooked.

**Author Contributions:** Conceptualization, J.R.; writing—original draft preparation, J.R. and L.M.S.; writing—review and editing, J.R. and L.M.S.; visualization, J.R. and L.M.S.; funding acquisition, J.R. All authors have read and agreed to the published version of the manuscript.

**Funding:** J.R. received funding from the German Research Foundation (DFG): grants RI 2920/2-1 and RI 2920/3-1.

**Conflicts of Interest:** The authors declare no conflict of interest.

# References

1. Fath, M.J.; Kolter, R. ABC transporters: Bacterial exporters. *Microbiol. Rev.* **1993**, *57*, 995–1017. [CrossRef] [PubMed]
2. Tanaka, K.J.; Song, S.; Mason, K.; Pinkett, H.W. Selective substrate uptake: The role of ATP-binding cassette (ABC) importers in pathogenesis. *Biochim. Biophys. Acta Biomembr.* **2018**, *1860*, 868–877. [CrossRef] [PubMed]
3. Gaballa, A.; Helmann, J.D. Identification of a Zinc-Specific Metalloregulatory Protein, Zur, Controlling Zinc Transport Operons in *Bacillus subtilis*. *J. Bacteriol.* **1998**, *180*, 5815–5821. [CrossRef]
4. Lazarevic, V.; Karamata, D. The tagGH operon of *Bacillus subtilis* 168 encodes a two-component ABC transporter involved in the metabolism of two wall teichoic acids. *Mol. Microbiol.* **1995**, *16*, 345–355. [CrossRef]
5. Chen, L.; Hou, W.-T.; Fan, T.; Liu, B.; Pan, T.; Li, Y.-H.; Jiang, Y.-L.; Wen, W.; Chen, Z.-P.; Sun, L.; et al. Cryo-electron Microscopy Structure and Transport Mechanism of a Wall Teichoic Acid ABC Transporter. *MBio* **2020**, *11*, e02749-19. [CrossRef] [PubMed]
6. Brown, S.; Santa Maria, J.P.; Walker, S. Wall Teichoic Acids of Gram-Positive Bacteria. *Annu. Rev. Microbiol.* **2013**, *67*, 313–336. [CrossRef]
7. Dawson, R.J.P.; Locher, K.P. Structure of a bacterial multidrug ABC transporter. *Nature* **2006**, *443*, 180–185. [CrossRef]
8. Locher, K.P. Mechanistic diversity in ATP-binding cassette (ABC) transporters. *Nat. Struct. Mol. Biol.* **2016**, *23*, 487–493. [CrossRef]
9. Steinfels, E.; Orelle, C.; Fantino, J.-R.; Dalmas, O.; Rigaud, J.-L.; Denizot, F.; Di Pietro, A.; Jault, J.-M. Characterization of YvcC (BmrA), a Multidrug ABC Transporter Constitutively Expressed in *Bacillus subtilis*. *Biochemistry* **2004**, *43*, 7491–7502. [CrossRef]
10. Davidson, A.L.; Dassa, E.; Orelle, C.; Chen, J. Structure, Function, and Evolution of Bacterial ATP-Binding Cassette Systems. *Microbiol. Mol. Biol. Rev.* **2008**, *72*, 317–364. [CrossRef]
11. Slotboom, D.J. Structural and mechanistic insights into prokaryotic energy-coupling factor transporters. *Nat. Rev. Microbiol.* **2014**, *12*, 79–87. [CrossRef] [PubMed]
12. Meisner, J.; Montero Llopis, P.; Sham, L.-T.; Garner, E.; Bernhardt, T.G.; Rudner, D.Z. FtsEX is required for CwlO peptidoglycan hydrolase activity during cell wall elongation in *Bacillus subtilis*. *Mol. Microbiol.* **2013**, *89*, 1069–1083. [CrossRef] [PubMed]
13. Dalmas, O.; Do Cao, M.-A.; Lugo, M.R.; Sharom, F.J.; Di Pietro, A.; Jault, J.-M. Time-Resolved Fluorescence Resonance Energy Transfer Shows that the Bacterial Multidrug ABC Half-Transporter BmrA Functions as a Homodimer. *Biochemistry* **2005**, *44*, 4312–4321. [CrossRef] [PubMed]
14. Mehmood, S.; Domene, C.; Forest, E.; Jault, J.-M. Dynamics of a bacterial multidrug ABC transporter in the inward- and outward-facing conformations. *Proc. Natl. Acad. Sci. USA* **2012**, *109*, 10832–10836. [CrossRef]
15. Lacabanne, D.; Orelle, C.; Lecoq, L.; Kunert, B.; Chuilon, C.; Wiegand, T.; Ravaud, S.; Jault, J.-M.; Meier, B.H.; Böckmann, A. Flexible-to-rigid transition is central for substrate transport in the ABC transporter BmrA from *Bacillus subtilis*. *Commun. Biol.* **2019**, *2*, 149. [CrossRef]
16. Orelle, C.; Gubellini, F.; Durand, A.; Marco, S.; Lévy, D.; Gros, P.; Di Pietro, A.; Jault, J.-M. Conformational Change Induced by ATP Binding in the Multidrug ATP-Binding Cassette Transporter BmrA. *Biochemistry* **2008**, *47*, 2404–2412. [CrossRef]
17. Wiegand, T.; Lacabanne, D.; Keller, K.; Cadalbert, R.; Lecoq, L.; Yulikov, M.; Terradot, L.; Jeschke, G.; Meier, B.H.; Böckmann, A. Solid-state NMR and EPR Spectroscopy of Mn 2+ -Substituted ATP-Fueled Protein Engines. *Angew. Chem. Int. Ed.* **2017**, *56*, 3369–3373. [CrossRef]
18. Fribourg, P.F.; Chami, M.; Sorzano, C.O.S.; Gubellini, F.; Marabini, R.; Marco, S.; Jault, J.-M.; Lévy, D. 3D Cryo-Electron Reconstruction of BmrA, a Bacterial Multidrug ABC Transporter in an Inward-Facing Conformation and in a Lipidic Environment. *J. Mol. Biol.* **2014**, *426*, 2059–2069. [CrossRef]
19. Krügel, H.; Licht, A.; Biedermann, G.; Petzold, A.; Lassak, J.; Hupfer, Y.; Schlott, B.; Hertweck, C.; Platzer, M.; Brantl, S.; et al. Cervimycin C resistance in *Bacillus subtilis* is due to a promoter up-mutation and increased mRNA stability of the constitutive ABC-transporter gene bmrA. *FEMS Microbiol. Lett.* **2010**, *313*, 155–163. [CrossRef]
20. Torres, C.; Galián, C.; Freiberg, C.; Fantino, J.-R.; Jault, J.-M. The YheI/YheH heterodimer from *Bacillus subtilis* is a multidrug ABC transporter. *Biochim. Biophys. Acta Biomembr.* **2009**, *1788*, 615–622. [CrossRef]
21. Mishra, S.; Verhalen, B.; Stein, R.A.; Wen, P.-C.; Tajkhorshid, E.; Mchaourab, H.S. Conformational dynamics of the nucleotide binding domains and the power stroke of a heterodimeric ABC transporter. *Elife* **2014**, *3*, e02740. [CrossRef] [PubMed]
22. Dezi, M.; Di Cicco, A.; Bassereau, P.; Levy, D. Detergent-mediated incorporation of transmembrane proteins in giant unilamellar vesicles with controlled physiological contents. *Proc. Natl. Acad. Sci. USA* **2013**, *110*, 7276–7281. [CrossRef] [PubMed]
23. Hutter, B.; Schaab, C.; Albrecht, S.; Borgmann, M.; Brunner, N.A.; Freiberg, C.; Ziegelbauer, K.; Rock, C.O.; Ivanov, I.; Loferer, H. Prediction of Mechanisms of Action of Antibacterial Compounds by Gene Expression Profiling. *Antimicrob. Agents Chemother.* **2004**, *48*, 2838–2844. [CrossRef] [PubMed]

24. Lin, J.T.; Connelly, M.B.; Amolo, C.; Otani, S.; Yaver, D.S. Global Transcriptional Response of *Bacillus subtilis* to Treatment with Subinhibitory Concentrations of Antibiotics That Inhibit Protein Synthesis. *Antimicrob. Agents Chemother.* **2005**, *49*, 1915–1926. [CrossRef]

25. Reilman, E.; Mars, R.A.T.; van Dijl, J.M.; Denham, E.L. The multidrug ABC transporter BmrC/BmrD of *Bacillus subtilis* is regulated via a ribosome-mediated transcriptional attenuation mechanism. *Nucleic Acids Res.* **2014**, *42*, 11393–11407. [CrossRef]

26. Chumsakul, O.; Takahashi, H.; Oshima, T.; Hishimoto, T.; Kanaya, S.; Ogasawara, N.; Ishikawa, S. Genome-wide binding profiles of the *Bacillus subtilis* transition state regulator AbrB and its homolog Abh reveals their interactive role in transcriptional regulation. *Nucleic Acids Res.* **2011**, *39*, 414–428. [CrossRef]

27. Nicolas, P.; Mader, U.; Dervyn, E.; Rochat, T.; Leduc, A.; Pigeonneau, N.; Bidnenko, E.; Marchadier, E.; Hoebeke, M.; Aymerich, S.; et al. Condition-Dependent Transcriptome Reveals High-Level Regulatory Architecture in *Bacillus subtilis*. *Science* **2012**, *335*, 1103–1106. [CrossRef]

28. Trach, K.A.; Hoch, J.A. Multisensory activation of the phosphorelay initiating sporulation in *Bacillus subtilis*: Identification and sequence of the protein kinase of the alternate pathway. *Mol. Microbiol.* **1993**, *8*, 69–79. [CrossRef]

29. Jiang, M.; Shao, W.; Perego, M.; Hoch, J.A. Multiple histidine kinases regulate entry into stationary phase and sporulation in *Bacillus subtilis*. *Mol. Microbiol.* **2000**, *38*, 535–542. [CrossRef]

30. Fukushima, S.; Yoshimura, M.; Chibazakura, T.; Sato, T.; Yoshikawa, H. The putative ABC transporter YheH/YheI is involved in the signalling pathway that activates KinA during sporulation initiation. *FEMS Microbiol. Lett.* **2006**, *256*, 90–97. [CrossRef]

31. Ohki, R.; Tateno, K.; Masuyama, W.; Moriya, S.; Kobayashi, K.; Ogasawara, N. The BceRS two-component regulatory system induces expression of the bacitracin transporter, BceAB, in *Bacillus subtilis*. *Mol. Microbiol.* **2003**, *49*, 1135–1144. [CrossRef] [PubMed]

32. Staroń, A.; Finkeisen, D.E.; Mascher, T. Peptide Antibiotic Sensing and Detoxification Modules of *Bacillus subtilis*. *Antimicrob. Agents Chemother.* **2011**, *55*, 515–525. [CrossRef] [PubMed]

33. Dintner, S.; Heermann, R.; Fang, C.; Jung, K.; Gebhard, S. A Sensory Complex Consisting of an ATP-binding Cassette Transporter and a Two-component Regulatory System Controls Bacitracin Resistance in *Bacillus subtilis*. *J. Biol. Chem.* **2014**, *289*, 27899–27910. [CrossRef] [PubMed]

34. Clemens, R.; Zaschke-Kriesche, J.; Khosa, S.; Smits, S.H.J. Insight into Two ABC Transporter Families Involved in Lantibiotic Resistance. *Front. Mol. Biosci.* **2018**, *4*, 91. [CrossRef] [PubMed]

35. Dintner, S.; Staron, A.; Berchtold, E.; Petri, T.; Mascher, T.; Gebhard, S. Coevolution of ABC Transporters and Two-Component Regulatory Systems as Resistance Modules against Antimicrobial Peptides in Firmicutes Bacteria. *J. Bacteriol.* **2011**, *193*, 3851–3862. [CrossRef] [PubMed]

36. Mascher, T. Intramembrane-sensing histidine kinases: A new family of cell envelope stress sensors in Firmicutes bacteria. *FEMS Microbiol. Lett.* **2006**, *264*, 133–144. [CrossRef]

37. Joseph, P.; Fichant, G.; Quentin, Y.; Denizot, F. Regulatory relationship of two-component and ABC transport systems and clustering of their genes in the *Bacillus/Clostridium* group, suggest a functional link between them. *J. Mol. Microbiol. Biotechnol.* **2002**, *4*, 503–513. [PubMed]

38. Bernard, R.; Guiseppi, A.; Chippaux, M.; Foglino, M.; Denizot, F. Resistance to Bacitracin in *Bacillus subtilis*: Unexpected Requirement of the BceAB ABC Transporter in the Control of Expression of Its Own Structural Genes. *J. Bacteriol.* **2007**, *189*, 8636–8642. [CrossRef]

39. Rietkötter, E.; Hoyer, D.; Mascher, T. Bacitracin sensing in *Bacillus subtilis*. *Mol. Microbiol.* **2008**, *68*, 768–785. [CrossRef]

40. Mascher, T.; Margulis, N.G.; Wang, T.; Ye, R.W.; Helmann, J.D. Cell wall stress responses in *Bacillus subtilis*: The regulatory network of the bacitracin stimulon. *Mol. Microbiol.* **2003**, *50*, 1591–1604. [CrossRef]

41. Koh, A.; Gibbon, M.J.; Van der Kamp, M.W.; Pudney, C.R.; Gebhard, S. Conformation control of the histidine kinase BceS of *Bacillus subtilis* by its cognate ABC-transporter facilitates need-based activation of antibiotic resistance. *Mol. Microbiol.* **2020**. [CrossRef] [PubMed]

42. Radeck, J.; Gebhard, S.; Orchard, P.S.; Kirchner, M.; Bauer, S.; Mascher, T.; Fritz, G. Anatomy of the bacitracin resistance network in B acillus subtilis. *Mol. Microbiol.* **2016**, *100*, 607–620. [CrossRef] [PubMed]

43. Fritz, G.; Dintner, S.; Treichel, N.S.; Radeck, J.; Gerland, U.; Mascher, T.; Gebhard, S. A New Way of Sensing: Need-Based Activation of Antibiotic Resistance by a Flux-Sensing Mechanism. *MBio* **2015**, *6*, e00975. [CrossRef] [PubMed]

44. Kingston, A.W.; Zhao, H.; Cook, G.M.; Helmann, J.D. Accumulation of heptaprenyl diphosphate sensitizes B acillus subtilis to bacitracin: Implications for the mechanism of resistance mediated by the BceAB transporter. *Mol. Microbiol.* **2014**, *93*, 37–49. [CrossRef] [PubMed]

45. Kobras, C.M.; Piepenbreier, H.; Emenegger, J.; Sim, A.; Fritz, G.; Gebhard, S. BceAB-Type Antibiotic Resistance Transporters Appear To Act by Target Protection of Cell Wall Synthesis. *Antimicrob. Agents Chemother.* **2019**, *64*, e02241-19. [CrossRef]

46. Jiang, X.; Geng, Y.; Ren, S.; Yu, T.; Li, Y.; Liu, G.; Wang, H.; Meng, H.; Shi, L. The VirAB-VirSR-AnrAB Multicomponent System Is Involved in Resistance of *Listeria monocytogenes* EGD-e to Cephalosporins, Bacitracin, Nisin, Benzalkonium Chloride, and Ethidium Bromide. *Appl. Environ. Microbiol.* **2019**, *85*, e01470-19. [CrossRef]

47. Grubaugh, D.; Regeimbal, J.M.; Ghosh, P.; Zhou, Y.; Lauer, P.; Dubensky, T.W.; Higgins, D.E. The VirAB ABC Transporter Is Required for VirR Regulation of *Listeria monocytogenes* Virulence and Resistance to Nisin. *Infect. Immun.* **2017**, *86*, e00901-17. [CrossRef]

48. Collins, B.; Curtis, N.; Cotter, P.D.; Hill, C.; Ross, R.P. The ABC Transporter AnrAB Contributes to the Innate Resistance of *Listeria monocytogenes* to Nisin, Bacitracin, and Various β-Lactam Antibiotics. *Antimicrob. Agents Chemother.* **2010**, *54*, 4416–4423. [CrossRef]

49. Kang, J.; Wiedmann, M.; Boor, K.J.; Bergholz, T.M. VirR-Mediated Resistance of *Listeria monocytogenes* against Food Antimicrobials and Cross-Protection Induced by Exposure to Organic Acid Salts. *Appl. Environ. Microbiol.* **2015**, *81*, 4553–4562. [CrossRef]

50. Zhu, X.; Long, F.; Chen, Y.; Knøchel, S.; She, Q.; Shi, X. A Putative ABC Transporter Is Involved in Negative Regulation of Biofilm Formation by *Listeria monocytogenes*. *Appl. Environ. Microbiol.* **2008**, *74*, 7675–7683. [CrossRef]

51. Mandin, P.; Fsihi, H.; Dussurget, O.; Vergassola, M.; Milohanic, E.; Toledo-Arana, A.; Lasa, I.; Johansson, J.; Cossart, P. VirR, a response regulator critical for *Listeria monocytogenes* virulence. *Mol. Microbiol.* **2005**, *57*, 1367–1380. [CrossRef] [PubMed]

52. Thedieck, K.; Hain, T.; Mohamed, W.; Tindall, B.J.; Nimtz, M.; Chakraborty, T.; Wehland, J.; Jänsch, L. The MprF protein is required for lysinylation of phospholipids in listerial membranes and confers resistance to cationic antimicrobial peptides (CAMPs) on *Listeria monocytogenes*. *Mol. Microbiol.* **2006**, *62*, 1325–1339. [CrossRef] [PubMed]

53. Staubitz, P.; Peschel, A. MprF-mediated lysinylation of phospholipids in *Bacillus subtilis*—Protection against bacteriocins in terrestrial habitats? *Microbiology* **2002**, *148*, 3331–3332. [CrossRef] [PubMed]

54. Heaton, M.P.; Neuhaus, F.C. Biosynthesis of D-alanyl-lipoteichoic acid: Cloning, nucleotide sequence, and expression of the Lactobacillus casei gene for the D-alanine-activating enzyme. *J. Bacteriol.* **1992**, *174*, 4707–4717. [CrossRef] [PubMed]

55. Kristian, S.A.; Dürr, M.; Van Strijp, J.A.G.; Neumeister, B.; Peschel, A. MprF-Mediated Lysinylation of Phospholipids in Staphylococcus aureus Leads to Protection against Oxygen-Independent Neutrophil Killing. *Infect. Immun.* **2003**, *71*, 546–549. [CrossRef]

56. Abachin, E.; Poyart, C.; Pellegrini, E.; Milohanic, E.; Fiedler, F.; Berche, P.; Trieu-Cuot, P. Formation of D-alanyl-lipoteichoic acid is required for adhesion and virulence of *Listeria monocytogenes*. *Mol. Microbiol.* **2002**, *43*, 1–14. [CrossRef]

57. Peschel, A.; Otto, M.; Jack, R.W.; Kalbacher, H.; Jung, G.; Götz, F. Inactivation of the dlt Operon in Staphylococcus aureus Confers Sensitivity to Defensins, Protegrins, and Other Antimicrobial Peptides. *J. Biol. Chem.* **1999**, *274*, 8405–8410. [CrossRef]

58. Chatterjee, S.S.; Hossain, H.; Otten, S.; Kuenne, C.; Kuchmina, K.; Machata, S.; Domann, E.; Chakraborty, T.; Hain, T. Intracellular Gene Expression Profile of *Listeria monocytogenes*. *Infect. Immun.* **2006**, *74*, 1323–1338. [CrossRef]

59. Camejo, A.; Buchrieser, C.; Couvé, E.; Carvalho, F.; Reis, O.; Ferreira, P.; Sousa, S.; Cossart, P.; Cabanes, D. In Vivo Transcriptional Profiling of *Listeria monocytogenes* and Mutagenesis Identify New Virulence Factors Involved in Infection. *PLoS Pathog.* **2009**, *5*, e1000449. [CrossRef]

60. Pazos, M.; Peters, K. Peptidoglycan. In *Subcellular Biochemistry*; Springer: Berlin/Heidelberg, Germany, 2019; pp. 127–168.

61. Egan, A.J.F.; Errington, J.; Vollmer, W. Regulation of peptidoglycan synthesis and remodelling. *Nat. Rev. Microbiol.* **2020**, *18*, 446–460. [CrossRef]

62. Nakagawa, J.; Tamaki, S.; Tomioka, S.; Matsuhashi, M. Functional biosynthesis of cell wall peptidoglycan by polymorphic bifunctional polypeptides. Penicillin-binding protein 1Bs of *Escherichia coli* with activities of transglycosylase and transpeptidase. *J. Biol. Chem.* **1984**, *259*, 13937–13946. [CrossRef]

63. Mayer, C. Bacterial Cell Wall Recycling. In *eLS*; John Wiley & Sons, Ltd.: Chichester, UK, 2012.

64. Mayer, C.; Kluj, R.M.; Mühleck, M.; Walter, A.; Unsleber, S.; Hottmann, I.; Borisova, M. Bacteria's different ways to recycle their own cell wall. *Int. J. Med. Microbiol.* **2019**, *309*, 151326. [CrossRef] [PubMed]

65. Benda, M.; Schulz, L.; Rismondo, J.; Stülke, J. The YtrBCDEF ABC transporter is involved in the control of social activities in *Bacillus subtilis*. *bioRxiv* **2020**. [CrossRef]

66. Yoshida, K.-I.; Fujita, Y.; Ehrlich, S.D. An Operon for a Putative ATP-Binding Cassette Transport System Involved in Acetoin Utilization ofBacillus subtilis. *J. Bacteriol.* **2000**, *182*, 5454–5461. [CrossRef] [PubMed]

67. Salzberg, L.I.; Luo, Y.; Hachmann, A.-B.; Mascher, T.; Helmann, J.D. The *Bacillus subtilis* GntR Family Repressor YtrA Responds to Cell Wall Antibiotics. *J. Bacteriol.* **2011**, *193*, 5793–5801. [CrossRef] [PubMed]

68. Wenzel, M.; Kohl, B.; Münch, D.; Raatschen, N.; Albada, H.B.; Hamoen, L.; Metzler-Nolte, N.; Sahl, H.-G.; Bandow, J.E. Proteomic Response of *Bacillus subtilis* to Lantibiotics Reflects Differences in Interaction with the Cytoplasmic Membrane. *Antimicrob. Agents Chemother.* **2012**, *56*, 5749–5757. [CrossRef]

69. Beckering, C.L.; Steil, L.; Weber, M.H.W.; Völker, U.; Marahiel, M.A. Genomewide Transcriptional Analysis of the Cold Shock Response in *Bacillus subtilis*. *J. Bacteriol.* **2002**, *184*, 6395–6402. [CrossRef]

70. Quentin, Y.; Fichant, G.; Denizot, F. Inventory, assembly and analysis of *Bacillus subtilis* ABC transport systems. *J. Mol. Biol.* **1999**, *287*, 467–484. [CrossRef]

71. Lopez, J.M.; Thoms, B. Beziehungen zwischen katabolischer Repression und Sporulation bei *Bacillus subtilis*. *Arch. Microbiol.* **1976**, *109*, 181–186. [CrossRef]

72. Huang, M.; Oppermann-Sanio, F.B.; Steinbüchel, A. Biochemical and Molecular Characterization of theBacillus subtilis Acetoin Catabolic Pathway. *J. Bacteriol.* **1999**, *181*, 3837–3841. [CrossRef]

73. Gardner, J.G.; Grundy, F.J.; Henkin, T.M.; Escalante-Semerena, J.C. Control of Acetyl-Coenzyme A Synthetase (AcsA) Activity by Acetylation/Deacetylation without NAD+ Involvement in *Bacillus subtilis*. *J. Bacteriol.* **2006**, *188*, 5460–5468. [CrossRef] [PubMed]

74. Gardner, J.G.; Escalante-Semerena, J.C. In *Bacillus subtilis*, the Sirtuin Protein Deacetylase, Encoded by the srtN Gene (Formerly yhdZ), and Functions Encoded by the acuABC Genes Control the Activity of Acetyl Coenzyme A Synthetase. *J. Bacteriol.* **2009**, *191*, 1749–1755. [CrossRef]

75. Grundy, F.J.; Waters, D.A.; Takova, T.Y.; Henkin, T.M. Identification of genes involved in utilization of acetate and acetoin in *Bacillus subtilis*. *Mol. Microbiol.* **1993**, *10*, 259–271. [CrossRef] [PubMed]
76. Thanh, T.N.; Jürgen, B.; Bauch, M.; Liebeke, M.; Lalk, M.; Ehrenreich, A.; Evers, S.; Maurer, K.-H.; Antelmann, H.; Ernst, F.; et al. Regulation of acetoin and 2,3-butanediol utilization in Bacillus licheniformis. *Appl. Microbiol. Biotechnol.* **2010**, *87*, 2227–2235. [CrossRef]
77. Grundy, F.J.; Turinsky, A.J.; Henkin, T.M. Catabolite regulation of *Bacillus subtilis* acetate and acetoin utilization genes by CcpA. *J. Bacteriol.* **1994**, *176*, 4527–4533. [CrossRef] [PubMed]
78. Koo, B.-M.; Kritikos, G.; Farelli, J.D.; Todor, H.; Tong, K.; Kimsey, H.; Wapinski, I.; Galardini, M.; Cabal, A.; Peters, J.M.; et al. Construction and Analysis of Two Genome-Scale Deletion Libraries for *Bacillus subtilis*. *Cell Syst.* **2017**, *4*, 291–305.e7. [CrossRef]
79. Toledo-Arana, A.; Dussurget, O.; Nikitas, G.; Sesto, N.; Guet-Revillet, H.; Balestrino, D.; Loh, E.; Gripenland, J.; Tiensuu, T.; Vaitkevicius, K.; et al. The Listeria transcriptional landscape from saprophytism to virulence. *Nature* **2009**, *459*, 950–956. [CrossRef]
80. Rismondo, J.; Schulz, L.M.; Yacoub, M.; Wadhawan, A.; Hoppert, M.; Dionne, M.S.; Gründling, A. EslB is required for cell wall biosynthesis and modification in *Listeria monocytogenes*. *J. Bacteriol.* **2020**. [CrossRef]
81. Burke, T.P.; Loukitcheva, A.; Zemansky, J.; Wheeler, R.; Boneca, I.G.; Portnoy, D.A. *Listeria monocytogenes* Is Resistant to Lysozyme through the Regulation, Not the Acquisition, of Cell Wall-Modifying Enzymes. *J. Bacteriol.* **2014**, *196*, 3756–3767. [CrossRef]
82. Durack, J.; Burke, T.P.; Portnoy, D.A. A prl Mutation in SecY Suppresses Secretion and Virulence Defects of *Listeria monocytogenes* secA2 Mutants. *J. Bacteriol.* **2015**, *197*, 932–942. [CrossRef]
83. Boneca, I.G.; Dussurget, O.; Cabanes, D.; Nahori, M.-A.; Sousa, S.; Lecuit, M.; Psylinakis, E.; Bouriotis, V.; Hugot, J.-P.; Giovannini, M.; et al. A critical role for peptidoglycan N-deacetylation in Listeria evasion from the host innate immune system. *Proc. Natl. Acad. Sci. USA* **2007**, *104*, 997–1002. [CrossRef] [PubMed]
84. Aubry, C.; Goulard, C.; Nahori, M.-A.; Cayet, N.; Decalf, J.; Sachse, M.; Boneca, I.G.; Cossart, P.; Dussurget, O. OatA, a Peptidoglycan O-Acetyltransferase Involved in *Listeria monocytogenes* Immune Escape, Is Critical for Virulence. *J. Infect. Dis.* **2011**, *204*, 731–740. [CrossRef] [PubMed]
85. Gueiros-Filho, F.J. A widely conserved bacterial cell division protein that promotes assembly of the tubulin-like protein FtsZ. *Genes Dev.* **2002**, *16*, 2544–2556. [CrossRef] [PubMed]
86. Sørensen, K.I.; Hove-Jensen, B. Ribose catabolism of Escherichia coli: Characterization of the rpiB gene encoding ribose phosphate isomerase B and of the rpiR gene, which is involved in regulation of rpiB expression. *J. Bacteriol.* **1996**, *178*, 1003–1011. [CrossRef] [PubMed]
87. Jaeger, T.; Mayer, C. The Transcriptional Factors MurR and Catabolite Activator Protein Regulate N-Acetylmuramic Acid Catabolism in Escherichia coli. *J. Bacteriol.* **2008**, *190*, 6598–6608. [CrossRef] [PubMed]
88. Kohler, P.R.A.; Choong, E.-L.; Rossbach, S. The RpiR-Like Repressor IolR Regulates Inositol Catabolism in Sinorhizobium meliloti. *J. Bacteriol.* **2011**, *193*, 5155–5163. [CrossRef] [PubMed]
89. Zhu, Y.; Nandakumar, R.; Sadykov, M.R.; Madayiputhiya, N.; Luong, T.T.; Gaupp, R.; Lee, C.Y.; Somerville, G.A. RpiR Homologues May Link Staphylococcus aureus RNAIII Synthesis and Pentose Phosphate Pathway Regulation. *J. Bacteriol.* **2011**, *193*, 6187–6196. [CrossRef]
90. Guzmán, K.; Campos, E.; Aguilera, L.; Toloza, L.; Giménez, R.; Aguilar, J.; Baldoma, L.; Badia, J. Characterization of the gene cluster involved in allantoate catabolism and its transcriptional regulation by the RpiR-type repressor HpxU in *Klebsiella pneumoniae*. *Int. Microbiol.* **2013**, *16*, 165–176. [CrossRef]
91. Aleksandrzak-Piekarczyk, T.; Stasiak-Różańska, L.; Cieśla, J.; Bardowski, J. ClaR—A novel key regulator of cellobiose and lactose metabolism in Lactococcus lactis IL1403. *Appl. Microbiol. Biotechnol.* **2015**, *99*, 337–347. [CrossRef]
92. Aleksandrzak-Piekarczyk, T.; Szatraj, K.; Kosiorek, K. GlaR (YugA)—A novel RpiR-family transcription activator of the Leloir pathway of galactose utilization in Lactococcus lactis IL 1403. *Microbiologyopen* **2019**, *8*, e00714. [CrossRef]
93. Van De Putte, P.; Van Dillewijn, J.; Rörsch, A. The selection of mutants of escherichia coli with impaired cell division at elevated temperature. *Mutat. Res. Mol. Mech. Mutagen.* **1964**, *1*, 121–128. [CrossRef]
94. Garti-Levi, S.; Hazan, R.; Kain, J.; Fujita, M.; Ben-Yehuda, S. The FtsEX ABC transporter directs cellular differentiation in *Bacillus subtilis*. *Mol. Microbiol.* **2008**, *69*, 1018–1028. [CrossRef] [PubMed]
95. Yamaguchi, H.; Furuhata, K.; Fukushima, T.; Yamamoto, H.; Sekiguchi, J. Characterization of a New *Bacillus subtilis* Peptidoglycan Hydrolase Gene, yvcE (Named cwlO), and the Enzymatic Properties of Its Encoded Protein. *J. Biosci. Bioeng.* **2004**, *98*, 174–181. [CrossRef]
96. Domínguez-Cuevas, P.; Porcelli, I.; Daniel, R.A.; Errington, J. Differentiated roles for MreB-actin isologues and autolytic enzymes in *Bacillus subtilis* morphogenesis. *Mol. Microbiol.* **2013**, *89*, 1084–1098. [CrossRef]
97. Liu, T.-Y.; Chu, S.-H.; Shaw, G.-C. Deletion of the cell wall peptidoglycan hydrolase gene *cwlO* or *lytE* severely impairs transformation efficiency in *Bacillus subtilis*. *J. Gen. Appl. Microbiol.* **2018**, *64*, 139–144. [CrossRef]
98. Hashimoto, M.; Ooiwa, S.; Sekiguchi, J. Synthetic Lethality of the lytE cwlO Genotype in *Bacillus subtilis* Is Caused by Lack of D,L-Endopeptidase Activity at the Lateral Cell Wall. *J. Bacteriol.* **2012**, *194*, 796–803. [CrossRef] [PubMed]
99. Brunet, Y.R.; Wang, X.; Rudner, D.Z. SweC and SweD are essential co-factors of the FtsEX-CwlO cell wall hydrolase complex in *Bacillus subtilis*. *PLoS Genet.* **2019**, *15*, e1008296. [CrossRef]
100. Crow, A.; Greene, N.P.; Kaplan, E.; Koronakis, V. Structure and mechanotransmission mechanism of the MacB ABC transporter superfamily. *Proc. Natl. Acad. Sci. USA* **2017**, *114*, 12572–12577. [CrossRef]

 *microorganisms*

*Review*

# Bacteriophages of Thermophilic 'Bacillus Group' Bacteria—A Review

Beata Łubkowska [1,2,*], Joanna Jeżewska-Frąckowiak [1], Ireneusz Sobolewski [1] and Piotr M. Skowron [1]

1 Department of Molecular Biotechnology, Faculty of Chemistry, University of Gdansk, Wita Stwosza 63, 80-308 Gdansk, Poland; j.jezewska-frackowiak@ug.edu.pl (J.J.-F.); ireneusz.sobolewski@ug.edu.pl (I.S.); piotr.skowron@ug.edu.pl (P.M.S.)
2 The High School of Health in Gdansk, Pelplinska 7, 80-335 Gdansk, Poland
* Correspondence: b.lubkowska@gmail.com

**Abstract:** Bacteriophages of thermophiles are of increasing interest owing to their important roles in many biogeochemical, ecological processes and in biotechnology applications, including emerging bionanotechnology. However, due to lack of in-depth investigation, they are underrepresented in the known prokaryotic virosphere. Therefore, there is a considerable potential for the discovery of novel bacteriophage-host systems in various environments: marine and terrestrial hot springs, compost piles, soil, industrial hot waters, among others. This review aims at providing a reference compendium of thermophages characterized thus far, which infect the species of thermophilic 'Bacillus group' bacteria, mostly from *Geobacillus* sp. We have listed 56 thermophages, out of which the majority belong to the *Siphoviridae* family, others belong to the *Myoviridae* and *Podoviridae* families and, apparently, a few belong to the *Sphaerolipoviridae*, *Tectiviridae* or *Corticoviridae* families. All of their genomes are composed of dsDNA, either linear, circular or circularly permuted. Fourteen genomes have been sequenced; their sizes vary greatly from 35,055 bp to an exceptionally large genome of 160,590 bp. We have also included our unpublished data on TP-84, which infects *Geobacillus stearothermophilus* (*G. stearothermophilus*). Since the TP-84 genome sequence shows essentially no similarity to any previously characterized bacteriophage, we have defined TP-84 as a new species in the newly proposed genus *Tp84virus* within the *Siphoviridae* family. The information summary presented here may be helpful in comparative deciphering of the molecular basis of the thermophages' biology, biotechnology and in analyzing the environmental aspects of the thermophages' effect on the thermophile community.

**Keywords:** bacteriophage; *Bacillus*; *Bacillus stearothermophilus*; *Geobacillus*; *Geobacillus stearothermophilus*; thermophile; bionanotechnology; probiotics; thermostable enzymes; TP-84

**Citation:** Łubkowska, B.; Jeżewska-Frąckowiak, J.; Sobolewski, I.; Skowron, P.M. Bacteriophages of Thermophilic 'Bacillus Group' Bacteria—A Review. *Microorganisms* **2021**, *9*, 1522. https://doi.org/10.3390/microorganisms9071522

Academic Editor: Imrich Barák

Received: 3 June 2021
Accepted: 12 July 2021
Published: 16 July 2021

**Publisher's Note:** MDPI stays neutral with regard to jurisdictional claims in published maps and institutional affiliations.



## 1. Introduction

Bacterial viruses (bacteriophages or phages) are the most abundant, genetically and evolutionary diversified biological entities in the biosphere [1]. They play key roles in the regulation and evolution of bacterial communities in essentially all ecosystems. They are found in all environments, including extreme ones, such as those associated with: very low or high temperatures, hot/cold deserts, hypersaline systems, low or high pH, such as soda lakes, high hydrodynamic pressures or containing poisonous elements or compounds, among others.

Here we focus on a narrow segment of such bacteriophages, meeting two criteria: (i) those living in a moderately high temperature segment of approximately 45–70 °C. Ecosystems providing such a condition are diverse and include: terrestrial hot springs, deep-sea hydrothermal vents areas, shallow ocean waters above geothermally active beds, human generated hot waters, hot deserts, compost piles, greenhouse soils, silage, rotting straw, river sludge, stable manure, digested sewage sludge, among others. These contain thermophilic bacterial communities and specific viruses that infect them (thermophilic bacteriophages or thermophages) [2,3] and (ii) thermophages infecting a diverse '*Bacillus* group' bacteria. A number of moderately thermophilic, spore-forming, aerobic bacteria with a growth range of 45–70 °C were classified into the genera *Aneurinibacillus*, *Alicyclobacillus*, *Sulfobacillus*, *Brevibacillus*, *Thermoactinomyces* and *Thermobacillus* [4–7]. However, genomes and proteomes have revealed that the majority of these thermophiles belong to the *Bacillus* sp. (*B.*) genetic groups 1 and 5 [8,9]. These findings further led to a more in-depth evaluation of thermophiles from group 5. In its course, it was determined, on the basis of high 16S rRNA sequence similarity (98.5–99.2%), that they form a phenotypically and phylogenetically coherent group of thermophilic bacilli. Thus, in 2001, the thermophiles belonging to *Bacillus* genetic group 5 were reclassified to be included into *Geobacillus* (*G.*) gen. nov. with the well-known *B. stearothermophilus* being assigned as the type strain *G. stearothermophilus* [9,10]. The *Geobacillus* name means earth or soil *Bacillus* and includes moderately thermophilic, aerobic or facultatively anaerobic, neutrophilic, motile, spore-forming rods [10]. The genus may also include some non-spore-forming bacteria, in which sporulation-related genes were inactivated by genome-integrated probacteriophage-like DNA segments [11]. The increasing number of discoveries of many novel thermophages, including those infecting the '*Bacillus* group', are providing a more complete understanding of thermophiles' biology, the mechanisms of biochemical adaptations needed for the life in high temperatures and the thermophilic host-bacteriophage evolution. From this perspective, especially interesting is the characterization of microbial and viral thermophilic communities in deep-sea hydrothermal vents fields, as this leads to deciphering the origin of ecosystems on the sea floor, the subsurface biosphere, the driving forces of evolution and the origin of life on the Earth [12,13]. Furthermore, it has been found that a horizontal gene transfer between thermophilic and mesophilic *Bacillus* species and bacteriophages takes place. This results in genetic mosaicism, as exemplified by the D6E thermophage, possibly leading to the contribution of substantial amounts of foreign DNAs by mobile DNA elements such as bacteriophages to entire microbial communities, including remote ones, such as the deep-sea vent community [14]. From a practical perspective, thermophilic *Bacillus* species are dominant bacterial workhorses in industrial fermentation processes, thus, these bacteria and their bacteriophages provide a novel source of genetic material and enzymes with great potential for application in biotechnology industry and science. Thus we provide an overview of the bacteriophages that infect thermophilic *Bacillus* sp. (many reclassified later as *Geobacillus* sp.), *Geobacillus* sp. and other related hosts that have been isolated from hot ecosystems around the world.

## 2. Environment Relation

Thermophilic bacilli, including *Geobacillus*, grow in essentially all environments, except those with ultra-harsh conditions, such as temperatures near or above boiling and extreme pH, among others. As mentioned above, geobacilli can be found in temperatures ranging from those typical for mesophiles (20–45 °C) up to those characteristic for moderate thermophiles (approximately 45–70 °C) in substantial numbers in ecological niches: geothermal areas, terrestrial hot springs, deep-sea hydrothermal vents areas, shallow ocean waters above geothermally active beds, human generated hot waters, hot deserts, compost piles, greenhouse soils, silage, rotting straw, river sludge, stable manure, digested sewage sludge, regardless of geographic location [15,16]. Especially rich sources of thermophilic *Geobacillus* bacteria, infected with bacteriophages, are present in humid organic matter, such as active compost piles and greenhouse soils due to active microbial metabolism and increased temperature, not reaching extreme values [2,17]. The *G. kaustophilus*-infecting thermophage GBK2 was found in a backyard compost pile after enrichment in liquid culture with *G. kaustophilus* (ATCC 8005) at 55 °C [18]. The minimum detectable growth temperature for *Geobacillus* isolates was described to be 45 °C [4], although in our hands, laboratory conditions allowed for a slow growth of. *G. stearothermophilus* strain 10 also in the mesophilic range. As typical soils rarely reach temperatures above 45 °C, this seems to be an evolutionary adaptation, allowing *Geobacillus* to grow extremely slowly, until more environmentally favorable growth temperatures occur. On the other hand, the temperatures in active organic matter easily exceed 45 °C. Very diverse conditions are present in hot springs: their temperatures range from 40 °C to 98 °C, and the bacteria capable of growing there, including geobacilli, are classified either as moderate thermophiles (40–71 °C) or hyperthermophiles (72–98 °C) [19]. Furthermore, the hot springs' pH values vary widely, depending on hot water dissolving components of various minerals—from pH 1 to 9—and are defined in three functional categories [20]: acidic (pH 1–5), approximately neutral (pH 6–7.5) and alkaline (pH > 7.5). This translates to a high diversity of '*Bacillus* group' hosts and their specific bacteriophages, which are able to infect most of the major taxonomic groups of *Bacillus* thermophiles [21], regardless of the type of ecosystem. For example, two thermophages—*Bacillus* virus BVW1 and *Geobacillus* virus GVE1—were isolated from deep-sea hydrothermal fields in the east Pacific [22]. Thermostability assays showed that GVE1, whose host is *Geobacillus* sp. E26323, isolated from the same location, was most stable at 60 °C [23]. During the cultivation of *Geobacillus* sp. strain E263, many uncharacterized bacteriophage plaques were observed, indicating the presence of a rich host-bacteriophage community. Another deep-sea hydrothermal fields thermophage GVE2 was found in geobacilli cultured at 65 °C [14,24]. Additionally, a *Geobacillus* thermophage D6E was characterized in deep-sea hydrothermal vent communities, where viruses play very important roles [14]. Yellowstone National Park is one of the best studied sites regarding the thermophages that infect the species of *Geobacillus Thermus* as well as *Archea*, among others. This is due to the fact that Yellowstone has one of the highest concentrations of hot springs in the world, which vary greatly in terms of temperature, pH and minerals content [10]. Bacteriophage GBSV1, infecting *Geobacillus* sp. 6k51 was isolated from a shallow (100 m depth) off-shore hot spring [25]. Similarly, the bacteriophage φOH2, which infects *G. kaustophilus* NBRC 102445(T), was isolated from the sediment of a hot spring [26] and bacteriophage AP45 and its host strain *Aeribacillus* sp. CEMTC656 were isolated from the Valley of Geysers, Kamchatka, Russia [27].

Regardless of the environmental diversity, the morphologies and compositions of the '*Bacillus* group' thermophilic bacteriophages are less diverse than that of the overall community of thermophages. In general, the thermophilic bacteriophages that have been found in all taxonomic groups of *Bacteria* and *Archaea*, exhibit diverse morphologies (e.g., tailed capsids, tail-less capsids, filamentous, lipid-containing), unique structures (e.g., very long-tailed *siphoviruses*), and extend through five taxonomic families: *Myoviridae*, *Siphoviridae*, *Inoviridae*, *Tectiviridae* and *Sphaerolipoviridae* [3]. Interestingly, in the diverse '*Bacillus* group' bacteria, the bacteriophages found thus far belong to a limited number of families: vastly dominating *Siphoviridae* (head–tail bacteriophages with non-contractile tails), *Myoviridae* (head–tail bacteriophages with contractile tails), *Podoviridae* (very short tail) and, possibly scarce, *Sphaerolipoviridae*, *Tectiviridae* or *Corticoviridae* members (icosahedral, with a lipid membrane) (Figure 1). All of them contain dsDNA genome, either linear or circular, or circularly permitted. On the whole, the ubiquity of thermophages and their impact on various trophic levels seems to have a powerful direct influence on diverse microbial communities that play a crucial role in biogeochemical cycles. All this, in turn, plays a role in altering those cycles [27–29]. As a result, thermophages can affect/stimulate the modifications of those cycles.

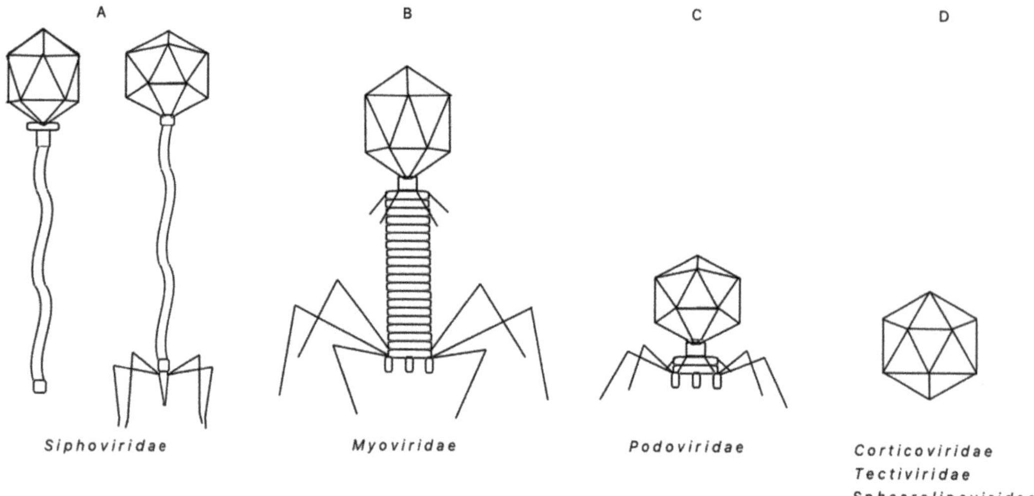


**Figure 1.** Schematic representation of morphological/structural forms of the *'Bacillus* group' thermophages. (**A**). *Siphoviridae;* (**B**). *Myoviridae;* (**C**). *Podoviridae;* (**D**). *Sphaerolipoviridae, Tectiviridae, Corticoviridae.*

### 3. *'Bacillus* Group' Bacteriophages—Chronological Review

The thermophilic *'Bacillus* group' bacteriophages, reviewed here, comprise 56 species, described in varying detail in original publications (Table 1). Although there have certainly been more such bacteriophages detected, they have not been sufficiently analyzed or have been just briefly mentioned in publications. They infect various *Geobacillus* (previously *Bacillus*) sp., *'B. stearothermophilus'*, *B. acidocaldarius*, *B. caldotenax*, *B. caldolyticus*, *G. kaustophilus*, *G. thermoglucosidasius*, *G. icigianus*, *B. pseudoalcaliphilus*, *B. bogoriensis*, *B. pseudofirmus*, *B. thermocatenulatus*, *B. cohnii*, undefined *Bacillus* sp. and *Aeribacillus* sp. As older species determinations are not always precise, we tend to retain the original author's nomenclature to avoid confusion and for the reader's convenience, when reading original publications. Some of these thermophages were discovered and analyzed as long ago as the 1950s, 1960s and 1970s, using the techniques available at the time. The data thus obtained is often incomplete or contains errors (such as the nearly 2-fold underestimation of TP-84's genome size) especially at the molecular level. Apparently, the earliest thorough/reliable report is that of the thermophage TP-84, discovered in 1952 and preliminarily characterized in the 1960s and 1970s [30]. Much later, its genome was sequenced, analyzed and its proteomics data was established by our group in 2018 [2] (Table 1). Below, we chronologically review the major features of each thermophage and its interactions with its corresponding host. Interestingly, all 56 bacteriophages contain dsDNA genomes. The hosts and bacteriophages were cultivated typically on rich, 'nutrient broth'-type media at approximately pH 7, unless otherwise noted.

**Table 1.** New (this work) experimentally confirmed functional CDSs of bacteriophage TP-84 genome—an update 1.

| CDS Name | CDS Length (bp) | Location in the Genome (bp) | CDS Arbitrary Orientation | Polypeptide Length (aa) | Predicted Polypeptide Molecular Weight (kDa) | Experimentally Determined Polypeptide Molecular Weight (kDa) | Predicted Isoelectric Point | Hypothetical Function (Analysis) | Confirmed by Proteomic Analysis |
|---|---|---|---|---|---|---|---|---|---|
| TP84_01 | 567 | 13–579 | + | 188 | 22.4 | 22.4 | 7.76 | terminase, small subunit | terminase, small subunit |
| TP84_02 | 1299 | 576–1874 | + | 432 | 50.0 | 50.1 | 7.74 | terminase, large subunit | terminase, large subunit |
| TP84_04 | 354 | 2586–2939 | + | 117 | 13.6 | ND | 4.36 | unknown | unknown |
| TP84_15 | 324 | 10358–10681 | + | 107 | 11.9 | ND | 6.26 | unknown | unknown |
| TP84_20 | 375 | 12687–13061 | + | 124 | 14.7 | 14.7 | 9.80 | unknown | unknown |
| TP84_26 | 2976 | 20179–23154 | + | 991 | 112.2 | 112.9 | 5.49 | glycosylase | glycosylase |
| TP84_27 | 432 | 23240–23671 | + | 143 | 15.5 | 15.5 | 6.20 | holin | holin |
| TP84_30 | 573 | 25329–25901 | + | 190 | 21.7 | 21.7 | 4.98 | unknown | unknown |
| TP84_33 | 291 | 26597–26887 | + | 96 | 11.0 | ND | 5.10 | unknown | unknown |
| TP84_41 | 273 | 28824–29096 | + | 90 | 10.6 | ND | 9.69 | unknown | unknown |
| TP84_43 | 309 | 29330–29639 | + | 102 | 11.0 | ND | 9.81 | unknown | unknown |
| TP84_48 | 2265 | 30527–32791 | + | 754 | 86 | ND | 7.09 | unknown | unknown |
| TP84_49 | 474 | 32871–33344 | + | 157 | 18.0 | ND | 5.85 | unknown | unknown |
| TP84_58 | 318 | 37942–38259 | + | 105 | 12.3 | ND | 4.92 | replicative helicase inhibitor | replicative helicase inhibitor |
| TP84_59 | 1305 | 38259–39563 | + | 434 | 49.1 | 49.2 | 5.53 | replicative DNA helicase | replicative DNA helicase |
| TP84_60 | 672 | 39632–40303 | + | 223 | 26.4 | 26.4 | 8.41 | unknown | unknown |
| TP84_61 | 435 | 40281–40715 | + | 144 | 16.7 | ND | 9.08 | HNH homing endonuclease | HNH homing endonuclease |
| TP84_70 | 330 | 43704–44033 | + | 109 | 12.7 | ND | 9.47 | unknown | unknown |
| TP84_73 | 264 | 44746–45009 | + | 87 | 10.4 | ND | 5.60 | unknown | unknown |

### 3.1. 'Thermophilic Lytic Principle' (Family/Genus Not Determined (ND), Host Putative G. stearothermophilus)

This is apparently the earliest discovered and reported thermophilic bacteriophage, with no assigned name, just mentioned as thermophilic lytic principle [31]. Its host, T60 bacteria was described as medium length, Gram-positive rods, isolated from milk, forming spores and growing between 20 °C and 60 °C, with the optimum at 45–52 °C, which suggests that the host belongs to *G. stearothermophilus*. The thermophilic lytic principle was capable of inoculating between series of cultures and was optimally lysing the host's cells at 52–55 °C; the lysis capability declining sharply at temperatures of 60 °C and with no lysis occurring at 62 °C [31].

### 3.2. 'Thermophilic Bacteriophage' (1) (Family/Genus ND, Host Thermophilic Bacillus sp.)

The bacteriophage, isolated by Adant in 1928 [32], had similar properties to the thermophilic lytic principle [30]. It was optimally active at 52 °C and was destroyed in 30 min. at 75 °C.

### 3.3. 'Thermophilic Bacteriophage' (2) (Family/Genus ND, Host 'Thermophilic Bacterium No. 10')

Scarce data was published concerning thermophilic bacteriophage [33]. It was found to infect a thermophilic bacterium no. 10, aerobic, Gram-positive, isolated from thermal springs in Yellowstone National Park, as being or closely resembling *B. stearothermophilus* [34]. The bacteria exhibited temperature growth range of 35–75 °C and the bacteriophage was propagated with stimulating adsorption $Ca^{2+}$ ions at 50–70 °C, with optimum at 65 °C, and a latent period of 80–90 min. The bacteriophage produced a halo around clear plaques, suggesting the existence of a diffusing envelope depolymerase. It exhibited substantial thermostability. At 2 h incubation at 95 °C in its growth media, significant phage titer remained, whereas at 100 °C after 30 min, some infective particles were still detectable. Transmission electron microscopy (TEM) images showing the bacteriophage morphology suggest that it belongs to the *Siphoviridae* family [33,35].

### 3.4. 'Thermophilic Bacteriophage' (3) (Family/Genus ND, Host Thermophilic Bacillus sp.)

One of the first isolated thermophilic bacteriophages included an undefined species of very high thermostability. As a host served gram-positive bacillus, spore-forming, strictly aerobic, capable of growth at over 70 °C, isolated from a compost pile in Tokyo, Japan [36]. The bacteriophage was capable of lysing a host's cells at between 55 °C and 70 °C, optimally at 65 °C, with a burst size of 205–220. Remarkably, that bacteriophage exhibited a very high thermostability. Suspended in its cultivation media, it was capable of surviving over 120 min at 100 °C with approximately a 1000-fold decrease in a titer. The bacteriophage TEM analysis has shown spherical particles of approximately 20 nm diameter with a tendency to aggregate. Aside from that, the results of the coding nucleic acid determination are questionable. The author claims, as based on chemical analysis, that the bacteriophage is composed of 74% nucleic acids, both DNA—42%, and RNA—32% [37,38].

### 3.5. TP-84 (Siphoviridae Family, Tp84virus Genus, Host B. stearothermophilus)

Early studies determined a number of microbiological and physical properties of bacteriophage TP-84. This was discovered in 1952 in greenhouse soil, infecting *B. stearothermophilus* strain 2184 (according to current classification—*G. stearothermophilus*) [30]. TP-84 exhibits high host specificity. Out of +24 related strains of *B. stearothermophilus*, two unclassified thermophilic *Bacillus* strains, T-27 and 194, also supported TP-84 infection as well as *G. stearothermophilus* strain 10 and 4 supporting TP-84 growth. The most effective host was *G. stearothermophilus* strain 10, thus it was further used for more detailed studies [39]. Here, we evaluated more *G. stearothermophilus* hosts that have been used by others for recombinant DNA studies in these bacteria, including: NUB3625 9A5 and NUB R [40]. These hosts supported TP-84 growth but much less effectively than *G. stearothermophilus* 10. The cultivation conditions, optimal for TP-84, included a rich media of pH 6.5, supplemented with fructose or glucose as a sporulation suppressor and essential $Ca^{2+}$ and $Mg^{2+}$ ions. TP-84 can reach a high titer of app. $10^{12}$. However, omitting the sugar addition decreased the titer approximately 1000-fold [39], while omitting divalent cations had a dramatic effect on final TP-84 yields, causing an approximate 200-fold decrease in titer [41]. The temperature growth range of 43–76 °C (optimal 55–60 °C) matched that of the hosts growth range. The latent period was 22–24 min. Clear plaques were surrounded by a turbid 'halo' ([41] Lubkowska et al., unpublished data), indicating diffusion of an enzyme, possibly glycosylase-depolymerase, degrading the bacterial envelopes. The bacteriophage has a hexagonal head, 53 nm in diameter with a 131 nm long 3–5 nm wide tail. It has a Svedberg coefficient of 436 and bands in CsCl at a density of 1.508 g/cm$^3$ (pH 8.5). The sedimentation and diffusion data allowed for calculation of the TP-84 particle molecular weight of approximately 50 MDa [39]. In our previous work, we characterized and sequenced the genome, and established the proteomics data on TP-84 [2]. TP-84 contains 47.7 kb dsDNA, 42% G+C genome. The bioinformatics analysis revealed the presence of 81 coding sequences (CDSs), coding for polypeptides of 4 kDa or larger. Interestingly, all CDSs are oriented in the same direction, pointing to a dominant transcription direction of one DNA strand only. Based on homology analysis, a hypothetical function could be assigned to 31 CDSs and a total of 33 biosyntheses of proteins was confirmed experimentally [2]. To further confirm the genome bioinformatics data, we analyzed: (i) purified TP-84 bacteriophage particles, subjected to SDS-PAGE, resolving individual proteins and (ii) purified TP-84 proteins from TP-84/*G. stearothermophilus* lysates, followed by LC-MS analysis (to be published elsewhere). Nineteen new TP-84-encoded proteins were found (Table 1) with at least two peptides sequence coverage. Those include: (i) proteins involved in DNA replication/packaging—TP84_01 (putative terminase, small subunit), TP84_02 (putative terminase, large subunit), TP84_59, replicative DNA helicase and TP84_58, replicative helicase inhibitor; (ii) TP84_27—holin, involved in cell membrane disintegration during host lysis; (iii) TP84_61—HNH homing endonuclease, which may be involved in recombination processes; (iv) TP84_26—glycosylase, possibly involved in disintegration of a very thick host's envelope. This is the largest protein of TP-84 (112 kDa) and, interestingly, it shows homology to a number of proteins from *Bacillus* and *Geobacillus* species, and one *Geobacillus* bacteriophage GBK2 protein (GenBank YP_009010491). No RNA or DNA polymerase-coding genes were found on the TP-84 genome, thus the bacteriophage relies on the host's transcription and replication enzymes. The TP-84 genome has a 'condensed' genome with CDSs, typically spaced by several-to-tens of bp only and often overlapping. The genome contains five putative promoter-like sequences resembling the host promoter consensus sequence with allowed 2-bp mismatches. Furthermore, ten putative rho-independent terminators were found. The TP-84 genome sequence shows essentially no similarity to any previously characterized bacteriophage, thus, TP-84 was established as a holotype of a new genus *Tp84virus* within the *Siphoviridae* family [2].

### 3.6. φμ-4 (Family/Genus Not Determined, Host B. stearothermophilus)

Bacteriophage φμ-4 was isolated from lysogenic *B. stearothermophilus* NU strain 4, and was propagated in *B. stearothermophilus* 10. It was cultivated at 50–65 °C with required $Ca^{2+}$ ions. The latent period at 65 °C was 35 min with a burst size of 175. TEM images have shown the bacteriophage particles size of 88–100 nm. The particles shape was 'roughly spherical' [42].

### 3.7. TP-1 (Putative Siphoviridae Family, Host B. stearothermophilus)

TP-1 bacteriophage was isolated as a mitomycin- or UV-induced lysogenic *B. stearothermophilus* 1503-4R, growing in the narrow temperature range of 50–65 °C, with optimum at 55 °C. $Ca^{2+}$ ions enhanced the bacteriophage propagation. TP-1 formed turbid plaques when titrated on *B. stearothermophilus* 4S. Furthermore, a clear-plaque mutant of TP-1C was found. Lysogen induction was rapid, causing a culture lysis within 45–60 min. Spontaneous induction was detected at a rate of 1 particle/app. $3 \times 10^6$ cells. Morphologically, TP-1 has a head of 65 nm and a long, flexible tail of 240 nm and 12 nm wide, which suggests that TP-1 belongs to the *Siphoviridae* family. The bacteriophage genome is composed of 42% G+C, dsDNA, with estimated molecular weight of $1.21 \times 10^7$ [43,44], which corresponds to approximately 18,516 bp. One of the authors [45] isolated a number of thermophilic bacteriophages, exemplified by TP-42, TP-56 and TP-68, but no characterization was provided.

### 3.8. ST1 (Siphoviridae Family, Genus ND, Host B. stearothermophilus)

Bacteriophage ST1 was propagated at 60 °C in *B. stearothermophilus* S13, forming clear plaques. TEM micrographs showed the head of ST1 as a regular icosahedral form of 50 nm in diameter, a tail of 100 nm length and 2 nm wide, ended with a terminal plate. The genetic material was dsDNA, containing 43% G+C [46].

### 3.9. Tφ3 (Tphi3) (Siphoviridae Family, Genus ND, Host G. stearothermophilus)

Tphi3 bacteriophage was isolated from a soil sample and infects the thermophilic *B. stearothermophilus* ATCC 8005. Along with Tphi3, there were two more bacteriophages isolated, apparently infecting *B. stearothermophilus*, not further characterized. Tphi3 is highly host-specific, as it does not infect *B. stearothermophilus* ATCC 7953, 7954, 10,149, 12,016. Optimal growth occurs at 60 °C with a latent period of 18 min and burst size of approximately 200. The Tphi3 exhibits high thermostability, up to 75 °C with a half-life of 12 min. Prolonged plates incubation resulted in a formation of a turbid 'halo' around clear plaques, suggesting a secretion of an enzyme, depolymerase, degrading the bacterial envelope. TEM imaging of Tphi3 shows that the bacteriophage has a 57 nm long regular icosahedron head. Edges of the head are 29 nm long, the tail is 125 nm long and 10 nm wide. The tail contains a stacked-ring structure. There are about 30 cross-strations that are spaced at 3.9-nm intervals along the tail and TEM images also suggest that the tail end contains small fibers. The buoyant density of the bacteriophage in a cesium chloride density gradient was 1.526 g/mL. $Ca^{2+}$ ions were required for efficient adsorption and/or growth [47]. Further reported Tphi3 DNA studies, indicated that its average genome length was 12.1 mm, which corresponds to app. 23.2 MDa or app. 35,700 bp with G+C content of 40.2% [48].

### 3.10. GH5, GH8 (Family/Genus ND, Host B. stearothermophilus)

A brief report by Humbert and Fields (1972) showed two bacteriophages, GH5 and GH8, with distinct features, both infecting *B. stearothermophilus* NCA1518 [49]. Both share the same plaque-forming temperature range of 42.5–67 °C. Bacteriophage ND8 was examined under TEM and exhibited an icosahedral head with a diameter of approximately 100 nm, a long tail of 330 nm and approximately 10 nm wide, thus, apparently belonging to the *Siphoviridae* family. The latent periods were 47 min (GH5) or 35 min (GH8), burst size 51 (GH5) or 72 (GH8), buoyant density 1.473 (GH5) or 1.506 (GH8), sensitivity to osmotic shock from 2 M sucrose (both), sensitivity to osmotic shock from 4 M NaCl (GH5 no, GH8 yes), extrapolated 1 min half-life 88 °C (GH5) or 76 °C (GH8), maximum half-life at pH 9.0 (GH5) or 7.0 (GH8) [49].

### 3.11. PhB1 (Siphoviridae Family, Genus ND, Host Thermophilic Bacillus sp.)

PhB1 was isolated from a farm soil sample, using a soil-isolated bacterium as a host, described by the authors as facultatively thermophilic *Bacillus* sp. strain B1, Gram-positive rods, aerobic, spore-forming, motile with temperature growth range approximately 30–60 °C and optimum between 45–50 °C. The bacteriophage was cultivated on the hosts, growing at 55 °C. Small plaques were obtained with PhB1 and had a clear center with a diameter less than one millimeter, surrounded by a turbid halo [50], which suggests the presence of secreted envelope depolymerases. It was determined that $Ca^{2+}$ ions and $Mg^{2+}$ ions were stabilizing the bacteriophage particles against heat inactivation. TEM showed a long-tailed particle with a probable octahedral head and small spikes at the tail tip.

### 3.12. D5–D8 (Family/Genus ND, Host B. stearothermophilus)

A report by Reanney and Marsch (1973) presented 4 bacteriophages: D5, D6, D7, D8 infecting thermophilic *B. stearothermophilus* NRS T91 and ATCC7953. The hosts were grown on rich media supplemented with glucose and $Ca^{2+}$ ions. Remarkably, the plaque-forming range for the bacteriophages extended down to the mesophilic region of 30–37 °C, with an optimum at 45 °C and reaching the upper limit at 55 °C. Bacteriophage proliferation in soil was greatest at 45 °C, while at 55 °C this was much less intense. The authors raised a hypothesis concerning the ecology of bacteriophages infecting the thermophilic soil organism *B. stearothermophilus*. The authors suggested that, in soil, these bacteria and their bacteriophages proliferate optimally at mesophilic rather than thermophilic temperatures. They proposed that, in the field, the vast majority of the metabolizing population of *B. stearothermophilus* are mesophilic in their biochemistry and that the 'thermophilic' strains, maintained in laboratories, represent a selection obtained by the isolation procedure of a minority of atypical, thermophilic cells [51].

### 3.13. φNS11 (PhiNS11) (Putative Sphaerolipoviridae, Tectiviridae or Corticoviridae Family, Host Acidophilic-Thermophilic B. acidocaldarius)

φNS11 is a 'double' extremophilic bacteriophage, as it infects both acidophilic and thermophilic bacterium, *A. acidocaldarius* strain TA6, growing in the Beppu hot springs in Japan. The bacteriophage was propagated in a pH range of 2 to 5 and exhibited optimum growth at 60 °C and pH 3.5. Above pH 6, it became unstable. φNS11 has a polyhedral shape, is tail-less and has a lipid-containing capsid with a diameter of 60–70 nm and spike-like structures. SDS-PAGE showed that φNS11 contains only six proteins. The genome was determined to be composed of dsDNA. The buoyant density of the bacteriophage was relatively low, 1.27 g/cm$^3$, similar to that of the lipid-containing phages PM2, PR4, φ6. Furthermore, it was highly sensitive to organic solvents. The fatty acid composition of the bacteriophage was slightly different from that of the host bacterium. However, o-cyclohexyl fatty acids typical of *B. acidocaldarius*, were also found in φNS11 as its main components [52]. These features point to classification of φNS11 as belonging to the *Sphaerolipoviridae*, *Tectiviridae* or *Corticoviridae* families.

### 3.14. JS001-JS027 Series (Sections 3.14–3.37 Below)

Sharp et al. (1986) have discovered and partially characterized 24 thermophilic bacteriophages, designated JS001-JS027, isolated from diversified sources: compost piles, soil, mud, sewage and river sludges, stable manure, silage and rotting straw, capable of infection of most of the major taxonomic groups of *Bacillus* thermophiles [21]. Various species/strains were used for the bacteriophages detection and propagation, including: '*B. caldotenax*' DSM 406, DSM 411, '*B. caldolyticus*' DSM 405, the *Bacillus* thermophile RS 93, *B. stearothermophilus* NCA 1503, NW 10, EP 240, EP 262, EP 136, DSM 2334, ATCC 12016 [53,54], streptomycin-resistant strains of: *B. stearothermophilus* NCA 1503, *Bacillus* thermophiles RS 239, RS 240, RS 241 and RS 242 [55]. The bacteriophages were uniformly detected/propagated in rich media of pH app. 7, and supplemented with $Ca^{2+}$ ions, at 55 °C. They exhibited significant differences in capsid morphology, comprising 4 groups: (i) comparatively short stubby tail and no evidence of a tail plate; (ii) icosahedral heads of approximately 40–65 nm in diameter with tails between 120 nm and 135 nm long and no evidence of a tail plate; (iii) icosahedral head of 85–90 nm in diameter and a long (400 nm) flexible tail with helical symmetry; (iv) cylindrical heads with rounded ends, with tails varying between 115 nm and 120 nm long. All the genomes were determined as dsDNA. A number of isolates were lost when chloroform was used to sterilize them, indicating vital lipids content or denaturation-sensitive capsid protein(s). Table 2 contains selected data, concerning JS001–JS027.

**Table 2.** Classification and major data describing '*Bacillus* group' thermophilic bacteriophages (chronologically as discovered).

| N° | Bacteriophage Species | Virus Family, Genus | GenBank Accession Number | Host (Used for Propagation) | Genome Type and Size [bp] | ORFs | G+C [%] | Original Discovery Reference | Isolate Location | Life Cycle | Growth Temperature (°C, Optimal/Range) and pH (Optimal/Range) |
|----|----|----|----|----|----|----|----|----|----|----|----|
| 1 | 'thermophilic lytic principle' | ND | ND | *B. stearothermophilus* T60 | ND | ND | ND | Koser, 1926 | sewage polluted river water (USA) | lytic | 52–60, app. 7 |
| 2 | 'thermophilic bacteriophage' | ND | ND | ND | ND | ND | ND | Adant, 1928 | ND | lytic | 52–55 |
| 3 | 'thermophilic bacteriophage' | ND | ND | *thermophilic bacterium* no. 10 | ND | ND | ND | White et al., 1954, 1955 | greenhouse soil (USA) | lytic | 65 (50–70), 7 |
| 4 | 'thermophilic bacteriophage' | ND | ND | *Bacillus* sp. | DNA, RNA(?) | ND | ND | Onodera; 1959; Onodera 1961 | compost (Japan) | lytic | 65 (55–70), 7.2 |
| 5 | TP-84 | *Siphoviridae, Tp84virus* | KY565347.1 | *B. stearothermophilus* strain 10 | circular, dsDNA, 47,718 | 81 | 54.5 | Saunders & Campbell, 1964 | greenhouse soil (USA) | lytic | 58 (43–76), 7.2 |
| 6 | φμ-4 | ND | ND | *B. stearothermophilus* | ND | ND | ND | Shafia & Thompson, 1964 | ND | lytic/lysogenic | 50–65, 7 |
| 7 | TP-1 | *Siphoviridae* (putative) | ND | *B. stearothermophilus* | dsDNA, app. 18,516 bp, (MW 12.1 Mda) | ND | 42 | Welker and Campbell, 1965 | ND | lysogenic/lytic | 55 (50–65), 7 |
| 8 | ST1 | *Siphoviridae Myoviridae* (putative) | ND | *B. stearothermophilus* strain S13 | dsDNA | ND | 43 | Carnevali & Donelli, 1968 | ND | lytic | 60, app. 7 |
| 9 | Tφ3 | *Siphoviridae* | ND | *B. stearothermophilus* ATCC 8005 S^R | ds DNA, app. 35,700 (MW app. 23.2 MDa) | ND | 40.2 | Egbert&Mitchel, 1967; Egbert, 1969 | soil (USA) | lytic | 60, 7.3 |
| 10 | GH5 | ND | ND | *B. stearothermophilus* NCA1518 | ND | ND | ND | Humbert & Fields, 1972 | greenhouse soil (USA) | lytic | 42.5–67, app. 7 |
| 11 | GH8 | *Siphoviridae* | ND | *B.stearothermophilus* NCA1518 | ND | ND | ND | Humbert & Fields, 1972 | greenhouse soli (USA) | lytic | 42.5–67, app. 7 |
| 12 | PhB1 | *Siphoviridae* | ND | *Bacillus* sp. strain B | ND | ND | ND | Junger & Edebo, 1972 | farm soil (Sweden) | lytic | 55, 7.3 |
| 13 | D5 | ND | ND | *B. stearothermophilus* NRS T91, ATCC7953 | ND | ND | ND | Reanney & Marsch, 1973 | ND | lytic | 45 (30–55), app. 7 |
| 14 | D6 | ND | ND | *B. stearothermophilus* NRS T91, ATCC7953 | ND | ND | ND | Reanney & Marsch, 1973 | ND | lytic | 45 (30–55), app. 7 |
| 15 | D7 | ND | ND | *B. stearothermophilus* NRS T91, ATCC7953 | ND | ND | ND | Reanney & Marsch, 1973 | ND | lytic | 45 (30–55), app. 7 |
| 16 | D8 | ND | ND | *B. stearothermophilus* NRS T91, ATCC7953 | ND | ND | ND | Reanney & Marsch, 1973 | ND | lytic | 45 (30–55), app. 7 |
| 17 | φNS11 | *Sphaerolipoviridae,* (putative) | ND | *B.acidocaldarius* TA6 | dsDNA | ND | ND | Sakaki & Oshima, 1976 | hot spring (Beppu, Japan) | lytic | 60, 3.5 (2–5) |

**Table 2.** *Cont.*

| N° | Bacteriophage Species | Virus Family, Genus | GenBank Accession Number | Host (Used for Propagation) | Genome Type and Size [bp] | ORFs | G+C [%] | Original Discovery Reference | Isolate Location | Life Cycle | Growth Temperature (°C, Optimal/Range) and pH (Optimal/Range) |
|---|---|---|---|---|---|---|---|---|---|---|---|
| 18 | JS001 | ND | ND | *B. caldotenax* | dsDNA | ND | ND | Sharp et al., 1986 | ND | lytic/lysogenic | 55 (50–70), 7.3 ± 0.2 |
| 19 | JS004 | ND | ND | *Bacillus thermophile* RS 239 | dsDNA | ND | ND | Sharp et al., 1986 | silage | lytic | 55 (50–70), 7.3 ± 0.2 |
| 20 | JS005 | ND | ND | *B. thermophile* RS 239 | dsDNA | ND | ND | Sharp et al., 1986 | rotting straw | lytic | 55 (50–70), 7.3 ± 0.2 |
| 21 | JS006 | ND | ND | *Bacillus thermophile* RS 239 | dsDNA | ND | ND | Sharp et al., 1986 | compost | lytic | 55 (50–70), 7.3 ± 0.2 |
| 22 | JS007 | ND | ND | *Bacillus thermophile* RS 240 | dsDNA | ND | ND | Sharp et al., 1986 | silage | lytic | 55 (50–70), 7.3 ± 0.2 |
| 23 | JS008 | ND | ND | *Bacillus thermophile* RS 241 | dsDNA | ND | ND | Sharp et al., 1986 | rotting straw | lytic | 55 (50–70), 7.3 ± 0.2 |
| 24 | JS009 | ND | ND | *Bacillus thermophile* RS 242 | dsDNA | ND | ND | Sharp et al., 1986 | stable manure | lytic | 55 (50–70), 7.3 ± 0.2 |
| 25 | JS010 | ND | ND | *Bacillus thermophile* RS 242 | dsDNA | ND | ND | Sharp et al., 1986 | compost | lytic | 55 (50–70), 7.3 ± 0.2 |
| 26 | JS011 | unclassified family | ND | *Bacillus thermophile* RS 239 | dsDNA | ND | ND | Sharp et al., 1986 | silage | lytic | 55 (50–70), 7.3 ± 0.2 |
| 27 | JS012 | ND | ND | *Bacillus thermophile* RS 239 | dsDNA | ND | ND | Sharp et al., 1986 | compost | lytic | 55 (50–70), 7.3 ± 0.2 |
| 28 | JS013 | ND | ND | *B.stearothermophilus* NCA 1503 | dsDNA | ND | ND | Sharp et al., 1986 | soil | lytic | 55 (50–70), 7.3 ± 0.2 |
| 29 | JS014 | ND | ND | *B.stearothermophilus* NCA 1503 | dsDNA | ND | ND | Sharp et al., 1986 | rotting straw | lytic | 55 (50–70), 7.3 ± 0.2 |
| 30 | JS015 | ND | ND | *B.stearothermophilus* NCA 1503 | dsDNA | ND | ND | Sharp et al., 1986 | compost | lytic | 55 (50–70), 7.3 ± 0.2 |
| 31 | JS017 | ND | ND | *B. caldotenax* | dsDNA | ND | ND | Sharp et al., 1986 | compost | lytic | 55 (50–70), 7.3 ± 0.2 |
| 32 | JS018 | ND | ND | *B. caldotenax* | dsDNA | ND | ND | Sharp et al., 1986 | rotting vegetation | lytic | 55 (50–70), 7.3 ± 0.2 |
| 33 | JS019 | ND | ND | *B. caldotenax* | dsDNA | ND | ND | Sharp et al., 1986 | rotting vegetation | lytic | 55 (50–70), 7.3 ± 0.2 |
| 34 | JS020 | ND | ND | *B.s caldotenax* | dsDNA | ND | ND | Sharp et al., 1986 | rotting vegetation | lytic | 55 (50–70), 7.3 ± 0.2 |
| 35 | JS021 | ND | ND | *B. caldotenax* | dsDNA | ND | ND | Sharp et al., 1986 | rotting vegetation | lytic | 55 (50–70), 7.3 ± 0.2 |
| 36 | JS022 | ND | ND | *B. caldotenax* | dsDNA | ND | ND | Sharp et al., 1986 | compost | lytic | 55 (50–70), 7.3 ± 0.2 |

**Table 2.** *Cont.*

| N° | Bacteriophage Species | Virus Family, Genus | GenBank Accession Number | Host (Used for Propagation) | Genome — Type and Size [bp] | ORFs | G+C [%] | Original Discovery Reference | Isolate Location | Life Cycle | Growth Temperature (°C, Optimal/Range) and pH (Optimal/Range) |
|---|---|---|---|---|---|---|---|---|---|---|---|
| 37 | JS023 | ND | ND | *B. caldotenax* | dsDNA | ND | ND | Sharp et al., 1986 | compost | lytic | 55 (50–70), 7.3 ± 0.2 |
| 38 | JS024 | ND | ND | *B. caldotenax* | dsDNA | ND | ND | Sharp et al., 1986 | compost | lytic | 55 (50–70), 7.3 ± 0.2 |
| 39 | JS025 | ND | ND | *B. caldotenax* | dsDNA | ND | ND | Sharp et al., 1986 | compost | lytic | 55 (50–70), 7.3 ± 0.2 |
| 40 | JS026 | ND | ND | *B. caldotenax* | dsDNA | ND | ND | Sharp et al., 1986 | compost | lytic | 55 (50–70), 7.3 ± 0.2 |
| 41 | JS027 | ND | ND | *Bacillus thermophile* RS 241 | dsDNA | ND | ND | Sharp et al., 1986 | compost | lytic | 55 (50–70), 7.3 ± 0.2 |
| 42 | BVW1 (W1) | *Siphoviridae* | ND | *Bacillus* sp. w13 | dsDNA, app. 18 kb | ND | ND | Liu et al., 2006 | deep-sea hydrothermal fields (West Pacific) | lytic | 60, 7.0 |
| 43 | GVE1 (E1) | *Siphoviridae* | ND | *Geobacillus* sp. E 26323 | dsDNA, app. 41 kb | ND | ND | Liu et al., 2006 | deep-sea hydrothermal fields (East Pacific) | lytic | 60, 7.0 |
| 44 | GVE2 (E2) | *Siphoviridae*, unclassified *Siphoviridae* | NC_009552 DQ453159 | *Geobacillus* sp. E 263 | linear, dsDNA, 40,863 | 62 | 44.8 | Liu & Zhang, 2008a,b | deep sea (China) | lysogenic | 65, 7.0 |
| 45 | GBSV1 | *Myoviridae*, *Svunavirus* | | *Geobacillus* sp. 6k512 | linear, dsDNA, 34,683 | 54 | 44.4 | Liu et al., 2009, 2010 | off shore hot spring, (Xiamen, China) | lytic | 65, 7.2 |
| 46 | BV1 | *Myoviridae*, *Svunavirus* | NC_009737.2, DQ840344 | *Geobacillus* sp. 6k512 | linear, dsDNA, 35,055 | 54 | 44.4 | Liu et al., 2009, 2010 | off shore hot spring, (Xiamen, China) | lytic | 65, 7.2 |
| 47 | D6E | *Myoviridae* | NC_019544 | *Geobacillus* sp. E 26323 | circular, dsDNA, 49,335 | 49 | 46 | Wang & Zhang, 2010 | deep-sea hydrothermal fields (East Pacific) | lytic | 65, 7.0 |
| 48 | φOH2 (phiOH2) | *Siphoviridae* | AB823818, NC_021784 | *G. kaustophilus* GBlys, *G. kaustophilus* NBRC 102445(T), lysogenic *G. kaustophilus* GBlys) | dsDNA, 38,099 | 60 | 45 | Doi et al., 2013 | hot spring sediment (Japan) | lytic/lysogenic | 55 |
| 49 | GBK2 | *Siphoviridae* | KJ159566 | *G. kaustophilus* | Circularly permuted, dsDNA, 39,078 | 62 | 43 | Marks & Hamilton, 2014 | compost (Cary, NC, USA) | lytic | 55, 7.3 |
| 50 | GVE3 (E3) | *Siphoviridae* | NC_029073, KP144388 | *G. thermoglucosidasius* | dsDNA 141,298 | 202 | 29.6 | Van Zyl et al., 2015 | ND | lytic/lysogenic | 60, 7.3 |

**Table 2.** *Cont.*

| N° | Bacteriophage Species | Virus Family, Genus | GenBank Accession Number | Host (Used for Propagation) | Genome | | | Original Discovery Reference | Isolate Location | Life Cycle | Growth Temperature (°C, Optimal/Range) and pH (Optimal/Range) |
|---|---|---|---|---|---|---|---|---|---|---|---|
| | | | | | Type and Size [bp] | ORFs | G+C [%] | | | | |
| 51 | AP45 | *Siphoviridae* | KX965989 | *Aeribacillus* sp. CEMTC656 | dsDNA 51,606 | 71 | 38.3 | Morozowa et al., 2019 | soil (Valley of Geysers, Kamchatka, Russia) | lytic/lysogenic | 55, 7.5 |
| 52 | vB_Bps-36 | ND | MH884513 | *B. pseudalcaliphilus* | dsDNA 50,485 | 68 | 41.1 | Akhwale et al., 2019 | Lake Elmenteita, (Kenya) | lytic/? | 30–40< 9< |
| 53 | vB_BpsM-61 | ND | MH884514 | *B. pseudofirmus* | dsDNA 48,160 | 75 | 43.5 | Akhwale et al., 2019 | Lake Elmenteita, (Kenya) | lytic/? | 30–40< 9< |
| 54 | vB_BboS-125 | ND | MH884509 | *B. bogoriensis* | dsDNA 58,528 | 81 | 48.6 | Akhwale et al., 2019 | Lake Elmenteita, (Kenya) | lytic/? | 30–40< 9< |
| 55 | vB_BcoS-136 | ND | MH884508 | *B. cohnii* | dsDNA 160,590 | 240 | 32.2 | Akhwale et al., 2019 | Lake Elmenteita, (Kenya) | lytic/? | 30–40< 9< |
| 56 | vB_BpsS-140 | ND | MH884512 | *B. pseudalcaliphilus* | dsDNA 55,091 | 68 | 39.8 | Akhwale et al., 2019 | Lake Elmenteita, (Kenya) | lytic/? | 30–40< 9< |

*3.15. JS001 (Family/Genus ND, Hosts Thermophilic Bacillus sp.)*

JS001 was obtained by induction of *B. stearothermophilus* NCA 1503 and propagated on *B. caldotenax*. Other susceptible hosts included: *B. stearothermophilus*: ATCC 12016, RS 93, EP 262; *Bacillus* thermophile: RS 15, RS 108, RS 125 [21].

*3.16. JS004 (Putative Podoviridae Family, Hosts Thermophilic Bacillus sp.)*

JS004 was isolated from a silage, propagated on *Bacillus* thermophile RS 239, yielding small 0.5–1 mm circular, hazy plaques. High thermal stability was observed up to 60 °C, then gradually declining, but still substantial at 70 °C. Other susceptible hosts included: *B. stearothermophilus* RS 93, *Bacillus* thermophile RS 241. The bacteriophage belongs to the group 1: it possesses an icosahedral head 55–60 nm long, a short cylindrical 15 nm stubby tail with collar and no evidence of a base plate, which suggests that JS004 belongs to the *Podoviridae* family [21].

*3.17. JS005 (Family/Genu ND, Hosts Thermophilic Bacillus sp.)*

JS005 was isolated from rotting straw, propagated on *Bacillus* thermophile RS 239 and yielding small 0.5–1 mm circular, slightly hazy plaques. The only other susceptible host was *Bacillus* thermophile RS 241 [21].

*3.18. JS006 (Family/Genus ND, Hosts Thermophilic Bacillus sp.)*

JS006 was isolated from compost, propagated on *Bacillus* thermophile RS 239 and yielding small 0.5–0.75 mm circular, clear plaques. High thermal stability was observed up to 60 °C, then gradually declining, but still substantial at 70 °C. Other susceptible hosts included: *B. stearothermophilus*: RS 93, NW 4S, EP 262, *Bacillus* thermophile RS 242 [21].

*3.19. JS007 (Family/Genus ND, Hosts Thermophilic Bacillus sp.)*

JS007 was isolated from silage, propagated on *Bacillus* thermophile RS 240, and yielding small 0.5–0.75 mm circular, clear plaques. High thermal stability was observed up to 60 °C, then gradually declining, but still substantial at 70 °C. Other susceptible hosts included: *B. stearothermophilus* RS 93, *and Bacillus* thermophile RS 239, RS 240, RS 241. The bacteriophage belongs to the group 3: it possesses an icosahedral head 80–85 nm, an exceptionally long flexible tail of 400–420 nm with cross-strations and a fine fiber at tip, which suggests that JS007 belongs to the *Siphoviridae* family [21].

*3.20. JS008 (Family/Genus ND, Hosts Thermophilic Bacillus sp.)*

JS008 was isolated from rotting straw, propagated on *Bacillus* thermophile RS 241, yielding 1–3.5 mm circular, hazy plaques. Another susceptible host was *Bacillus* thermophile RS 239 [21].

*3.21. JS009 (Family/Genus ND, Hosts Thermophilic Bacillus sp.)*

JS009 was isolated from stable manure, propagated on *Bacillus* thermophile RS 242 and yielding small 0.5 mm circular plaques with a hazy center. Other susceptible hosts included: *B. stearothermophilus* RS 93, *Bacillus* thermophile RS 241 [21].

*3.22. JS010 (Family/Genus ND, Hosts Thermophilic Bacillus sp.)*

JS010 was isolated from compost, propagated on *Bacillus* thermophile RS 242 and yielding 3 mm, irregular, hazy plaques with clear areas at the edge. Other susceptible hosts included: *B. stearothermophilus* RS93, *Bacillus* thermophile RS 241 [21].

*3.23. JS011 (Family/Genus ND, Hosts Thermophilic Bacillus sp.)*

J7S011 was isolated from silage, propagated on *Bacillus* thermophile RS 239 and yielding 3 mm, irregular, hazy plaques with clear areas at the edge [21].

### 3.24. JS012 (Family/Genus ND, Hosts Thermophilic Bacillus sp.)

JS012 was isolated from compost, propagated on *Bacillus* thermophile RS 239 and yielding 0.5–2 mm, irregular, clear plaques [21].

### 3.25. JS013 (Family/Genus ND, Hosts Thermophilic Bacillus sp.)

JS013 was isolated from soil, propagated on *B. stearothermophilus* NCA 1503 and yielding 0.75–1 mm, circular, hazy plaques [21].

### 3.26. JS014 (Putative Siphoviridae, Hosts Thermophilic Bacillus sp.)

JS014 was isolated from rotting straw, propagated on *B. stearothermophilus* NCA 1503 and yielding small 0.5–1 mm circular, slightly hazy plaques. High thermal stability was observed up to 60 °C, then gradually declining, but still substantial at 70 °C. Another susceptible host was *B. stearothermophilus* NW 10. The bacteriophage belongs to the group 3: it possesses icosahedral head 85–90 nm, an exceptionally long flexible tail of 400 nm with cross-strations, which suggests that JS007 belongs to the *Siphoviridae* family [21].

### 3.27. JS015 (Family/Genus ND, Hosts Thermophilic Bacillus sp.)

JS015 was isolated from compost, propagated on *B. stearothermophilus* NCA 1503 and yielding 0.5–2 mm, circular, slightly hazy plaques. Another susceptible host was *B. stearothermophilus* NW 10 [21].

### 3.28. JS017 (Putative Myoviridae Family, Hosts Thermophilic Bacillus sp.)

JS017 was isolated from compost, propagated on *B. caldotenax* and yielding small 1–1.5 mm circular, turbid/hazy center plaques. High thermal stability was observed up to 60 °C, then gradually declining, but still substantial at 70 °C. Other susceptible hosts included: *B. caldotenax* DSM 406, *B. caldovelox* DSM 411, *B. caldolyticus* DSM 405, *B. stearothermophilus*: ATCC 12016, NW 10, RS 93, EP 262, EP 240, ATCC 8005, *B. thermocatenulatus* DSM 730. The JS017 belongs to the group 4: it possesses a cylindrical head with round ends 80 × 40 nm, a rigid tail of 115 nm with cross-strations and triangular base plate, which suggests that JS017 belongs to the *Myoviridae* family. Interestingly, a 1–5% portion of the bacteriophage population had heads of twice the normal length and/or less frequently longer tails [21].

### 3.29. JS018 (Putative Myoviridae Family, Hosts Thermophilic B. caldotenax)

JS018 was isolated from rotting vegetation, propagated on *B. stearothermophilus* NCA 1503 and yielding 1 mm, circular, clear center plaques. Other susceptible hosts included: *B. caldotenax* DSM 406, *B. caldovelox* DSM 411; *B. caldolyticus* DSM 405, *B. stearothermophilus*: RS 93, EP 240 [21].

### 3.30. JS019 (Putative Myoviridae Family, Hosts Thermophilic B. caldotenax)

JS019 was isolated from rotting vegetation, propagated on *B. caldotenax* and yielding small 1–1.5 mm circular, clear center, hazy at edges plaques. High thermal stability was observed up to 60 °C, then gradually declining, but still substantial at 70 °C. The bacterio-phage exhibited broad specificity. Other susceptible hosts included: *B. caldotenax* DSM 406, *B. caldovelox* DSM 411; *B. caldolyticus* DSM 405, *B. stearothermophilus*: RS 93, EP 240. The JS019 belongs to the group 2: it possesses an icosahedral head of 50 nm and rigid tail of 130, which suggests that JS019 belongs to the *Myoviridae* family [21].

### 3.31. JS020 (Family/Genus ND, Hosts Thermophilic B. caldotenax)

JS020 was isolated from rotting vegetation, propagated on *B. caldotenax* and yielding 1–2 mm, circular, clear center, hazy outer halo 4–8 mm plaques. This suggest that the bacteriophage produces very active depolymerase, degrading the host's envelope. Other susceptible hosts included: *B. caldotenax* DSM 406, *B. caldovelox* DSM 411, *B. caldolyticus* DSM 405, *B. stearothermophilus*: RS 93, EP 240 [21].

### 3.32. JS021 (Family/Genus ND, Hosts Thermophilic B. caldotenax)

JS021 was isolated from rotting vegetation, propagated on *B. caldotenax* and yielding 2–3 mm, circular, clear centre, hazy outer halo plaques. This suggests that the bacteriophage produces an envelope depolymerase. Other susceptible hosts included: *B. caldotenax* DSM 406, *B. caldovelox* DSM 411; *B. caldolyticus* DSM 405, *B. stearothermophilus*: RS 93, EP 240, ATCC 12016 [21].

### 3.33. JS022 (Putative Myoviridae Family, Hosts Thermophilic B. caldotenax)

JS022 was isolated from compost, propagated on *B. caldotenax* and yielding 1–2 mm, circular, clear centre plaques. Other susceptible hosts included: *B. caldotenax* DSM 406, *B. caldovelox* DSM 411; *B. caldolyticus* DSM 405, *B. stearothermophilus*: RS 93, EP 240. The JS022 belongs to the group 2: it possesses an icosahedral head 60–65 nm long and a rigid tail of 130 nm, which suggests that JS022 belongs to the *Myoviridae* family [21].

### 3.34. JS023 (Family/Genus ND, Hosts Thermophilic B. caldotenax)

JS023 was isolated from compost, propagated on *B. caldotenax* and yielding 0.5–1.5 mm, circular, clear centre, hazy edge plaques. Other susceptible hosts included: *B. caldotenax* DSM 406, *B. caldovelox* DSM 411; *B. caldolyticus* DSM 405, *B. stearothermophilus*: RS 93, EP 136 [21].

### 3.35. JS024 (Putative Siphoviridae or Myoviridae Family, Hosts Thermophilic B. caldotenax)

JS024 was isolated from compost, propagated on *B. caldotenax* and yielding 0.5–1.5 mm, circular, clear center, hazy edge plaques with a less hazy outer edge. Other susceptible hosts included: *B. caldotenax* DSM 406, *B. caldovelox* DSM 411; *B. caldolyticus* DSM 405, *B. stearothermophilus*: RS 93, EP 240, EP 136. The JS024 belongs to the group 2: it possesses an icosahedral head 60 nm long and a tail of limited flexibility 120 nm, which suggests that JS024 belongs to either the *Siphoviridae* or *Myoviridae* family [21].

### 3.36. JS025 (Putative Myoviridae Family, Hosts Thermophilic B. caldotenax)

JS025 was isolated from compost, propagated on *B. caldotenax* and yielding 1–1.5 mm, circular, hazy plaques. Other susceptible hosts included: *B. caldotenax* DSM 406, *B. caldovelox* DSM 411; *B. caldolyticus* DSM 405, *B. stearothermophilus*: RS 93, EP 240. The JS025 belongs to the group 2: it possesses an icosahedral head 45–50 nm long and a rigid tail of 120 nm, which suggests that JS025 belongs to the *Myoviridae* family [21].

### 3.37. JS026 (Putative Siphoviridae or Myoviridae Family, Hosts Thermophilic B. caldotenax)

JS026 was isolated from compost, propagated on *B. caldotenax* and yielding 1–1.5 mm, circular, hazy plaques. Other susceptible hosts included: *B. caldotenax* DSM 406, *B. caldovelox* DSM 411; *B. caldolyticus* DSM 405, *B. stearothermophilus*: RS 93, ATCC 12016. The JS026 belongs to the group 4: it possesses a cylindrical head with round ends 80 × 45–50 nm and a limited flexibility tail of 120 nm, which suggests that JS026 belongs to either the *Siphoviridae* or *Myoviridae* family [21].

### 3.38. JS027 (Putative Podoviridae Family, Hosts Thermophilic Bacillus sp.)

JS027 was isolated from compost, propagated on *Bacillus* thermophile RS241 and yielding 0.75 mm, circular, clear plaques. Another susceptible host is *Bacillus* thermophile RS239. The JS027 belongs to the group 1: it possesses an icosahedral head 55–60 nm long and a very short, stubby tail with no evidence of a tail plate, which suggests that JS027 belongs to the *Podoviridae* family [21].

### 3.39. W1 (BVW1) (Siphoviridae Family, Genus ND, Host Thermophilic Bacillus sp.)

W1 (BVW1) was isolated from a deep-sea hydrothermal field, located in the west Pacific (19_ 24′08″ N, 148_44′79″ E, depth of 5060 m), along with its host *Bacillus* sp. W13 (GenBank AY583457), which is aerobic, rod-shaped and spore-forming, growing within a temperature range of 45–85 °C with an optimum at 65–70 °C. The host range was narrow, as the W1 infects only one thermophilic strain out of 10 tested. The bacteriophage propagates most effectively at 60 °C. Morphologically, W1 has a hexagonal head of 70 nm in diameter and a very long tail of 300 nm and 15 nm in width. The bacteriophage genome is composed of dsDNA with an estimated size of 18 kb. As determined, using SDS-PAGE of purified virions, six major proteins were detected with estimated molecular weights: 32, 45, 50, 57, 60 and 70 kDa [22].

### 3.40. GVE1 (E1) (Siphoviridae Family, Genu NDs, Host Geobacillus sp.)

GVE1 was isolated from a deep-sea hydrothermal field, located in the east Pacific (12_42′29″ N, 104_02′01″ W, depth of 3083 m), along with its host *Geobacillus* sp. E26323 (GenBank DQ225186), which is aerobic, rod-shaped and spore-forming, growing in rich media of pH 7, within a temperature range of 45–85 °C and an optimum at 65–70 °C. The host range is narrow, as the GVE1 infects only one thermophilic strain out of 10 tested. The bacteriophage propagates most effectively at 60 °C. GVE1 possesses a large hexagonal head of 130 nm in diameter, a tail of 180 nm and 30 nm in width. The bacteriophage genome is composed of dsDNA with an estimated size of 41 kb. As determined on SDS-PAGE of purified virions, six major proteins were detected with estimated molecular weights: 34, 37, 43, 60, 66 and 100 kDa [22].

### 3.41. GVE2 (E2) (Siphoviridae Family, Genus ND, Hosts Geobacillus sp.)

Siphovirus GVE2 (also known as E2) is one of the very few deeply studied *Geobacillus* thermophages and was isolated from the same location as GVE1 [22]. The authors in the publications concerning GVE2, refer to their previous publication concerning the discovery of GVE1, thus, it is possible that the bacteriophages are identical or closely related.

GVE2 is a virulent *Siphoviridae* bacteriophage infecting deep-sea thermophilic *Geobacillus* sp. E263 (CGMCC1.7046), capable of growth at 45–80 °C, with optimum at 60–65 °C, which was selected for GVE2 propagation. The bacteriophage genome is a 40,863 bp, 44.8% G+C, linear dsDNA with 62 ORFs [22]. Its genome shows sequence similarity to the genomes of several *Geobacillus* sp. bacteriophages. Six GVE2 proteins were characterized in the laboratory. Transcriptomic analysis has shown that 74.2%, 46 of the ORFs, were transcribed and potentially expressed as proteins. The six structural proteins of purified GVE2 virions were also determined by mass spectrometry and, for one of those, novel portal protein VP411 protein, its coding gene was cloned/expressed in *Escherichia coli* (*E. coli*). The recombinant VP411 was characterised in details and confirmed by immune electron microscopy using gold-labelled secondary antibody, as located in the portal-neck region of the GVE2 [24]. Subsequently, the endolysin coding gene was cloned in *E. coli*, expressed and the protein purified and characterised [56]. Further, the holin-endolysin system was studied and it was shown that GFP–endolysin fusion aggregated in GVE2-infected *Geobacillus* sp. and revealed that the endolysin and holin interacted directly as well, as with the host *Geobacillus* protein ABC transporter system, which apparently participates in the regulation of the lysis process [57].

### 3.42. GBSV1 (Myoviridae Family, Svunavirus Genus, Hosts Geobacillus sp.)

GBSV1 was isolated from an off-shore location in shallow waters at the depth of 100 m. It was propagated in thermophilic *Geobacillus* sp. 6k51 (GenBank DQ141699), which is an aerobic rod-shaped bacterium growing at 45–85 °C [25]. It was cultivated and infected with GBSV1 at 65 °C, supplemented with $Ca^{2+}$ and $Mg^{2+}$ ions at pH app. 7. The 16S rDNA analysis revealed that this host shared 99.93% sequence identity with that of *G. toebii* mc-14 (GenBank EU214628). TEM analysis of GBSV1 showed a typical *Myoviridae* morphology with a 60 nm hexagonal head and 80–150 nm contractile tail. The GBSV1 genome analysis is composed of linear, 44.4% G+C dsDNA of 34,683 bp, coding for 54 putative ORFs. Thirty-three of them exhibit sequence similarities to genes from seven species of *Geobacillus* or *Bacillus* and to the bacteriophages that infect them. This may indicate their biological role of upmost importance in promoting genetic exchanges among *Bacillus* and/or *Geobacillus* bacteria: short-term adaptation, long term evolution of bacteria and population genetics. Twenty-two ORFs were bioinformatically annotated and five structural proteins of the purified GBSV1 were identified by proteomic analyses. The ORFs are grouped into functional clusters on the genome. Interestingly, seven of the GBSV1 ORFs exhibit sequence similarities with the genes from the bacteria relevant to human diseases. The authors hypothesize that thermophages may be the potential evolutionary connection between thermophiles and human pathogens. The GBSV1 shares over 90% proteome similarity with BV1, as based on genomes analysis. Two groups of genes exhibit high similarity between GBSV1, BV1 and phiOH2: (i) lysis-related proteins—a probacteriophage anti-repressor and lysin with >95% similarity, and (ii) bacteriophage genome maintenance proteins and integration—integrase, >95%, recombinase—>65% similarity [25]. These similarities indicate common, possibly interchangeable, infection modules across thermophilic host species of distinct bacterial phyla [3].

### 3.43. BV1 (Myoviridae, Family, Svunavirus Genus, Hosts Geobacillus sp.)

BV1 is highly similar to GBSV1 (see previous section) and the characteristics are essentially the same, with minor differences: a hexagonal head is 60 nm in diameter and a tail 90nm in length and 10 nm in width, the genome is 44.8% G+C, 35,055 bp. It contains 54 ORFs, classified into eight functional groups. SDS-PAGE analysis of purified BV1 revealed that it contains seven major polypeptides with molecular weights of app.: 100, 72, 42, 33, 31, 28 and 26 kDa. BLAST analysis revealed that 38 ORFs exhibited similarities to proteins present in databases. Interestingly, a diverse pattern of similarities suggests that BV1 could be involved in gene transfer between the *Bacillus* and *Geobacillus* species [58].

### 3.44. D6E (Myoviridae Family, Genus ND, Host Geobacillus sp.)

D6E was isolated from a deep-sea hydrothermal field (12°42′29″ N, 104°02′01″ W, depth of 3083 m) at the same location as GVE1, along with the same host *Geobacillus* sp. E26323 and was propagated at 65 °C. TEM results have shown that D6E is a myovirus with an icosahedral head (60 nm), a contractile tail (16 nm in width and 60 nm in length), decorated with a tail fiber (4 nm in width and 60 nm in length). The bacteriophage genome of 49,335 bp is composed of 46% G+C dsDNA, possibly circular, as based on restriction and PCR analysis. It contains 49 ORFs, with 30 of them having bioinformatically assigned homologies/function, grouped in functional clusters: (i) DNA packaging and capsid assembly, (ii) tail assembly, (iii) lysis and lysogeny, (iv) DNA replication and transcription. D6E genome does not code for any tRNA genes. Comparative genomics and proteomics analyses have indicated an extensive mosaicism of D6E genome with other mesophilic and thermophilic bacteriophages, including GVE1. The authors conclude, that there were at least three recombination events during GVE1 evolution: (i) the overall gene organization, (ii) recombination between gene boundaries and (iii) mutations in genes. As determined on SDS-PAGE and mass spectrometry 10 virion-associated proteins biosynthesis were confirmed [14].

### 3.45. φOH2 (phiOH2) (Family/Genus ND, Host G. kaustophilus)

Bacteriophage φOH2 is closely grouped with *Bacillus* virus 1 BV1 thermophage [3]. It was found in the sediment of a hot spring as lytic, infecting factor for *G. kaustophilus* NBRC102445. The genome of bacteriophage φOH2 is integrated into the lysogenic host genome *G. kaustophilus* GBlys, thus being a prophage. The host was cultivated at 55 °C. Its genome of 3,541,481 bp consists of 216 sequence contigs with OH2 prophage genome (38,099 bp, 45% G = C, 60 ORFs), situated astride contigs GBL0024 and GBL0110 and flanked by the host genes encoding GroES, GroEL. The host transposase gene is located next to the OH2 genome. A putative integrase gene is located at the terminus of the OH2 sequence. There are no excisionase gene homologues, and the attachment sites located next to the integrase gene are dissimilar to those coded by bacteriophages [59]. BLAST analysis show the presence of completely or partially homologous φOH2 sequences in many '*Bacillus* group' bacteria and bacteriophages, which indicates intense horizontal gene transfer within the group.

### 3.46. GBK2 (Siphoviridae Family, Genus ND, Host G. kaustophilus)

GBK2 bacteriophage was extracted out of the compost pile using a liquid culture of *G. kaustophilus* ATCC8005, cultivated at 55 °C. GBK2 has a circularly permuted dsDNA genome of 39,078 bp and 43% G+C content. The genome contains seven copies of a 17-bp inverted DNA repeat, which forms a potential origin of replication for GBK2. Bioinformatics analysis of the genome reveals 62 putative ORFs. Twenty-five of them show homology to the known proteins and 37 are proteins with undetermined functions. The ORFs are grouped into functional clusters: (i) DNA packaging, (ii) structural proteins, (iii) lysis, (iv) unknown, (v) DNA replication/metabolism. Overall, the genome exhibits a limited DNA sequence similarity to any other sequenced bacteriophages [1]. It is intriguing that GBK2 is more closely related to the bacteriophages that infect mesophilic hosts than to those that infect thermophilic *Geobacillus* sp. or other related thermophiles species. The overall genomic arrangement of GBK2 is quite similar to that of mesophilic bacteriophage SPP1, infecting *B. subtilis* [22]. Fourteen ORFs are coding proteins that are homologous to SPP1 bacteriophage and most of them are involved in DNA replication, while two ORFs are predicted to encode capsid structural proteins. DNA-related genes, include those coding for: exonuclease/recombinase, recombinase of RecT-type, helicase, helicase loader and three DNA metabolism enzymes [18]. Furthermore, in SPP1 genome there are three genes, non-essential for growth, of unknown function. Nevertheless, homologs of these genes are found in GBK2 bacteriophage. It was concluded that they may play a significant role in the thermophage life cycles under certain conditions [60]. These functional and sequence similarities point to a common origin of SPP1 and GBK2 bacteriophages or a massive horizontal gene transfer.

### 3.47. GVE3 (Siphoviridae Family, Genus ND, Host G. thermoglucosidasius)

GVE3 bacteriophage is a B1 morphotype group *siphovirus* that infects, very specifically, the thermophilic bacteria *G. thermoglucosidasius* as the only susceptible host, propagated at 60 °C. The bacteriophage has an isometric head of 90–100 nm diameter and a non-contractile tail of 210 nm. The plaques had characteristic "bulls-eye" morphology, suggesting host lysogeny [61] with several bacterial colonies growing inside plaques. Once isolated, the bacteria from those colonies were resistant to GVE3 infection. The lysogenic host genome sequencing revealed that the GVE3 genome had inserted into the specific location, attB site, with a 23-bp sequence that was duplicated upon insertion. The bacteriophage GVE3 encodes a putative pyrimidine nucleoside phosphorylase, situated downstream of GVE3 resolvase gene. The attP site is located within the gene. Apparently, the host's pyrimidine nucleoside phosphorylase is inactivated upon the GVE3 genome insertion.

The presence of a GVE3-encoded pyrimidine nucleoside phosphorylase suggests an obligatory requirement for its activity: upon GVE3 genome integration the host relies on the bacteriophage enzyme. There is an alternative explanation: if the enzyme is not essential for either the GVE3 or the host, since the mutation in the pyrimidine nucleoside phosphorylase-coding gene is non-lethal, this may suggest that GVE3 is a specialized transducing thermophage [62]. Interestingly, the genome size is 141,298 bp comprising the largest known among the bacteriophages that infect *Geobacillus* sp., and has a unique DNA sequence, with no close relatives. A large number of 202 putative ORFs were bioinformatically detected and, out of those, putative functions to 62 of them were assigned. As observed in a number of other members of the *Siphoviridae*, the GVE3 genome displays the functional clusters [62]. Remarkably, GVE3 resembles GBK2 in that it is apparently most closely related to the mesophilic *B. anthracis* bacteriophage vB_BanS_Tsamsa, rather than the *Geobacillus*-infecting bacteriophages. Tetranucleotide usage deviation analysis confirms this finding, indicating that the GVE3 genome sequence correlates mostly with *B. anthracis* and *B. cereus* sequences, rather than *Geobacillus* sp. sequences [62,63], which again points to widespread genetic recombination among meso- and thermophilic 'Bacillus group' bacteriophages. The analysis also suggests that there exist a different 'natural' host, unidentified *Geobacillus* sp., with tetranucleotide usage deviation patterns more similar to the two *Bacillus* species that might be the "natural" hosts for GVE3. Alternatively, the possibility exists that GVE3 has evolved from a mesophilic counterpart and that the correlation to mesophilic *Bacillus* species is an evolutionary link to its heritage. Similar findings have been shown for GBK2 [64]. The GVE3 genome sequence has a much lower G+C content (29.6%) than its host *G. thermoglucosidasius* (44%) [62], which is also observed for most bacteriophage–host pairs. It is suggested that its base composition bias might have resulted from a competition for metabolic resources and/or reflects an adaptation to thermophily [65]. The genome GC skew analysis has suggested a location of a replication terminus. A search for direct and inverted repeats of 7–30 bp resulted in finding as many as 582 repeats, including two invert repeats (TATTTTT/TAATTAT) predicted to be part of the origin of the replication. The large genome segment, DNA replication, modification and repair, contains tens of genes, not all of them with assigned functions. These make the GVE3 enzymatic machinery remarkably independent, including three subunits diverging hosts DNA polymerase to the bacteriophage replication and three DNA modification methyltransferases, protecting the genome from restriction, among others [62].

### 3.48. AP45 (Siphoviridae Family, Genus ND, Host Aeribacillus sp.)

AP45 bacteriophage was isolated from a soil sample collected in the Valley of Geysers, Kamchatka, Russia. It was propagated in thermophilic *Aeribacillus* sp. strain CEMTC 656, a Gram-positive, aerobic, rod-shaped, non-motile, forming sub-terminally located endospores. The optimum growth temperature was at 50–65 °C in nutrient broth at pH 6.5, containing 0.3 M NaCl. AP45 was propagated at 55 °C. The host range is very narrow, out of 13 tested *Geobacillus* sp., only *G. icigianus* CMBI G1w1, also isolated from the Valley of Geysers, was supporting infection. AP45 exhibited high thermostability and a latent period of approximately 200 min, with a burst size of approximately 40 bacteriophage particles/infected cell. The bacteriophage produces both a 'halo' around clear plaque centers and 'bulls-eye' morphology, indicating secretion of a cell envelope depolymerase and suggesting formations of host lysogeny. The colonies isolated from the plaques proved to be resistant to AP45 infection, however, PCR using primers to AP45 has failed, thus, it is not clear whether other resistance mechanisms take place. TEM analysis of AP45 indicated *Siphoviridae* morphology with an icosahedral head of 60 nm with a diameter and a tail approximately 160 nm in length. The AP45 genome analysis has shown that it is composed of 38.3% G+C dsDNA of 51,606 bp, containing 71 ORFs, 40 of them have had putative functions predicted bioinformatically. Furthermore, ORFs were divided into three functional clusters: (i) nucleic acids metabolism and regulation of the viral life cycle; (ii) structural proteins; (iii) lysis of bacterial cells. No genes encoding DNA and RNA polymerases were identified, thus, AP45 apparently uses host polymerases. The AP45 genome exhibited limited similarity to other bacteriophage sequences. The highest value of 36% was detected for thermophilic *Geobacillus* myovirus D6E. Interestingly, the majority of putative AP45 proteins exhibited a higher similarity to proteins from bacteria belonging to the *Bacillaceae* family, than to bacteriophages. Moreover, over half of the putative AP45 ORFs were highly similar to probacteriophage sequences of the *Aeribacillus pallidus* strain 8m3. It was suggested that the AP45 bacteriophage and revealed probacteriophages might be members of a new genus within the *Siphoviridae* family [27].

### 3.49. Series of 5 Bacteriophages from Haloalkaline Lake Elmenteita of Kenyan Rift Valley (Sections 3.49–3.54 Below)

All bacteriophages were isolated from Lake Elmenteita [66], situated at 0°27′ S 36°15′ E on the floor of the Kenyan Rift Valley, at 1776 m above sea level without a direct water outlet. This area has a hot, dry and semi-arid climate, which results in rapid evaporation rates. The alkalinity of the water is high alkaline with pH above 9 and with a high concentration of carbonates, chlorides and sulphates [67]. The water temperature ranges between 30 °C and 40 °C. Comparison of genome sizes, coding density, number of ORFs, GC% and gene organizations revealed that the bacteriophages had no relationship with each other as well as no significant similarities in the GenBank® sequence data. Furthermore, there is evidence of genes mosaicism to described genes with conserved domains [66]. The bacterial hosts were not described in details. Nevertheless these were extremophilic bacteria with regard to salinity and alkalinity from the 'Bacillus group', possibly also entering moderate thermophilic range. No bacteriophages characteristics are available at the time of this review submission, except bacteriophages genomes bioinformatics analysis. The authors cite ongoing publication, which will contain descriptions of the bacteriophages [66].

### 3.50. vB_BpsS-36 (Family/Genus ND, Host B. pseudalcaliphilus)

The vB_BpsS-36 genome is composed of 50,485 bp dsDNA with 41.1% GC content. It codes for 68 ORFs distributed on both forward and reverse strands and six transcription terminators with coding density of 91.6%. On the DNA level, the bacteriophage shows essentially no homologies to other bacteriophages, while on the amino acids level there exists weak similarities to short regions of *Bacillus* bacteriophages phages Tsamsa, Riggi and CampHawk. A conserved replication factor was detected in the genome, suggesting that vB_BpsS-36 employs DNA polymerase III subunit alpha for DNA replication. Furthermore, vB_BpsS-36 codes for replication proteins—domains DnaB and DnaG [66].

### 3.51. vB_BpsM-61 (Family/Genus ND, Host B. pseudofirmus)

The vB_BpsM-61 genome is composed of 48,160 bp dsDNA with 43.5% GC content. It codes for 75 ORFs distributed on both forward and reverse strands and eight transcription terminators with coding density of 93.0%. The bacteriophage exhibits weak similarities to *Bacillus* bacteriophage PM1 and *Geobacillus* phage GBK2 in cluster coding for structural components. The vB_BpsM-61 genome is also coding for putative proteins: replicative DnaC, helix destabilizing Ssb, dUTPase, Holliday junction resolvase [66].

### 3.52. vB_BboS-125 (Family/Genus ND, Host B. bogoriensis)

The vB_BboS-125 genome is composed of 58,528 bp dsDNA with 48.6% GC content. It codes for 81 ORFs distributed on forward strands only and six transcription terminators with coding density of 92.2%. The bacteriophage exhibits weak similarities to other bacteriophage proteins, more pronounced to structural components and DNA packaging located in genomic regions in *Exiguobacterium* sp. AB2 and *Brevibacillus* sp. CF112, apparently being not-annotated prophage segments. In addition, the vB_BboS-125 shows similarities to genes with conserved domains of the replication cluster, helicase and primase, present in the genome of *Brevibacillus* sp. CF112. Furthermore, the vB_BboS employs a host's DNA polymerase I for replication. Interestingly, the bacteriophage terminase did not cluster with previously described bacteriophages, but formed a distinct phyletic line, which suggests that vB_BboS-125 belongs to a new genus of bacteriophages [66].

### 3.53. vB_BcoS-136 (Family/Genus ND, Host B. cohnii)

The vB_BcoS-136 has the largest genome of an entire group of eight bacteriophages found in haloalkaline in Lake Elmenteita, 160,590 bp dsDNA with low GC content of 32.2%. It codes for 240 ORFs distributed on both forward and reverse strands, 15 transcription terminators and 17 tRNA with a coding density of 93.0%. A very high number of tRNA coding genes suggest substantial differences in codon usage between the bacteriophage and its host, thus, needed for efficient translation and vB_BcoS-136 propagation. Furthermore, it may suggest that the bacteriophage relatively recently has adopted its host. The bacteriophage exhibits very low similarities to *Bacillus* bacteriophage Tsamsa. BcoS-136 terminase is dissimilar to other bacteriophage's terminases with an already known DNA packaging strategy, but groups in a distinct phyletic line, thus pointing to the possibility of the existence of another new genus. vB_BcoS-136 utilizes a host's DNA polymerase III subunit alpha and codes for other DNA replication/DNA metabolism putative proteins: DNA ligase, DNA gyrase, ribonuclease HI and integrase. The latter suggests that vB_BcoS-136 can also enter lysogenic mode [66].

### 3.54. vB_BpsS-140 (Family/Genus ND, Host B. pseudalcaliphilus)

The vB_BpsS-140 genome is composed of 55,091 bp dsDNA with 39.8% GC content. It codes for 68 ORFs distributed on forward strands only and four transcription terminators with coding density of 91%. The vB_BpsS-140 encoded genes for proteins similar to homologs of a terminase and some structural proteins of *Bacillus* bacteriophages IEBH and 250, respectively. Homology comparison of the large terminase subunit shows that vB_BpsS-140 clusters with P22 bacteriophage, which is established as a model bacteriophage that follows headful packaging strategy [68]. As this protein is considered the most universally conserved gene sequence among bacteriophages [69], it is used to get insight into evolutionary relationships among distant bacteriophages [70]. Interestingly, the bacteriophage codes for atypical endopeptidase endolysin, are absent in other bacteriophages published by the authors [66].

## 4. Scientific, Biotechnology, Environmental and Medical Potential—A Short Note

In general, bacteriophages, and their corresponding enzymatic activities, offer great potential for the development of agents for the control of human pathogens [71]. Bacteriophage bionanostructures are used in science, industry and medicine, as exemplified by bioimaging diagnostics, ultrasensitive biomarker detection, targeted drug and gene delivery, improved tissue formation, directed stem cell differentiation, new generation vaccines and nanotherapeutics for targeted disease treatment, among others [71]. These applications may be substantially improved if based on robust, thermostable bacteriophage bionanoparticles. The study of *Geobacillus* host bacteria, their bacteriophages and their mutual interaction, presents great potential for gaining knowledge about the functioning of high temperature ecosystems, including the thermophages crucial role as regulators of bacterial populations. No less importantly, this knowledge is likely to help improve the known processes or to develop new processes in biotechnology. The list of viral genome sequences reported thus far in the National Center for Biotechnology Information Genome database, as of June 2021, includes 10,607 genomes; 3746 of them being bacterial hosts, with only four hosts listed under 'Geobacillus'. This proportion shows that there are significant gaps in the understanding of the genetics of the 'Bacillus group' thermophages as well as of the type and dynamics of their biological processes at a local and global scale.

Even though thermophiles and extreme thermophiles are presently attracting substantial attention, the area remains in large part a 'terra incognita' in terms of approaching understanding of mesophilic host-bacteriophage systems (such as *E. coli*/bacteriophage lambda). Nevertheless, they can provide model systems for deciphering the molecular biology and biochemistry of adaptations towards life at high temperatures and their effects on atypical ecosystems, and can affect many biogeochemical and ecological processes.

The scientific interest in thermophilic bacilli, especially *Geobacillus* sp., is fully justified, due to their potential applications in biotechnology and environment remediation processes. Nevertheless, thermophilic bacteriophages can cause huge economical losses, upon undesired infections of industrial fermentations. Bacteriophage-sensitive processes are shown in the following examples.

As *Geobacillus* sp. possesses the ubiquitous capability to metabolize various organic compounds, including hydrocarbons, the fermentation of C-5 and C-6 sugars is being used to biosynthesize ethanol for biofuel and organic acids production [72]. Even very 'difficult' substrates, such as cellulose, the most abundant organic carbon reservoir on Earth, are effectively degraded by *Geobacillus* sp. cellulases.

More thermostable enzymes, coded by *Geobacillus* sp., are also of interest to the biotechnology industry, such as: proteases, amylases, polysaccharides, lipases, glycanases and herbicides metabolizing enzymes for agricultural biotechnology [10,73–76].

Certain *Geobacillus* sp. are also capable of the destruction of quorum sensing in some Gram-negative bacteria [10]. Furthermore, some strains of *G. thermoleovorans* are secreting large bacteriocins that cause lysis of various other bacteria, including pathogenic *Salmonella typhimurium* [77]. These capabilities could be a basis for future therapeutic methods. Thus, cataloguing bacteriophages and studying their interactions with their hosts is also important for practical purposes. Furthermore, bacteriophages can affect the human microbiome, including facultatively thermophilic 'Bacillus group' species. Several clades from genus *Bacillus* of the *Bacillaceae* family involve bio-safe species, regarded as bimodal probiotic microbiobiota of humans and animals [78]. These are strains generally derived from environmental locations such as soil, marsh and plants, but are also found in animal's and human's skin, gastrointestinal tract or respiratory system. Several *Bacillus* species are typically encountered in traditional fermented foods, especially those of Asian or African origin that are prepared prevalently from protein rich soybeans or locust beans. The outstanding potential for the survival of environmental probiotics results from their ability to form durable spores [79] capable of enduring harsh conditions of draught, UV exposure, extreme temperatures and pH values, the presence of organic and inorganic chemicals, salts, detergents and heavy metal ions. During the past decade, probiotic *Bacillus* species, including *B. subtilis*, *B. coagulans*, *B. clausii* and *B. licheniformis*, are of growing interest and seem to be promising models for biotechnological applications, involving industrial, agricultural, medical or, typically, scientific applications [80,81]. Pathogenic activity cases of 'Bacillus group' are extremely rare [82], thus, no phage therapy against these bacteria has been developed for human use. However, for animal protection purposes, successful phage therapy has been used, employing FBL1 bacteriophage preparation for combating the mortality of the white Pacific shrimp, *Litopenaeus vannamei*, infected by *B. licheniformis*. Some strains of these bacteria are capable of extending their upper temperature growth limit up to approximately 50 °C, thus stepping into the thermophilic range. The FBL1 not only impaired its host growth, but also prevented the formation of biofilm [83]. Besides biofilm degradation/prevention, there are other molecular mechanisms that can be exploited to control bacterial population. The bacteriophage SPO1 of *B. subtilis* is coding for Gp44 protein, which is part of an unusual molecular mimicry strategy inhibiting host RNA polymerase. Gp44 is composed of a DNA binding motif, a flexible DNA mimic domain and a random-coiled domain. Such composition allows selective binding to bacterial RNA polymerase through β and β' subunits, blocking bacterial growth. This a non-specific mechanism that SPO1 employs to target different bacterial transcription apparatus, regardless of the structural variations of RNA polymerases. This opens up the possibility for the generation of genetically modified SPO1 for the purpose of targeted gene therapy [84,85].

It is known that these industrial and medical biotechnology applications can be dramatically affected by undesired thermophage infections. On the other hand, thermophages are important in biotechnology, not only because of their potential to cause big economic losses but also because they also code for a number of valuable enzymes, which are often more robust than those of their hosts due to the rapid nature of a high temperature thermophage infection requirements. Some of these enzymes, with a big potential for novel applications, are the less recognized endolysins-lysozymes and depolymerases-glycosidases of thermophage origin [1,86,87]. These enzymes could be used for treating industrial installations overgrown with various bacteria as well as in biofilm removal for medical usage [1,86,87]. Since the critical initial step in a bacteriophage infection is the adsorption to its compatible receptor on the surface of the bacterial cell, the presence of exopolysaccharide envelopes can effectively block this stage. Some bacteriophages are producing a 'halo' around clear plaque centers, indicating a diffusing factor, hydrolyzing and allowing for the penetration of the extracellular exopolysaccharide matrix—a major bacterial biofilm component [88–90]. A number of thermophages produce thermostable depolymerases-glycosylases, endolysins (lysozymes) and holins, involved in actions against external bacterial layers, capsule, cell wall, cell membrane, protecting the internal cell cytoplasmatic environment [64]. The thermophage TP-84 codes for all three types of enzymes [2]. The combination of three such types of thermostable, robust enzymes in a cocktail for biofilms removal may become a powerful tool in industrial, environmental and medical applications. The ability of bacteria to form biofilms is an important element of their pathogenicity, and biofilm destruction is of high interest in medicine as a potential alternative or supplement to existing therapies [91]. TP-84 is a bacteriophage that deprives *G. stearothermophilus* bacteria of their capsule, thus facilitating the permeation of bacteriophages into deeper biofilm layers and lysis of the susceptible bacterial cells. Considering the rather broad specificity of TP-84 lytic enzymes, if applied to pathogen biofilms, it may prevent their formation and contribute to eradication of the biofilm bacteria [91]. Furthermore, thermophages often produce other useful enzymes, for example, GVE3 is encoding DNA ligase, DNA polymerase, among others [92]. Additionally, there are works being conducted to develop a GVE3 thermophage-based system for the introduction of novel or engineered metabolic and biosynthetic pathways [61]. Some of the lysogenic thermophages contain integrases, which are of growing importance for the genetic engineering of living eukaryotic cells, as they can mediate precise site-specific recombination between two different sequences [93,94]. An application unexplored thus far would be the construction of a robust, thermostable phage display system for directed evolution *in vitro*. Such systems have made a major breakthrough in molecular biology and biotechnology, so that the inventors were honored with the 2018 Nobel Prize.

**Author Contributions:** Conceptualization, B.Ł. and P.M.S.; methodology, B.Ł., P.M.S., I.S., J.J.-F.; resources, B.Ł. and P.M.S.; writing—original draft preparation B.Ł. and P.M.S.; writing—review and editing, B.Ł., I.S., J.J.-F. and P.M.S.; supervision P.M.S.; project administration, B.Ł. and P.M.S.; funding acquisition P.M.S. All authors have read and agreed to the published version of the manuscript.

**Funding:** This work was financially supported by the National Center for Research and Development (Poland) grant TECHMATSTRATEG2/410747/11/NCBR/2019 to Piotr Skowron.

**Institutional Review Board Statement:** Not applicable.

**Informed Consent Statement:** Not applicable.

**Data Availability Statement:** Brief data presented in the Table 1 will be published in details elsewhere.

**Acknowledgments:** Agnieszka-Zylicz-Stachula is appreciated for valuable suggestions. Jakub Zaleski is appreciated for English corrections.

**Conflicts of Interest:** The authors declare no conflict of interest.

## References

1. Shapiro, J.W.; Putonti, C. Gene co-occurrence networks reflect bacteriophage ecology and evolution. *mBio* **2018**, *9*, e01870-17. [CrossRef] [PubMed]
2. Skowron, P.M.; Kropinski, A.M.; Zebrowska, J.; Janus, L.; Szemiako, K.; Czajkowska, E.; Maciejewska, N.; Skowron, M.; Łoś, J.; Łoś, M.; et al. Sequence, genome organization, annotation and proteomics of the thermophilic, 47.7-kb *Geobacillus stearothermophilus* bacteriophage TP-84 and its classification in the new *Tp84virus* genus. *PLoS ONE* **2018**, *13*, e0195449. [CrossRef]
3. Zablocki, O.; van Zyl, L.; Trindade, M. Biogeography and taxonomic overview of terrestrial hot spring thermophilic bacteriophages. *Extremophiles* **2018**, *22*, 827–837. [CrossRef]
4. Wisotzkey, J.D.; Jurtshuk, P., Jr.; Fox, G.E.; Deinhard, G.; Poralla, K. Comparative sequence analyses on the 16S rRNA (rDNA) of *Bacillus acidocaldarius*, *Bacillus acidoterrestris*, and *Bacillus cycloheptanicus* and proposal for creation of a new genus, *Alicyclobacillus* gen. nov. *Int. J. Syst. Bacteriol.* **1992**, *42*, 263–269. [CrossRef]
5. Dufresne, S.; Bousquet, J.; Boissinot, M.; Guay, R. *Sulfobacillus disulfidooxidans* sp. nov., a new acidophilic, disulfide-oxidizing, gram-positive, spore-forming bacterium. *Int. J. Syst. Bacteriol.* **1996**, *46*, 1056–1064. [CrossRef] [PubMed]
6. Heyndrickx, M.; Lebbe, L.; Vancanneyt, M.; Kersters, K.; De Vos, P.; Logan, N.A.; Forsyth, G.; Nazli, S.; Ali, N.; Berkeley, R.C.W. A polyphasic reassessment of the genus *Aneurinibacillus*, reclassification of *Bacillus thermoaerophilus* (Meier-Stauffer et al. 1996) as Aneurinibacillus thermoaeropilus comb. nov., and emended descriptions of A. aneurinilyticus corrig., A migulanus, and A. thermoaerophilus. *Int. J. Syst. Bacteriol.* **1997**, *47*, 808–817.
7. Touzel, J.P.; O'Donohue, M.; Debeire, P.; Samain, E.; Breton, C. *Thermobacillus xylanilyticus* gen. nov., sp. nov., a new aerobic thermophilic xylan-degrading bacterium isolated from farm soil. *Int. J. Syst. Evol. Microbiol.* **2000**, *50*, 315–320. [CrossRef] [PubMed]

8.  Rainey, F.A.; Fritze, D.; Stackebrandt, E. The phylogenetic diversity of thermophilic members of the genus *Bacillus* as revealed by 16S rDNA analysis. *FEMS Microbiol. Lett.* **1994**, *115*, 205–211. [CrossRef] [PubMed]

9.  Nazina, T.N.; Tourova, T.P.; Poltaraus, A.B.; Novikova, E.V.; Grigoryan, A.A.; Ivanova, A.E.; Lysenko, A.M.; Petrunyaka, V.V.; Osipov, G.A.; Belyaev, S.S.; et al. Taxonomic study of aerobic thermophilic bacilli: Descriptions of *Geobacillus subterraneus* gen. nov., sp. nov. and *Geobacillus uzenensis* sp. nov. from petroleum reservoirs and transfer of *Bacillus stearothermophilus*, *Bacillus thermocatenulatus*, *Bacillus thermoleovorans*, *Bacillus kaustophilus*, *Bacillus thermoglucosidasius* and *Bacillus thermodenitrificans* to *Geobacillus* as the new combinations *G. stearothermophilus*, *G. thermocatenulatus*, *G. thermoleovorans*, *G. kaustophilus*, *G. thermoglucosidasius* and *G. thermodenitrificans*. *Int. J. Syst. Evol. Microbiol.* **2001**, *51*, 433–446. [CrossRef]

10. McMullan, G.; Christie, J.M.; Rahman, T.J.; Banat, I.M.; Ternan, N.G.; Marchant, R. Habitat, applications and genomics of the aerobic, thermophilic genus *Geobacillus*. *Biochem. Soc. Trans.* **2004**, *32*, 214–217. [CrossRef]

11. Abe, K.; Yoshinari, A.; Aoyagi, T.; Hirota, Y.; Iwamoto, K.; Sato, T. Regulated DNA rearrangement during sporulation in *Bacillus weihenstephanensis* KBAB4. *Mol. Microbiol.* **2013**, *90*, 415–427. [CrossRef] [PubMed]

12. Roger, W.; Hendrix, R.W. Bacteriophages: Evolution of the mMajority. *Theor. Popul. Biol.* **2002**, *61*, 471–480.

13. Reysenbach, A.-L.; Shock, E. Merging genomes with geochemistry in hydrothermal ecosystems. *Science* **2002**, *296*, 1077–1082. [CrossRef] [PubMed]

14. Wang, Y.; Zhang, X. Genome analysis of deep-sea thermophilic bacteriophage D6E. *Appl. Environ. Microbiol.* **2010**, *76*, 7861–7866. [CrossRef] [PubMed]

15. Sharp, R.J.; Riley, P.W.; White, D. *Thermophilic Bacilli*; Kristjansson, J.K., Ed.; CRC Press: Boca Raton, FL, USA, 1992; pp. 19–50.

16. Maugeri, T.L.; Gugliandolo, C.; Caccamo, D.; Stackebrandt, E. Three novel halotolerant and thermophilic *Geobacillus* strains from shallow marine vents. *Syst. Appl. Microbiol.* **2002**, *25*, 450–455. [CrossRef]

17. Blanc, M.; Marilley, L.; Beffa, T.; Aragno, M. Rapid identification of heterotrophic, thermophilic, spore-form-ing bacteria isolated from hot composts. *Int. J. Syst. Bacteriol.* **1997**, *47*, 1246–1248. [CrossRef]

18. Marks, T.J.; Hamilton, P.T. Characterization of a thermophilic bacteriophage of *Geobacillus kaustophilus*. *Arch. Virol.* **2014**, *159*, 2771–2775. [CrossRef]

19. Bell, E.M. *Life at Extremes: Environments, Organisms, and Strategies for Survival*; CABI: Wallingford/Oxfordshire, UK, 2012.

20. Pentecost, A. High temperature ecosystems and their chemical interactions with their environment. In *Ciba Foundation Symposium 202-Evolution of Hydrothermal Ecosystems on Earth (And Mars?): Ciba Foundation Symposium 202*; Bock, G.R., Goode, J.A., Eds.; John Wiley & Sons Ltd.: New York, NY, USA, 1996; pp. 99–111. [CrossRef]

21. Sharp, R.J.; Ahmad, S.I.; Munster, A.; Dowsett, B.; Atkinson, T. The isolation and characterization of bacterio-bacteriophages infecting obligately thermophilic strains of *Bacillus*. *J. Gen. Microbiol.* **1986**, *132*, 1709–1722. [CrossRef]

22. Liu, B.; Wu, S.; Song, Q.; Zhang, X.; Xie, L. Two novel bacteriophages of thermophilic bacteria isolated from deep-sea hydrothermal fields. *Curr. Microbiol.* **2006**, *53*, 163–166. [CrossRef]

23. Nagayoshi, Y.; Kumagae, K.; Mori, K.; Tashiro, K.; Nakamura, A.; Fujino, Y.; Hiromasa, Y.; Iwamoto, T.; Kuhara, S.; Ohshima, T.; et al. Physiological properties and genome structure of the hyperthermophilic filamentous bacteriophage φOH3 which infects *Thermus thermophilus* HB8. *Front. Microbiol.* **2016**, *7*, 50. [CrossRef]

24. Liu, B.; Zhang, X. Deep-sea thermophilic *Geobacillus* bacteriophage GVE2 transcriptional profile and proteomic characterization of virions. *Appl. Microbiol. Biotechnol.* **2008**, *80*, 697–707. [CrossRef]

25. Liu, B.; Zhou, F.; Wu, S.; Xu, Y.; Zhang, X. Genomic and proteomic characterization of a thermophilic Geobacillus bacteriophage GBSV1. *Res. Microbiol.* **2009**, *16*, 166–171. [CrossRef] [PubMed]

26. Doi, K.; Mori, K.; Martono, H.; Nagayoshi, Y.; Fujino, Y.; Tashiro, K.; Kuhara, S.; Ohshima, T. Draft genome sequence of *Geobacillus kaustophilus* GBlys, a lysogenic strain with bacteriophage φOH2. *Genome Announc.* **2013**, *1*, e00634-13. [CrossRef]

27. Morozova, V.; Bokovaya, O.; Kozlova, Y.; Kurilshikov, A.; Babkin, I.; Tupikin, A.; Bondar, A.; Ryabchikova, E.; Brayanskaya, A.; Peltek, S.; et al. A novel thermophilic *Aeribacillus* bacteriophage AP45 isolated from the Valley of Geysers, Kamchatka: Genome analysis suggests the existence of a new genus within the *Siphoviridae* family. *Extremophiles* **2019**, *23*, 599–612. [CrossRef]

28. Wei, D.; Zhang, X. Identification and characterization of a single-stranded DNA-binding protein from thermophilic bacteriophage GVE2. *Virus Genes* **2008**, *36*, 273–278. [CrossRef] [PubMed]

29. Brister, J.R.; Ako-Adjei, D.; Bao, Y.; Blinkova, O. NCBI viral genomes resource. *Nucleic Acids Res.* **2015**, *43*, D571–D577. [CrossRef]

30. Saunders, G.F.; Campbell, L.L.; Postgate, J.R. Abstract. In Proceedings of the 148th National Meeting of the American Chemical Society, Chicago, IL, USA, 30 August–4 September 1964.

31. Koser, S.A. Action of the bacteriophage on a thermophilic *Bacillus*. *Exp. Biol. Med.* **1926**, *24*, 109–111. [CrossRef]

32. Adant, M. Les bacteriophages des microbes thermophiles. *Compt. Rend. Soc. Biol.* **1928**, *99*, 1244–1245.

33. White, R.; Georgi, C.E.; Militzer, W. Heat studies on a thermophilic bacteriophage. *Proc. Soc. Exp. Biol. Med.* **1954**, *85*, 137–139. [CrossRef]

34. Marsh, C.L.; Larsen, D.H. Characterization of some thermophilic bacteria from the Hot Springs of Yellowstone National Park. *J. Bacteriol.* **1953**, *65*, 193–197. [CrossRef] [PubMed]

35. White, R.; Georgi, C.E.; Militzer, W.E. Characteristics of a thermophilic bacteriophage. *Proc. Soc. Exp. Biol. Med.* **1955**, *88*, 373–377. [CrossRef] [PubMed]

36. Hirano, J.; Mutoh, Y.; Kitamura, M.; Asai, S.; Nakajima, I.; Takamiya, A. Effect of change of temperature upon the viability of a thermophilic bacterium. *J. Gen. Appl. Microbiol.* **1958**, *4*, 188–199. [CrossRef]

37. Onodera, N. *Electron-Microscopy* **1959**, *7*, 2–3, 19. (In Japanese)
38. Onodera, N. On some characteristics of a newly isolated thermophilic bacteriophage and consideration of its thermostability. *J. Electron. Microsc. (Tokyo)* **1961**, *10*, 91–102. [CrossRef]
39. Saunders, G.F.; Campbell, L.L. Characterization of a thermophilic bacteriophage for *Bacillus stearothermophilus*. *J. Bacteriol.* **1966**, *91*, 340–348. [CrossRef]
40. Zeigler, D.R. The genus *Geobacillus*. In *Bacillus Genetic Stock Center, Catalog of Strains*, 7th ed.; Department of Biochemistry, The Ohio State University: Columbus, OH, USA, 2001; Volume 3, pp. 1–25.
41. Epstein, I.; Campbell, L. Production and purification of the thermophilic bacteriophage TP-84. *Appl. Microbiol.* **1975**, *29*, 219–223. [CrossRef]
42. Shafiai, F.; Thompson, T.L. Isolation and preliminary characterization of bacteriophage φμ-4. *J. Bacteriol.* **1964**, *87*, 999–1002. [CrossRef]
43. Walker, E.; Campbell, L. Induction and properties of a temperate bacteriophage from *Bacillus stearothermophilus*. *J. Bacteriol.* **1965**, *89*, 175–184. [CrossRef]
44. Walker, E.; Campbell, L. Biochemical changes in lysogenic *Bacillus stearothermophilus* after bacteriophage induction. *J. Bacteriol.* **1965**, *90*, 1129–1137. [CrossRef]
45. Welker, E. Transduction in *Bacillus stearothermophilus*. *J. Bacteriol.* **1988**, *170*, 3761–3764. [CrossRef]
46. Carnevali, F.; Donelli, G. Some properties of a thermophilic phage DNA. *Arch. Biochem. Biophys.* **1968**, *125*, 376–377. [CrossRef]
47. Egbert, L.N.; Mitchell, H.K. Characteristics of Tφ3, a bacteriophage for *Bacillus stearothermophilus*. *J. Virol.* **1967**, *1*, 610–616. [CrossRef]
48. Egbert, L.N. Characteristics of the deoxyribonucleic acid of T phi 3, a bacteriophage for *Bacillus stearothermophilus*. *J. Virol.* **1969**, *3*, 528–532. [CrossRef]
49. Humbert, R.D.; Fields, M.L. Study of two bacteriophages of *Bacillus stearothermophilus* strain NCA1518. *J. Virol.* **1972**, *9*, 397–398. [CrossRef] [PubMed]
50. Jurgen, C.L.; Edero, M. The influence of inorganic ions on the heat stability of a thermophilic bacteriophage. *Physiol. Plant.* **1972**, *27*, 182–186.
51. Reanney, D.C.; Marsh, S.C.N. The ecology of viruses attacking *Bacillus stearothermophilus* in soil. *Soil Biol. Biochem.* **1973**, *5*, 399–406. [CrossRef]
52. Sakaki, Y.; Oshima, T. A new lipid-containing phage infecting acidophilic thermophilic bacteria. *Virology* **1976**, *75*, 256–259. [CrossRef]
53. Sharp, R.J.; Bown, K.J.; Atkinson, A. Phenotypic and genotypic characterization of some thermophilic species of *Bacillus*. *J. Gen. Microbiol.* **1980**, *117*, 201–210. [CrossRef]
54. Wolf, J.; Sharp, R.J. Taxonomic and related aspects of thermophiles within the Genus *Bacillus*. In *The Aerobic Endospore-Forming Bacteria: Classification and Identification*; Berkeley, R.C.W., Goodfellow, M., Eds.; Academic Press: London, UK, 1981; pp. 251–296.
55. Sharp, R.J.; Woodrow, R. Numerical taxonomy of *Bacillus* thermophiles. In *Abstracts of the XHIth International Congress of Microbiology Boston USA*; Nottingham Trent University: Nottingham, UK, 1982; pp. 61–62.
56. Wang, Y.; Zhang, X. Characterization of a novel portal protein from deep-sea thermophilic bacteriophage GVE2. *Gene* **2008**, *421*, 61–66. [CrossRef] [PubMed]
57. Jin, M.; Ye, T.; Zhang, X. Roles of bacteriophage GVE2 endolysin in host lysis at high temperatures. *Microbiology* **2013**, *159*, 1597–1605. [CrossRef] [PubMed]
58. Liu, B.; Wu, S.; Xie, L. Complete genome sequence and proteomic analysis of a thermophilic bacteriophage BV1. *Acta Oceanol. Sin.* **2010**, *29*, 84–89. [CrossRef]
59. Fujimoto, D.F.; Higginbotham, R.H.; Sterba, K.M.; Maleki, S.J.; Segall, A.M.; Smeltzer, M.S.; Hurlburt, B.K. *Staphylococcus aureus* SarA is a regulatory protein responsive to redox and pH that can support bacteriophage lambda integrase-mediated excision/recombination. *Mol. Microbiol.* **2009**, *74*, 1445–1458. [CrossRef]
60. Alonso, J.C.; Lüder, G.; Stiege, A.C.; Chai, S.; Weise, F.; Trautner, T.A. The complete nucleotide sequence and functional organization of *Bacillus subtilis* bacteriophage SPP1. *Gene* **1997**, *204*, 201–212. [CrossRef]
61. Levine, M.; Truesdall, S.; Ramakrishan, T.; Bronson, M.J. Dual control of lysogeny by bacteriophage P22: An antirepressor locus and its controlling elements. *J. Mol. Biol.* **1975**, *91*, 421–438. [CrossRef]
62. van Zyl, L.J.; Sunda, F.; Taylor, M.P.; Cowan, D.; Trindade, M.I. Identification and characterization of a novel *Geobacillus thermoglucosidasius* bacteriophage, GVE3. *Arch. Virol.* **2015**, *160*, 2269–2282. [CrossRef]
63. van Zyl, L.J.; Taylor, M.P.; Trindade, M. Engineering resistance to bacteriophage GVE3 in *Geobacillus thermoglucosidasius*. *Appl. Microbiol. Biotechnol.* **2016**, *100*, 1833–1841. [CrossRef]
64. Maszewska, A. Bacteriophage associated polysaccharide depolymerases—Characteristics and application. *Postepy Hig. Med. Dosw.* **2015**, *69*, 690–702. [CrossRef] [PubMed]
65. Rocha, E.P.C.; Danchin, A. Base composition bias might result from competition for metabolic resources. *TRENDS Genet.* **2002**, *18*, 291–294. [CrossRef]
66. Akhwale, J.K.; Rohde, M.; Rohde, C.; Bunk, B.; Sproer, C.; Klenk, H.-P.; Boga, H.I.; Wittmann, J. Comparative genomic analysis of eight novel haloalkaliphilic bacteriophages from Lake Elmenteita, Kenya. *PLoS ONE* **2019**, *14*, e0212102. [CrossRef]

67. Mwirichia, R.; Cousin, S.; Muigai, A.W.; Boga, H.I.; Stackebrandt, E. Archaeal diversity in the haloalkaline Lake Elmenteita in Kenya. *Curr. Microbiol.* **2010**, *60*, 47–52. [CrossRef] [PubMed]

68. Streisinger, G.; Emrich, J.; Stahl, M.M. Chromosome structure in phage T4, iii. Terminal redundancy and length determination. *Proc. Natl. Acad. Sci. USA* **1967**, *57*, 292–295. [CrossRef] [PubMed]

69. Casjens, S. Prophages and bacterial genomics: What have we learned so far? *Mol. Microbiol.* **2003**, *49*, 277–300. [CrossRef] [PubMed]

70. Wittmann, J.; Dreiseikelmann, B.; Rohde, C.; Rohde, M.; Sikorski, J. Isolation and characterization of numerous novel phages targeting diverse strains of the ubiquitous and opportunistic pathogen *Achromobacter xylosoxidans*. *PLoS ONE* **2014**, *9*, e86935. [CrossRef] [PubMed]

71. Sunderland, K.S.; Yang, M.; Mao, C. Phage-Enabled Nanomedicine: From Probes to Therapeutics in Precision Medicine. *Angew Chem. Int. Ed. Engl.* **2017**, *56*, 1964–1992. [CrossRef] [PubMed]

72. Bustard, M.T.; Whiting, S.; Cowan, D.A.; Wright, P.C. Biodegradation of high-concentration isopropanol by a solvent-tolerant thermophile, *Bacillus pallidus*. *Extremophiles* **2002**, *6*, 319–323. [CrossRef]

73. Taylor, M.P.; Eley, K.L.; Martin, S.; Tuffin, M.I.; Burton, S.G.; Cowan, D.A. Thermophilic ethanologenesis: Future prospects for second-generation bioethanol production. *Trends Biotechnol.* **2009**, *27*, 398–405. [CrossRef]

74. Cuebasg, M.; Sannino, D.; Bini, B. Isolation and characterization of an arsenic resistant *Geobacillus kaustophilus* strain from geothermal soils. *J. Basic Microbiol.* **2011**, *51*, 364–371. [CrossRef]

75. Feng, L.; Wang, W.; Cheng, J.; Ren, Y.; Zhao, G.; Gao, C.; Tang, Y.; Lui, X.; Han, W.; Peng, X.; et al. Genome and proteome of long-chain alkane degrading *Geobacillus thermodenitrificans* NG80-2 isolated from a deep-subsurface oil reservoir. *Proc. Natl. Acad. Sci. USA* **2007**, *104*, 5602–5607. [CrossRef]

76. Moriello, V.S.; Lama, L.; Poli, A.; Gugliandolo, C.; Maugeri, T.L.; Gamacorta, A.; Nicolaus, B. Production of exopolysaccharides from a thermophilic microorganism isolated from a marine hot spring in flegrean areas. *J. Ind. Microbiol. Biotechnol.* **2003**, *30*, 95–101. [CrossRef]

77. Novotny, J.F.; Perry, J.J. Characterization of bacteriocins from two strains of Bacillus thermoleovorans, a thermophilic hydrocarbon-utilizing species. *Appl. Environ. Microbiol.* **1992**, *58*, 2393–2396. [CrossRef]

78. Jeżewska-Frąckowiak, J.; Seroczyńska, K.; Banaszczyk, J.; Jedrzejczak, G.; Żylicz-Stachula, A.; Skowron, P.M. The promises and risks of probiotic *Bacillus* species. *Acta Biochim. Pol.* **2018**, *65*, 509–519. [CrossRef]

79. Luo, Y.; Korza, G.; De Marco, A.M.; Kuipers, O.P.; Li, Y.Q.; Setlow, P. Properties of spores of *Bacillus subtilis* with or without a transposon that decreases spore germination and increases spore wet heat resistance. *J. Appl. Microbiol.* **2021**. [CrossRef]

80. Kuebutornye, F.K.A.; Abarike, E.D.; Lu, Y. A review on the application of *Bacillus* as probiotics in aquaculture. *Fish Shellfish Immunol.* **2019**, *87*, 820–828. [CrossRef]

81. Mu, Y.; Cong, Y. *Bacillus coagulans* and its applications in medicine. *Benef. Microbes* **2019**, *10*, 679–688. [CrossRef]

82. Logan, N.A. *Bacillus* and relatives in foodborne illness. *J. Appl. Microbiol.* **2012**, *112*, 417–429. [CrossRef] [PubMed]

83. Rehman, S.; Ali, Z.; Khan, M.; Bostan, N.; Naseem, S. The dawn of phage therapy. *Rev. Med. Virol.* **2019**, *29*, e2041. [CrossRef]

84. Wang, Z.; Wang, H.; Mulvenna, N.; Sanz-Hernandez, M.; Zhang, P.; Li, Y.; Ma, J.; Wang, Y.; Matthews, S.; Wigneshweraraj, S.; et al. A Bacteriophage DNA Mimic Protein Employs a Non-specific Strategy to Inhibit the Bacterial RNA Polymerase. *Front. Microbiol.* **2021**, *12*, 692512. [CrossRef] [PubMed]

85. Mulvenna, N.; Hantke, I.; Burchell, L.; Nicod, S.; Bell, D.; Turgay, K.; Wigneshweraraj, S. Xenogeneic modulation of the ClpCP protease of *Bacillus subtilis* by a phage-encoded adaptor-like protein. *J. Biol. Chem.* **2019**, *294*, 17501–17511. [CrossRef] [PubMed]

86. Gil, J.F.; Mesa, V.; Estrada-Ortiz, N.; Lopez-Obando, M.; Gómez, A.; Plácido, J. Viruses in Extreme Environments, Current Overview, and Biotechnological Potential. *Viruses* **2021**, *13*, 81. [CrossRef] [PubMed]

87. Maat, D.S.; Biggs, T.; Evans, C.; Van Bleijswijk, J.D.L.; Van Der Wel, N.N.; Dutilh, B.E.; Brussaard, C.P.D. Characterization and temperature dependence of arctic micromonas polaris viruses. *Viruses* **2017**, *9*, 134. [CrossRef] [PubMed]

88. Hughes, K.A.; Sutherland, I.W.; Clark, J.; Jones, M.V. Bacteriophage and associated polysaccharide depolymerases—Novel tools for study of bacterial biofilms. *J. Appl. Microbiol.* **1998**, *85*, 583–590. [CrossRef]

89. Hughes, K.A.; Sutherland, I.W.; Clark, J.; Jones, M.V. Biofilm susceptibility to bacteriophage attack: The role of bacteriophageborne polysaccharide depolymerase. *Microbiology* **1998**, *144*, 3039–3047. [CrossRef]

90. Donlan, R.M. Preventing biofilms of clinically relevant organisms using bacteriophage. *Trends Microbiol.* **2009**, *17*, 66–72. [CrossRef]

91. Parasion, S.; Kwiatek, M.; Gryko, R.; Mizak, L.; Malm, A. Bacteriophages as an alternative strategy for fighting biofilm development. *Pol. J. Microbiol.* **2014**, *63*, 137–145. [CrossRef] [PubMed]

92. Szekera, K.; Zhou, X.; Schwab, T.; Casanueva, A.; Cowan, D.; Mikhailopulo, I.A.; Neubauer, P. Comparative investigations on thermostable pyrimidine nucleoside phosphorylases from *Geobacillus thermoglucosidasius* and *Thermus thermophiles*. *J. Mol. Cat. B Enzym.* **2012**, *84*, 27–34. [CrossRef]

93. Julien, B. Characterization of the integrase gene and attachment site for the *Myxococcus xanthus* bacteriophage Mx9. *J. Bacteriol.* **2003**, *185*, 6325–6330. [CrossRef] [PubMed]

94. Piazzolla, D.; Calì, S.; Spoldi, E.; Forti, F.; Sala, C.; Magnoni, F.; Dehò, G.; Ghisotti, D. Expression of phage P4 integrase is regulated negatively by both Int and Vis. *J. Gen. Virol.* **2006**, *87*, 2423–2431. [CrossRef]

 *microorganisms*

**MDPI**

*Article*

# Establishment Genes Present on pLS20 Family of Conjugative Plasmids Are Regulated in Two Different Ways

Jorge Val-Calvo [1], Andrés Miguel-Arribas [1], Fernando Freire [2], David Abia [2], Ling Juan Wu [3,*] and Wilfried J.J. Meijer [1,*]

1   Centro de Biología Molecular "Severo Ochoa" (CSIC-UAM), Instituto de Biología Molecular "Eladio Viñuela" (CSIC), C. Nicolás Cabrera 1, Universidad Autónoma, Canto Blanco, 28049 Madrid, Spain; val.calvo.jorge@gmail.com (J.V.-C.); amiguelarr@gmail.com (A.M.-A.)
2   Centro de Biología Molecular "Severo Ochoa" (CSIC-UAM), Bioinformatics Facility, C. Nicolás Cabrera 1,Universidad Autónoma, Canto Blanco, 28049 Madrid, Spain; nangdoide@gmail.com (F.F.); dabia@cbm.csic.es (D.A.)
3   Centre for Bacterial Cell Biology, Biosciences Institute, Newcastle University, Richardson Road, Newcastle Upon Tyne NE4AX, UK
*   Correspondence: l.j.wu@newcastle.ac.uk (L.J.W.); wmeijer@cbm.csic.es (W.J.J.M.); Tel.: +34-91-196-4539 (W.J.J.M.)

**Citation:** Val-Calvo, J.; Miguel-Arribas, A.; Freire, F.; Abia, D.; Wu, L.J.; Meijer, W.J. Establishment Genes Present on pLS20 Family of Conjugative Plasmids Are Regulated in Two Different Ways. *Microorganisms* **2021**, *9*, 2465. https://doi.org/10.3390/microorganisms9122465

Academic Editor: Imrich Barák

Received: 26 October 2021
Accepted: 25 November 2021
Published: 29 November 2021

**Publisher's Note:** MDPI stays neutral with regard to jurisdictional claims in published maps and institutional affiliations.



**Abstract:** During conjugation, a conjugative DNA element is transferred from a donor to a recipient cell via a connecting channel. Conjugation has clinical relevance because it is the major route for spreading antibiotic resistance and virulence genes. The conjugation process can be divided into different steps. The initial steps carried out in the donor cell culminate in the transfer of a single DNA strand (ssDNA) of the conjugative element into the recipient cell. However, stable settlement of the conjugative element in the new host requires at least two additional events: conversion of the transferred ssDNA into double-stranded DNA and inhibition of the hosts' defence mechanisms to prevent degradation of the transferred DNA. The genes involved in this late step are historically referred to as establishment genes. The defence mechanisms of the host must be inactivated rapidly and—importantly—transiently, because prolonged inactivation would make the cell vulnerable to the attack of other foreign DNA, such as those of phages. Therefore, expression of the establishment genes in the recipient cell has to be rapid but transient. Here, we studied regulation of the establishment genes present on the four clades of the pLS20 family of conjugative plasmids harboured by different *Bacillus* species. Evidence is presented that two fundamentally different mechanisms regulate the establishment genes present on these plasmids. Identification of the regulatory sequences were critical in revealing the establishment regulons. Remarkably, whereas the conjugation genes involved in the early steps of the conjugation process are conserved and are located in a single large operon, the establishment genes are highly variable and organised in multiple operons. We propose that the mosaical distribution of establishment genes in multiple operons is directly related to the variability of defence genes encoded by the host bacterial chromosomes.

**Keywords:** conjugation; transcriptional regulation; Gram-positive bacteria; antibiotic resistance; establishment genes; *Bacillus subtilis*; plasmid; pLS20

## 1. Introduction

Conjugation is a horizontal gene transfer (HGT) route by which a DNA element is transferred from a donor to a recipient cell via a connecting channel. Conjugation occurs at a large scale in both Gram-negative (G−) and Gram-positive (G+) bacteria. Conjugative elements can be located in a bacterial chromosome or plasmid, and are named integrated conjugative elements (ICEs) or conjugative plasmid, respectively. The basic principles of the conjugation processes are conserved in G+ and G− bacteria. Conjugation can be divided into five discernible steps. The first step involves selection of and attachment to a recipient cell by the donor cell. The second step comprises the synthesis of a membrane-embedded DNA translocation machinery that is a T4-type secretion system (T4SS). In the third step, the conjugative DNA is processed in the donor cell in order to generate the single-stranded DNA (ssDNA) that is subsequently transferred through the translocation channel into the recipient cell. The final two steps take place in the recipient cell and are referred to as the establishment steps. Firstly, the transferred ssDNA must be circularised and converted into double-stranded plasmid DNA. Secondly, defence mechanisms that protect the cell against incoming foreign DNA must be inactivated. Most conjugation studies address aspects of the first three steps, but very little is known about the establishment step, particularly the way by which conjugative elements inhibit defence systems of the recipient cell.

One well-known type of bacterial defence mechanism is the restriction–modification (RM) systems, which encode a restriction endonuclease (REase) and a methyltransferase (MTase). The MTase methylates specific short DNA sequences, and thereby prevents these sequences from being recognised and digested by the cognate REase of the RM system. The REase will digest any foreign DNA entering a cell that is not properly methylated (including conjugative DNA) [1–3]. This system can be considered an innate defence mechanism. Another well-known system, which can be considered as an adaptive defence mechanism, is the CRISPR–Cas system [4]. Additional defence systems have been discovered in recent years [5]. For a conjugation event to be successful, the defence systems have to be evaded or inactivated, implying that conjugative elements encode inhibitors of bacterial defence systems. Although this topic has been studied little, it is known for more than two decades that many conjugative plasmids contain an anti-restriction gene whose encoded product is able to inhibit a restriction enzyme [6–9], i.e., anti-restriction genes are typical establishment genes. Importantly, inactivation of the bacterial defence systems must occur temporarily, because prolonged inactivation would make the cell vulnerable to entry of other foreign DNA, e.g., phage DNA. This implies that anti-restriction genes and other establishment genes must be regulated in a special way, such that they are expressed rapidly and transiently upon arrival of the plasmid in the recipient cell. In previous work, we showed that the conjugative plasmid p576 from *Bacillus pumilis* NRS576 contains an anti-restriction gene, and also encodes a transcriptional regulator, named $Reg_{p576}$, that efficiently represses the promoter of the anti-restriction gene [10]. While the DNA, but not the repressor protein, is transferred during conjugation, the anti-restriction gene will be expressed in the recipient cell soon after it enters the cell until sufficient repressor protein is produced to repress the promoter again. $Reg_{p576}$ also regulates its own expression and it represses two other promoters controlling the expression of four genes, which presumably also play a role in establishment. In other words, plasmid p576 contains an establishment regulon whose genes are regulated by $Reg_{p576}$.

Plasmid p576 shares similarity with the conjugative plasmid pLS20 from *Bacillus subtilis*. Although pLS20 is among the best-studied conjugative plasmids in G+ bacteria, its establishment genes have not been studied so far. Very recently, we have found that pLS20 is the prototype of a new family of conjugative plasmids that includes p576 [11]. The family has over 30 members and the plasmids are hosted by *Bacillus* species that are distributed worldwide. In this work, we studied the establishment regulons present on the pLS20 family of plasmids. We found that all plasmids contain an establishment regulon and that these regulons contain at least one gene that interferes with RM systems. Interestingly, these establishment genes are regulated by two fundamentally different systems, prototyped by the systems present on p576 and pLS20, respectively. Furthermore, whereas the conjugation genes are highly conserved between members of the pLS20 family of plasmids, there is a high level of variation in the establishment genes. We discuss the possibility that the diversity of the establishment genes is directly related to the defence genes encoded by the bacterial genome.

## 2. Materials and Methods

### 2.1. Bacterial Strains, Plasmids, and Oligonucleotides

Bacterial strains were grown in lysogeny broth (LB) or LB agar plates (LB with 1.5% agar [12]). Where appropriate, the following antibiotics were added: ampicillin (100 µg/mL) for *E. coli*; or spectinomycin (100 µg/mL) for *B. subtilis*. The strains, plasmids, and oligonucleotides used are listed in supplementary Tables S1–S3, respectively. All oligonucleotides were purchased from Integrated DNA Technologies (Belgium).

### 2.2. Construction of Plasmids and Strains

Standard molecular methods were used to manipulate DNA [13]. The correctness of all constructs was verified by sequence analysis. All enzymes were purchased from New England Biolabs, USA. *E. coli* transformation was carried out using standard methods [13]. Competent *B. subtilis* 168 cells were prepared as described before [14]. Transformants were selected on LB agar plates supplemented with appropriate antibiotics.

The following strategy was used to construct derivatives of *B. subtilis* 168 that contain at their *amyE* locus a cassette in which the wild-type or derivative of the EGeRS1-B region of pLS20 is placed between the inducible IPTG $P_{spank}$ promoter and the superfolder *gfp* gene (for simplicity, named here *gfp*). First, the intergenic region between pLS20cat genes *82c* and *85*, encompassing EGeRS1-B, or a subregion, was amplified by PCR using pairs of primers listed in supplemental Table S3. The amplified product was purified, digested with *SalI*, and cloned into the unique *SalI* site that is present on the *amyE* integration vector pAND101 in between the $P_{spank}$ promoter and the *gfp* gene [15]. The ligation mixtures were used to transform competent *E. coli* XL1-Blue cells. Recombinant plasmids were identified by colony PCR. The orientation of the insert was determined by PCR using primer pDR111_U in combination with one of the two oligonucleotides used to amplify the corresponding fragment. The names of the pAND101 derivatives are listed in supplemental Table S2. Constructed plasmids were then used to transform competent *B. subtilis* 168 cells, and spectinomycin-resistant transformants were tested for the loss of amylase activity to identify clones resulting from double crossover events at *amyE*. The resulting strains, which are listed in supplemental Table S1, contain the configuration $P_{spank}$-[fragment X]-*gfp*.

### 2.3. Flow Cytometry

Promoter strength quantification using fluorescence and flow cytometry was carried out as described before [16]. In short, overnight grown samples in LB medium at 37 °C were diluted to an $OD_{600} \approx 0.025$ in fresh 37 °C LB supplemented with or without IPTG (1 mM), and grown again until $OD_{600} \approx 0.8$–1. For each culture, a 2 mL sample was pelleted, washed twice with 2 mL of filtered 1xPBS, and resuspended in 1 mL of filtered 1xPBS. Fluorescence value is expressed as the mean value of the Geomean value of 100,000 cells measured in three independent experiments.

### 2.4. In Silico Analyses

The homologous genes of the studied plasmids were initially identified using the Get_Homologues or Blast tools. The InterProScan program was used to detect conserved domains or classify proteins into families already described. The homology relationships detected were represented on the sequences of the plasmids, giving rise to a comparative genetic map. Genetic maps were made using the GenomeDiagram module (Biopython project), and subsequently edited with Inkscape software to manually annotate and represent the homology relationships on the plasmid maps.

The EGeRS1 regions were identified as duplications in the plasmids on dot plot graphs (plasmid sequence against itself) or by direct observation of the sequences. The ViennaRNA Web Services toolkit was used to analyse the secondary structures of the EGeRS1 regions, either in RNA or ssDNA [17]. The RNAfold program was used to predict secondary structures of each EGeRS region, while the RNAalifold program [18] was used to predict conserved secondary structures in MSA of the EGeRS regions. The parameters used in both programs were the default values, except selecting the Turner model (2004) for RNA and Matthews model (2004) for ssDNA. The alignments used as input to the RNAalifold program were performed with the Pro-Coffee algorithm with standard parameters [19]. Presentation of the alignments was prepared using the Esprit3.0 program [20], while the arc diagrams were made using the R-CHIE web server [21]. The software used to generate the phylogenetic tree was IQ-TREE [22]. The IQ-TREE software was allowed to determine the substitution model to be used [23], and the ultrafast bootstrap statistical method was applied (1000 replicates). The phylogenetic trees were drawn using the FigTree v1.4.4 software (http://tree.bio.ed.ac.uk/, accessed on 7 September 2021, Andrew Rambout research group, Edinburgh, UK). The phylogenetic tree was unrooted, although a root was subsequently added, located in the midpoint. The sequences of the EGeRS regions of plasmids pLS20, pBatNRS213, pBamB1895, pBglSRCM103574, pBliYNP2†, and pBspNMCC4† were aligned by the Pro-Coffee algorithm. The likelihood tree was made from 30 sequences with 1240 base pairs. The model selected by ModelFinder was K2P + R2. The tree with the highest log likelihood ($-6291.1512$) is shown.

### 3. Results

*3.1. pLS20 Contains an Anti-Restriction Gene Similar to the One on Plasmid p576, but Both Genes Are Preceded by Very Different Sequences*

In previous work, we found that the *B. pumilus* plasmid p576 contains an anti-restriction gene, $ardC_{p576}$, which is a typical establishment gene [10]. The upstream sequences hold the key for the transient expression of $ardC_{p576}$ after the plasmid has entered a recipient cell. Sequences sharing between 50 and 68% identity to those upstream of $ardC_{p576}$ are also present upstream of some other p576 genes/operons, and these genes were found to be regulated in an identical or similar way as $ardC_{p576}$, strongly indicating that these genes also play a role in establishment of the plasmid in the recipient cell. In other words, the conserved upstream sequences of the typical establishment gene $ardC_{p576}$ formed the crux in identifying the establishment regulon of p576 [10]. A schematic view of the regulatory mechanism of the establishment operons on plasmid p576 is shown in supplemental Figure S1.

We wondered if we could use the same strategy to identify the establishment regulon of plasmid pLS20 from *B. subtilis* that is related to p576 [24,25]. Like p576, pLS20 contains an anti-restriction gene (gene *82c*, according to our annotation, and renamed here $ardC_{pLS20}$) whose encoded product shares 51% identity with $ArdC_{p576}$ (supplemental Figure S2). Like $ardC_{p576}$ of p576, $ardC_{pLS20}$ would be expressed only transiently after conjugative transfer of pLS20 into the recipient cell. However, the sequences upstream of $ardC_{pLS20}$ are very different from those upstream of the p576 establishment genes: the region is about fivefold larger, and it contains multiple inverted repeated sequences that are absent in p576 (Figure 1A,C). The presence of completely different sequences upstream of $ardC_{pLS20}$ and upstream of $ardC_{p576}$ and other p576 establishment genes strongly suggests that they are regulated in different ways.

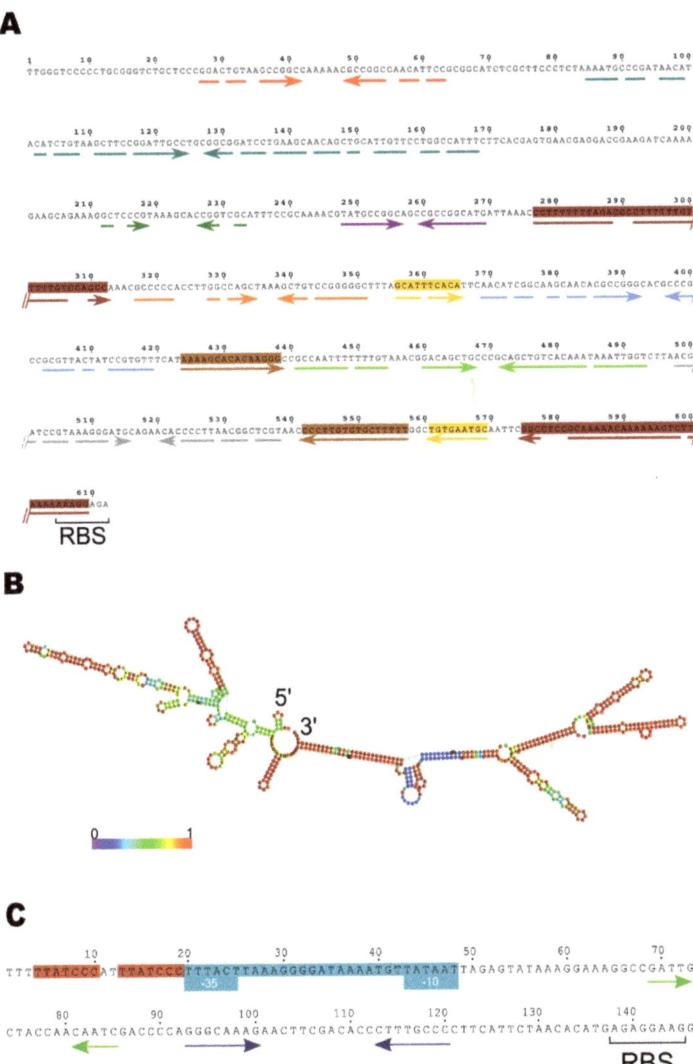

**Figure 1.** The region upstream of *ardC$_{pLS20}$*, which is different from the upstream region of *ardC$_{p576}$*, contains multiple inverted repeated sequences predicted to generate stable secondary structures in ssDNA or RNA. (**A**) The 600 bp region immediately upstream of the ribosome binding site (RBS) of *ardC$_{pLS20}$* contains multiple inverted repeated sequences. Inverted repeated sequences are indicated with coloured arrows shown above the sequence. The inverted repeated sequences that are separated by sequences of more than 100 bp, which themselves can contain other inverted repeated sequences, are boxed. (**B**) Centroid secondary structure predicted to be formed according to the RNAfold web server (rna.tbi.univie.ac.at//cgibin/RNAWebSuite/RNAfold.cgi) when the upstream region is in its single-stranded form. A very similar secondary structure is predicted when the region is transcribed into RNA. The sequence immediately upstream of the RBS is predicted to form a branch of ~300 nt with four ramifications (positioned at the right side of the image). Upstream of this branch, another branch of ~300 nt would be formed, with two and three ramifications predicted to be formed with low and high probability, respectively. Colours indicate the degree of probability that the predicted

secondary structure is formed. Red nucleotides reflect very high probabilities. Levels of decreasing probabilities are indicated by nucleotides in orange, yellow, green, light blue, and dark blue colours. (**C**) The 140 bp region of plasmid p576 located upstream of $ardC_{p576}$. The promoter sequences of $P_{ardCp576}$ are shown on a blue background and the −35 and −10 boxes are indicated. The $Reg_{p576}$ operator is shown on a red background. The two short inverted repeats, indicated with blue and green convergent arrows, most likely function to enhance the half-life time of the RNA molecule [10].

### 3.2. Sequences Highly Similar to Those Upstream of $ardC_{pLS20}$ Are also Present Upstream of Four Operons on pLS20: Identification of the Establishment Regulon of pLS20

Further analysis of the pLS20 revealed that sequences sharing an identity ranging between 62 and 95% to the region upstream of $ardC_{pLS20}$ were present at four other positions on pLS20, all confined within a region spanning about 20% of the pLS20 genome (see Figure 2A). An alignment of these five conserved sequences, ranging from 420 to 750 bp, is shown in Figure 2B. Interestingly, all five sequences are located immediately upstream of a likely ribosomal binding site (RBS), and each is followed by a putative operon (see Figure 2A). To distinguish between them, we refer to these five operons as operon A to E, which are composed of the following pLS20 genes: *79c-78c-77c-76c* (operon A); genes *82c-81c* (operon B, gene *82c* = $ardC_{pLS20}$); genes *90c-89c-88c-87c-86c* (operon C); genes *6c-5c-4c-3c-2c* (operon D); and genes *9c-8c-7c* (operon E) (see Figure 2A).

By analogy with the situation on plasmid p576, it is likely that the upstream sequences are crucial for regulating the expression of the genes present in the downstream operons of pLS20. In other words, the 19 genes from these five operons probably form the establishment regulon of pLS20. We refer to the conserved sequences upstream of the putative establishment operons present on pLS20 and p576 as establishment gene regulatory sequence (EGeRS) type 1 (or EGeRS1) and EGeRS type 2 (or EGeRS2), respectively. To discriminate between the five EGeRS1 regions present on pLS20, their EGeRS1 names are extended with the letter corresponding to the downstream operon; hence, the region upstream of $ardC_{pLS20}$ is named EGeRS1-B. Although the EGeRS1 sequences are 62 to 95% identical to each other, there are some noteworthy differences. For instance, the 420 bp EGeRS1-A is smaller than the approximate 610 bp long regions of EGeRS1-B, D, and E, because it lacks the 180 bp sequence located at the 5′ side of the EGeRS1 regions in B, D, and E. Finally, EGeRS1-C has a size of 751 bp due to an internal duplication of 131 bp plus 9 extra bp (Figure 2B).

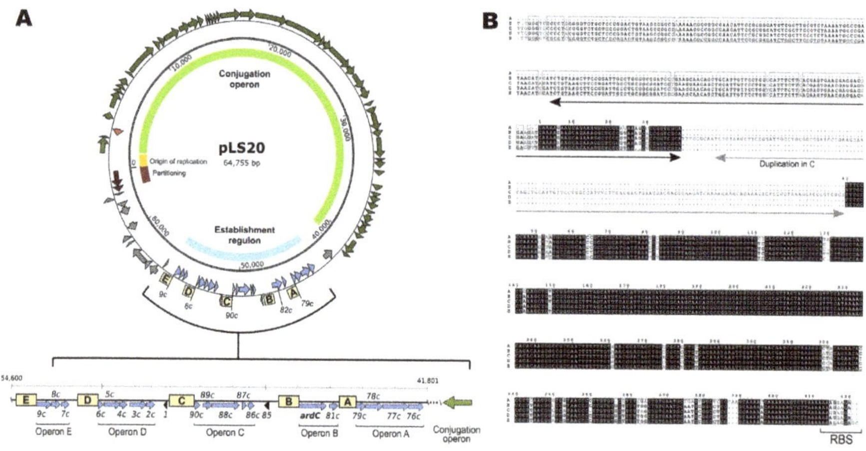

**Figure 2.** A conserved region of about 600 bp is present upstream of five putative operons on pLS20. (**A**) Circular map of pLS20. The different operons/modules are indicated with thick coloured lines on the inner circle: green, conjugation operon;

yellow, origin of replication; brown, partitioning genes; and blue, establishment genes. Genes are indicated with wide arrows on the outer ring. Green- and blue-coloured arrows represent conjugation and establishment genes, respectively. Other genes are coloured grey, except the gene encoding the repressor of the conjugation genes, which is shown in red. The blow-up shown in the lower panel shows the organisation of the five establishment operons encompassing genes *76c* to *9c*, corresponding to plasmid positions 41,801–54,600 of pLS20cat (accession number AB615352). Positions of repeated sequences are indicated with yellow boxes and numbered A to E. Genes are indicated with wide arrows. The green arrow indicates the position and orientation of the conjugation genes. (**B**) Alignment of the five conserved regions located upstream of the putative operons on pLS20. Note that region "A" is shorter due to the absence of about 195 bp at the $5'$ end that is conserved in the other four regions. Region "C" is longer due to an internal repeat of 131 bp (underlined with a grey arrow; the black arrow indicates the duplicated segment). Each sequence is located just upstream of the RBS of the first gene of the operon. Residues conserved in all five sequences are shown in a black background, while residues conserved in at least four sequences are boxed.

### 3.3. Features of the EGeRS1 Sequences

Two conspicuous features characterise the EGeRS1 regions. First, their GC content is significantly higher than the mean GC content of pLS20 (51.4 versus 37.7%, respectively). Second, they contain multiple inverted repeated sequences (Figure 1A). When present in ssDNA or RNA, these sequences are predicted to form complex secondary structures. Since the DNA is transferred into the recipient cell as ssDNA, it is very possible that secondary structures form temporally soon after transfer into the recipient cell. According to the RNAfold web server, EGeRS1 sequences are predicted to form very stable secondary structures, with calculated free energies of $-152.1/-71$ kcal/mol (EGeRS1-A; RNA/DNA) or around $-220/-110$ kcal/mol (other EGeRS1s; RNA/DNA). Figure 1B shows an example of the predicted secondary structures formed in the ssDNA of EGeRS1-B. The RNAalifold web server of the RNAWebServer was used to predict a consensus structure of the five EGeRS1 sequences (see supplementary Figure S3). The predicted structure shows that the region of about 325 nt located immediately upstream of the RBSs, which is conserved in all five EGeRS1 sequences, would form a branch containing four ramifications. Except for EGeRS1-A, which lacks the 180 bp at the $5'$ end, the other EGeRS1 sequences were predicted to generate an additional branch. In the case of the EGeRS1 sequences B, D, and E, this branch would form two stem-loops. Due to the 131 bp internal duplication in EGeRS1-C, these upstream secondary structures are more complex. The predicted secondary structures may be important for function, although, at present, we do not have evidence that the regulatory mechanism involving the EGeRS1 regions depends on ssDNA or RNA.

### 3.4. Functional Analysis of pLS20 EGeRS1-B

- EGeRS1-B does not contain a constitutive promoter

The EGeRS2 sequences of p576, i.e., the sequences preceding the p576 establishment genes, contain a strong promoter [10]. We tested whether EGeRS1 sequences also contain a promoter by constructing strains containing a transcriptional fusion of (parts of) the EGeRS1-B region with a *gfp* reporter gene (see Materials and Methods) and measuring fluorescence levels by flow cytometry. Figure 3A shows a schematic presentation of the different EGeRS1-B regions fused to *gfp*. In short, fragments were cloned in between the IPTG-inducible promoter $P_{spank}$ and the *gfp* reporter gene present on the *amyE* integration vector pAND101 [15], and the resulting cassette "$P_{spank}$-[fragment X]-*gfp*" was subsequently placed at the *amyE* locus of the *B. subtilis* chromosome. Since the $P_{spank}$ promoter is tightly repressed in the absence of the inductor [16], any fluorescence detected in the absence of the inducer IPTG is due to promoter activity originating from within the cloned fragment. These strains can also be exploited to study possible effects of the cloned insert on upstream initiated transcription by comparing fluorescence levels in the absence and presence of IPTG (see below). The results of the cytometry analysis are presented in Figure 3B. As expected, very low fluorescence levels (~2 arbitrary units (AU)) were obtained for the negative control strain AND101 when cells were grown in the absence of IPTG (see Figure 3B,

AND101, green bar), demonstrating that the P*spank* promoter is tightly repressed under these conditions. Importantly, very low fluorescence levels, similar to those obtained with the negative control strain AND101, were also obtained for strain JV62A containing the entire EGeRS1-B region fused to *gfp* in its native orientation. This result suggested that either the cloned EGeRS1-B region did not contain a constitutive promoter, or the activity of the promoter was inhibited or masked by some DNA sequences that were also present in the cloned fragment. To test the latter possibility, strains containing progressive deletions at either the 5′ or the 3′ end were constructed (Figure 3A, strains JV63A to JV65A, and JV66 to JV69). However, low fluorescence levels were also obtained for these strains (Figure 3B). Next, we constructed strains JV62B–JV65B that contained the EGeRS1-B region in the reverse orientation compared to strains JV62A–JV65A (see Figure 3A) to eliminate the possibility that the promoter activity might have been obscured by a convergently oriented promoter also present in the cloned fragment. Again, no promoter activity was detected in these strains. We then considered the possibility that a promoter was located further upstream of EGeRS1-B. Therefore, the regions cloned in strains JV66 to JV69 were extended at their 5′ ends to include the 338 bp sequence upstream of EGeRS1-B, which ended in the divergently oriented gene *85*. However, promoter activity was still not observed for strains JV66–JV69, indicating that the EGeRS1-B upstream region did not contain a constitutive promoter. Moreover, contrary to the situation on p576, manual inspection of these sequences did not result in the identification of sequences sharing clear similarities to the consensus sequences of $\sigma^A$-dependent promoters (not shown). Together, these results suggest that EGeRS1-B and sequences upstream do not contain a constitutive $\sigma^A$-dependent promoter that would transcribe the downstream-located *ardC*$_{pLS20}$ establishment gene, unlike the situation in p576, where the upstream region of *ardC*$_{p576}$ contains a strong promoter [10]. However, we could not exclude the possibility that the EGeRS1-B region contains a promoter that can be induced under certain conditions.

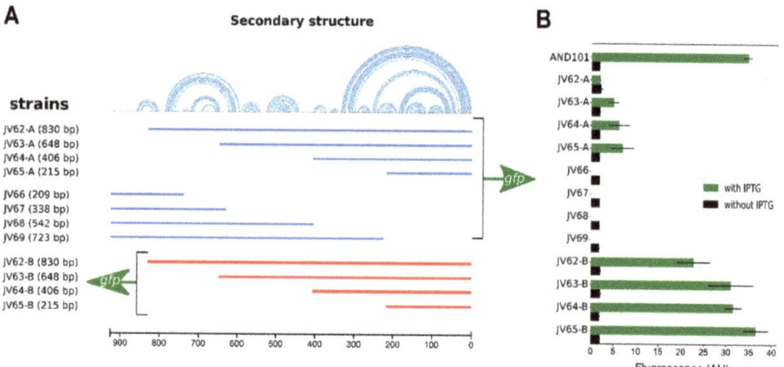



**Figure 3.** EGeRS1-B located upstream of *ardC*$_{pLS20}$ does not contain a regular promoter, but its presence prevents readthrough from upstream promoters. (**A**) Schematic overview of the *gfp* fusions generated. 5′ deletion fragments were cloned in both orientations between the IPTG-inducible promoter P*spank* and the *gfp* reporter gene in vector pAND101. 3′ deletion fragments were cloned in the native orientation. Names of the *B. subtilis* strains containing these *gfp* fusions at the chromosomal *amyE* locus are shown at the left; names ending with letter "A" and "B" correspond to strains containing the cloned fragment in the native and reverse orientations, respectively. The secondary structures predicted to form in ssDNA or RNA are indicated with arcs above the lines representing the cloned regions. (**B**) Fluorescence levels of cells taken from late exponentially growing cultures (OD$_{600}$ = 1) of strains grown in the absence (black bars) or presence of 1 mM IPTG (green bars). Strain AND101 (P*spank*-*gfp*) served as a control. Error bars indicate standard deviations. Each experiment was performed at least three times.

- EGeRS1-B sequence prevents readthrough of upstream promoters

We next considered the possibility that the presence of EGeRS1-B might interfere with progression of transcripts initiated at upstream promoters by determining the levels of fluorescence produced by the upstream-located $P_{spank}$ promoter. As expected, under the conditions where $P_{spank}$ was activated (i.e., strains grown in the presence of 1 mM IPTG), high fluorescence levels were obtained for the control strain AND101 (Figure 3B, green bar). Interestingly, very low fluorescence levels were obtained for strains JV62A–JV65A containing the EGeRS1-B region in its native orientation. In the case of JV62A containing the entire EGeRS1-B region, only background levels of fluorescence were obtained. Strain JV65A, which contained only the 215 bp 3′ region of EGeRS1-B, also produced very low levels of fluorescence. Thus, the presence of the ~3′ half of the EGeRS1-B region, which is conserved in all five EGeRS1 regions, was sufficient for interfering with transcripts starting at the upstream $P_{spank}$ promoter and preventing the expression of the downstream *gfp* gene. This probably indicates that, in its native setting, the EGeRS1-B region prevents expression of the $ardC_{pLS20}$ gene resulting from readthrough of upstream-located promoter(s), thereby contributing to the strict control of genes $ardC_{pLS20}$ and gene *81c*, which is likely to be important for proper functioning and fitness of the host cell. We also tested the effect of the EGeRS1-B sequences when present in the reverse orientation. As shown in Figure 3B, moderate to high fluorescence levels were obtained for strains JV62B–JV65B; therefore, when present in the reverse orientation, the EGeRS1-B sequences did not block or majorly interfere with transcription progression.

In summary, the above results suggest that, contrary to EGeRS2 regions present on plasmid p576, the EGeRS1 regions on pLS20 do not contain a constitutive $\sigma^{A}$-dependent promoter responsible for the expression of the downstream-located establishment genes. However, it seems that the presence of the EGeRS1-B region prevents read through of upstream promoter(s), thereby contributing to proper expression of the $ardC_{pLS20}$ establishment gene.

### 3.5. Establishment Genes Present on pLS20 Family Plasmids Are Regulated by One of the Two Different Mechanisms

Recently, we showed that pLS20 is the prototype of a family of related plasmids present in different *Bacillus* species. To gain insights into their phylogenetic relationships, two maximum likelihood trees were constructed. One of these was based on the replication region and the other on a concatenated sequence of nine conserved orthologous genes. This analysis resulted in two very similar trees in which these plasmids were divided into four clades [11]. pLS20 belongs to clade I, together with 23 other plasmids, and p576 belongs to clade II, together with six other plasmids. Clade III comprises three plasmids, and clade IV is constituted by a single member.

Since the establishment genes/operons of pLS20 and p576 are preceded by completely different sequences, we wondered whether the upstream (regulatory) sequences of the establishment genes/operons in the other pLS20 family plasmids were similar to those present on pLS20 or p576, or perhaps would be preceded by different sequences. This analysis revealed that all pLS20 family plasmids contain multiple copies of sequences that were >60% identical to either EGeRS1 of pLS20 or EGeRS2 of p576. Moreover, as in p576 and pLS20, the conserved sequences were located upstream of the putative establishment genes/operons (see below). These results strongly indicate that the establishment genes present on the pLS20 family of plasmids are regulated by one of the two different mechanisms that are exemplified by those present on pLS20 and p576.

Besides p576, EGeRS2 sequences are present on six other plasmids that all belong to clade II. All the other plasmids of the pLS20 family belonging to clades I, III, and IV contain EGeRS1-like sequences.

### 3.6. Analysis of the EGeRS1 Sequences

Clade I, III, and IV plasmids each contain multiple EGeRS1 sequences. To see how similar these sequences were, and whether their similarities correlated with the different clades, a phylogenetic tree was constructed for the EGeRS1 sequences present on the reference plasmids of clade I (pLS20 (clade IA), pBatNRS213 (clade IB), and pBamB1895 (clade IC)), clade III (pBglSRCM103574 (clade IIIA), and pBliYNP2† (clade IIIB)), and clade IV (pBspNMCC4†). The complete sequences of the reference plasmids share less than 80% sequence identity between them. During our analysis, we discovered that EGeRS1 sequences are also present on at least 10 putative conjugative but non-pLS20 family plasmids harboured by bacilli. One of these, the 102 kb plasmid pBS72 from *B. subtilis* strain 72 (accession number KX711616.1), has been studied in some detail [26]. For instance, the replication and partitioning modules of pBS72 are different from those present on the pLS20 family plasmids [27]. pBS72 contains four EGeRS1 sequences and, as observed for the pLS20 family plasmids of clades I, III, and IV, these EGeRS1 sequences precede putative operons. We included the EGeRS1 sequences of pBS72 in the phylogenetic analysis. The phylogenetic tree (Figure 4) shows that the EGeRS1 sequences can be divided into two groups. The first group comprises EGeRS1 sequences that are present on pBS72 and plasmids belonging to clade I of the pLS20 family of plasmids. The second group is composed of EGeRS1 sequences that are present on plasmids of clade III and IV. In addition, the tree shows that, in almost all cases, EGeRS1 sequences of the same plasmid cluster together. An exception may be the EGeRS1-A sequence of pLS20 (clade IA), which clusters together with an EGeRS1 sequence present on pBatNRS213 (clade IB) and pBamB1895 (clade IC).

Several conclusions can be drawn from this analysis. First, EGeRS1 sequences are not limited to members of the pLS20 family of plasmids. Second, the EGeRS1 sequences can be divided into two phylogenetic groups. Third, the EGeRS1 sequences present on pBS72 and those present on clade I plasmids of the pLS20 family belong to the same group.

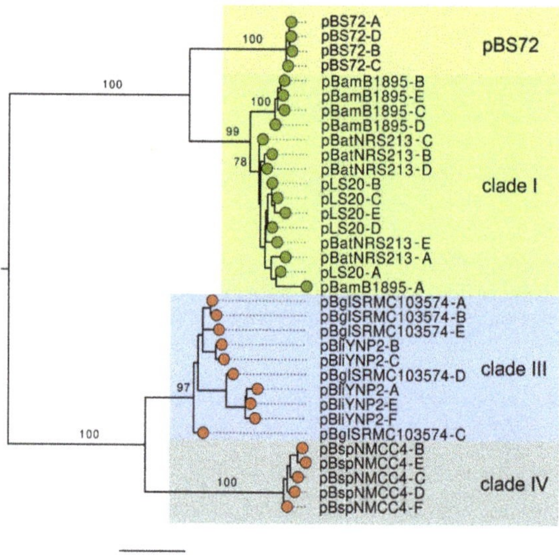


**Figure 4.** Phylogenetic relatedness of EGeRS1 sequences present on the reference plasmids belonging to pLS20 family clades I, III, IV, and plasmid pBS72. Phylogenetic trees were constructed for the 30 EGeRS1 sequences that are distributed on the pLS20 family reference plasmids of clade IA (pLS20, A to E), clade IB (pBatNRS213, A to E), clade IC (pBamB1895, A to E), clade IIIA (pBglSRCM103574, A to E), clade IIIB (pBliYNP2, A to C and E- F), clade IV (pBspNMCC4, B to F), and plasmid pBS72 A to D. Note that the letters used to differentiate EGeRS1 sequences are arbitrary and do not reflect the relatedness between the EGeRS1 sequences from different plasmids. Plasmids pBliYNP2† and pBspNMCC4† are deposited in the database as a single large contig whose endpoints correspond to an EGeRS1 region. Consequently, one EGeRS1 sequence on each of these two plasmids is incomplete,

and these two incomplete EGeRS1 sequences were not included in the analysis. The trees were constructed using the maximum likelihood method and Kimura 2-parameter model + R2. Statistical evidence for each branch is provided by bootstraps analysis (1000 replicates, indicated as percentages). The roots are located at the midpoint.

### 3.7. Analysis of the EGeRS2 Sequences

We have previously shown that the four EGeRS2 sequences on plasmid p576 contain a strong promoter [10]. In all four cases, a dual heptamer sequence separated by two bp (5′-TTATCCCnnTTATCCC-3′), named dual motifs (DMs), is located immediately upstream of the −35 box of the promoter. In two of the four EGeRS2 regions, an additional DM is located just downstream of the transcription start site. One of the four EGeRS2 regions is located upstream of $reg_{p576}$, which encodes the repressor that binds to the DMs with high affinity and co-operativity. $Reg_{p576}$ represses its own promoter less strictly than the other promoters. Consequently, low levels of $Reg_{p576}$ ensure tight repression of the three other promoters in donor cells and forms the core of the regulatory system [10].

EGeRS2 sequences are also present on the other six clade II plasmids. Thus, as for p576, all clade II plasmids contain a homologue of gene $reg_{p576}$ and promoters being flanked at one or both sides by Reg operators (see supplemental Figures S4 and S5). Like $Reg_{p576}$ [10], the other Reg homologues are predicted to be ribbon–helix–helix (RHH) DNA binding proteins. Together, these data strongly indicate that the establishment genes present on clade II pLS20 family plasmids are regulated in the same or similar way as those described for p576.

### 3.8. (Putative) Establishment Genes Regulated by EGeRS2

The establishment genes located downstream of the EGeRS2 sequences on p576 are transiently expressed after the plasmid is transferred into the recipient cell [10]. Most likely, this also applies to the establishment genes located downstream of the EGeRS2 sequences present on the other six clade II plasmids of the pLS20 family. Figure 5 shows linear genetic maps of the regions encompassing the putative establishment genes of clade II plasmids of the pLS20 family that are controlled by EGeRS2 sequences. We made an inventory of the genes on these plasmids for two reasons: to determine whether the establishment genes were conserved between plasmids, and to gain insights into the possible function of these genes. Features of the (putative) establishment genes regulated by EGeRS2 and their distribution on the pLS20 clade II plasmids are given in supplementary Table S4 and Figure 6, respectively. As expected, all seven plasmids contain a *reg* gene, which is essential for the transient regulation of the establishment genes. In p576, *reg* forms part of a bicistronic operon, together with gene *26c*, whose function is unknown but seems not to be required for the regulation of the Reg-regulated promoters [10]. This is supported by the observation that p576 gene *26c* is not conserved in all the other plasmids. All seven plasmids also contain an anti-restriction gene (*ardC*). An orthologue of p576 gene *23c* of unknown function is present in five of the seven plasmids. Remarkably, another gene replaces gene *23c* in plasmid pBalRIT380‡. Gene replacement also appears on plasmid pBpuPs115; an unrelated gene replaces the conserved gene *20c*. Finally, whereas homologues of p576 genes *18c–20c*, *23c*, and *26c* are present on several pLS20 clade II plasmids, other genes are present on only one of these plasmids.

In summary, the seven closely related clade II pLS20 family plasmids contain similarly organised establishment regulons in which only two establishment genes, *ardC* and *reg*, are conserved in all members. Other establishment genes are conserved in some plasmids, and some plasmids contain an establishment gene that is not present on any of the other six plasmids. Except for *ardC* and *reg*, the function of the other putative establishment genes is unknown.

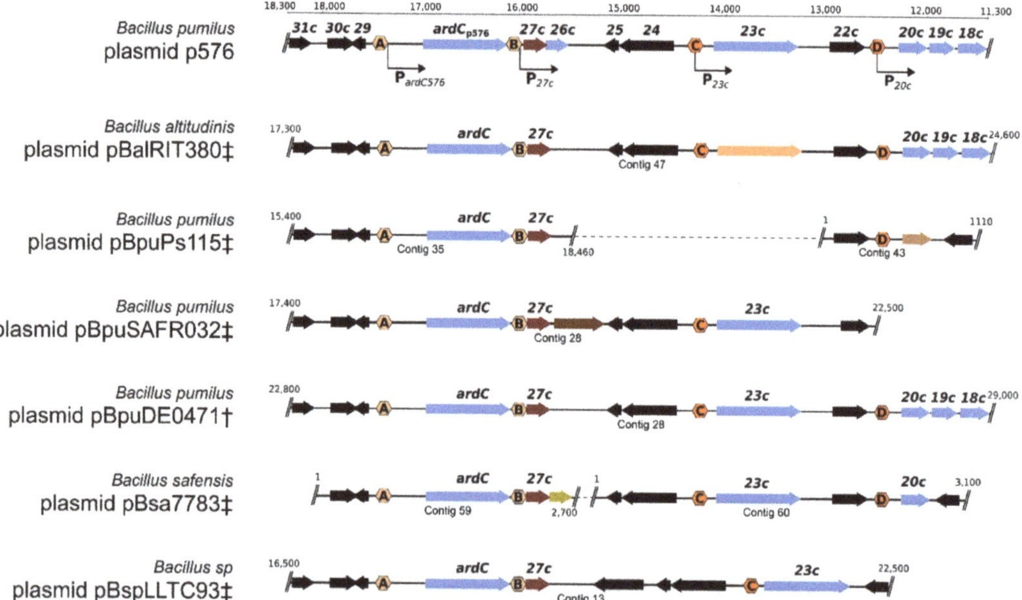



**Figure 5.** Putative establishment regulons present on clade II pLS20 family plasmids. Schematic view of the genetic organisation of the putative establishment regulon present on plasmid p576 and other clade II pLS20 family plasmids regulated by EGeRS2. The region shown for p576 corresponds to positions 11,300–18,400 according to accession number NZ_LR026977. Putative promoters are indicated with bent arrows. Genes are indicated with wide arrows. The positions of the repressor gene *reg* are indicated with red arrows. The positions of the (putative) promoters containing Reg operators (i.e., EGeRS2 regions) are indicated with hexagonal boxes labelled with a letter; orange and purple hexagonals indicate promoters flanked on one and both sides by a Reg operator, respectively. Genes controlled by EGeRS2 sequences other than *reg* are shown in blue, yellow, orange, or brown. Black arrows indicate genes that are not controlled by EGeRS2 sequences. Putative establishment regulons of the other clade II plasmids are approximately aligned with that of p576. The yellow, orange, and brown arrows correspond to putative establishment genes for which no homolog is present on p576. Note that p576 gene *23c* is not conserved in plasmid pBalRIT380. The only two genes for which a function can be attributed are the *ardC* and the *reg* genes. Plasmids for which the entire sequence is not known are indicated with a dagger symbol. In those cases where the plasmid sequence was present in the database as one large linear contig that probably corresponds to most of the plasmid sequence, the name was extended with a single dagger. Names are extended with a double dagger symbol when the plasmid was reconstructed form two or several contigs [11].

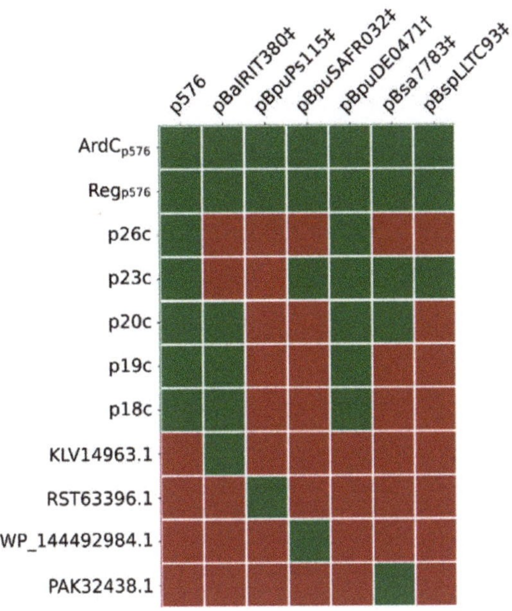

**Figure 6.** Conserved putative establishment proteins encoded by clade II plasmids of the pLS20 family. Homologous proteins indicated with green boxes were identified using the COG or OMCL algorithms of the tool "get homologues" with coverage and e-value thresholds of 75% and 1e–10, respectively.

### 3.9. (Putative) Establishment Genes Present on pLS20 Family Plasmids Regulated by EGeRS1

We also made an inventory of the putative establishment genes regulated by EGeRS1. For this inventory, we analysed the pLS20 family reference plasmids pLS20 (clade IA), pBatNRS213 (clade IB), pBamB1895 (clade IC), pBglSRMC103574 (clade IIIA), pBliYNP2† (clade IIIB), pBspNMCC4 (clade IV), and the non-pLS20 family plasmid pBS72. Figure 7 shows a schematic overview of the putative establishment regulons present on these plasmids. Features of these putative establishment genes and their distribution on the different plasmids are given in supplementary Table S5 and Figure 8, respectively. The number of establishment genes per plasmid varies between 14 and 20, summing up to a total of 118 genes on these seven plasmids. Of these, 91 genes can be grouped into 31 homologous protein clusters, with at least two members in each cluster (see supplementary Table S5), while the remaining 27 are singletons. Although there is a large variety in the putative establishment genes present on these seven plasmids, Figure 7 shows that each plasmid contains at least one gene coding for a protein able to interfere with the RM defence system of the host, being either an anti-restriction or a methylase gene (a C5-MTase (IPR001525) or an N-6 adenine-specific MTase (IPR003356)). The function of only a few of the establishment genes could be deduced based on similarity of the encoded proteins with proteins of known functions. Examples of these are genes encoding a putative SNase-like thermonuclease (present on four plasmids), and a winged-helix DNA-binding protein (present on one plasmid). Proteins encoded by three other genes contain regions that share similarity to a conserved domain of unknown function. However, as observed for those regulated by EGeRS2, the function of most of the putative establishment genes regulated by EGeRS1 is unknown.

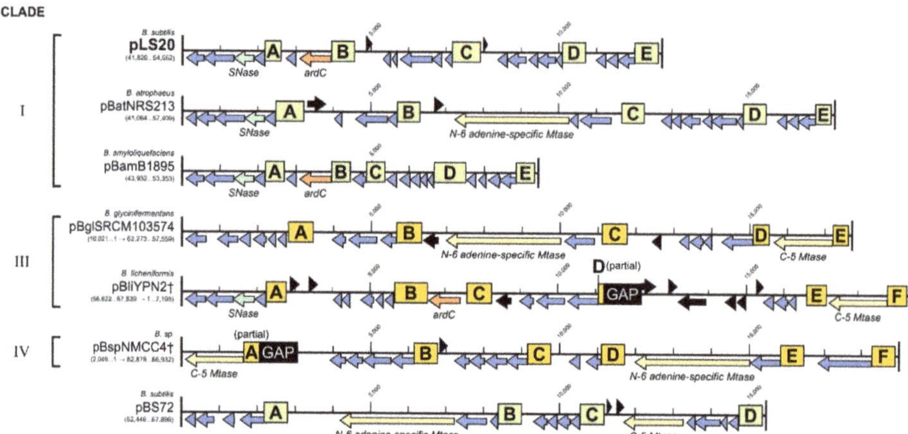

**Figure 7.** Putative establishment regulons present on plasmid pBS72 and on the pLS20 family reference plasmids of clades I, III, and IV. EGeRS1 regions are boxed and labelled with letters (A to D/F). Green and yellow-coloured boxes are used to indicate EGeRS1 sequences that belong to the same phylogenetic clade (see Figure 4). In each plasmid, the EGeRS1 sequences precede putative operons of two to six genes. Genes controlled by EGeRS1 sequences are shown in blue, except the homologues of gene *ardC* (highlighted in orange), genes encoding putative MTases (highlighted in yellow), and homologue genes for SNase (in light green). Black arrows indicate genes that are not controlled by EGeRS2 sequences.

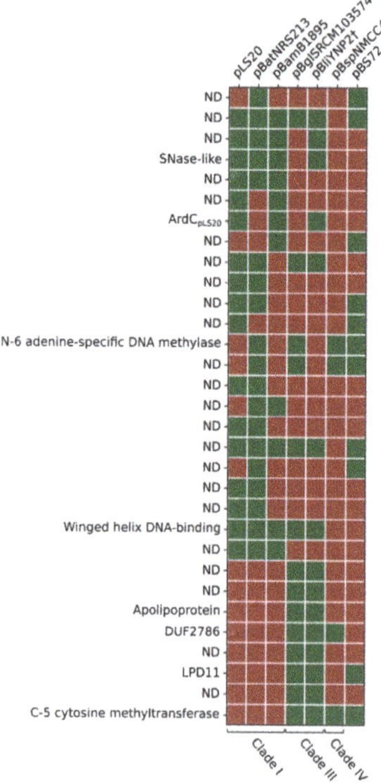

**Figure 8.** Conserved establishment proteins encoded by pBS72 and plasmids belonging to clades I, III and IV of the pLS20 family. Homologous proteins indicated with green boxes were identified using the COG or OMCL algorithms of the tool "get homologues" with coverage and e-value thresholds of 75% and 1e–10, respectively.

## 4. Discussion

Establishment genes play roles in at least two functions: converting ssDNA into dsDNA and protecting the transferred DNA against degradation by defence mechanisms of the recipient cell. Although some establishment genes are expressed in the donor cell, and the synthesised proteins are then transported into the recipient cell together with the DNA; most establishment genes are expressed only in the recipient cell. These latter genes are expressed rapidly and transiently upon arrival of the plasmid in the recipient cell. Transient expression is particularly important for the establishment proteins that inhibit the defence mechanisms, because prolonged expression makes the cell vulnerable to other foreign DNA. Here, we analysed the regulatory sequences of the establishment genes present on the pLS20 family of plasmids, which, in turn, have resulted in the determination of the establishment regulons present on these plasmids.

### 4.1. Establishment Gene Regulatory Sequences Present on pLS20 Family of Plasmids

Our analyses reveal that the establishment genes present on the pLS20 family of plasmids are regulated in two fundamentally different ways. One of these mechanisms is based upon a regulatory protein that controls its own promoter, as well as promoters regulating the expression of establishment genes, as has been shown for the clade II plasmid p576. The repressor protein of p576 strictly represses the activity of the strong promoters located upstream of the establishment genes in the donor cells. During conjugation, the DNA but not the repressor protein is transferred, resulting in transient high-level expression of the establishment genes until sufficient repressor has been synthesised to repress the establishment promoters [10]. This mechanism is similar to the regulation of anti-CRISPR genes present on bacteriophages. Several bacteriophages encode anti-CRISPR proteins that inhibit the CRISPR–Cas defence systems of the host, and a gene encoding a repressor often follows an anti-CRISPR gene. Upon infection, both the anti-CRISPR and the repressor protein are expressed at high levels; the repressor protein subsequently curtails expression by repressing the promoter located upstream of both genes [28–30]. Therefore, as for the establishment genes on the conjugative plasmids, this results in transient high expression of anti-CRISPR genes. However, the repressors encoded by the clade II pLS20 family of conjugative plasmids belong to the ribbon–helix–helix family, while those encoded by phages are helix–turn–helix family proteins.

Here, we show that all clade II members of the pLS20 family of plasmids use the repressor-based mechanism, but the establishment genes present on plasmids belonging to clade I, III, and IV are regulated in a fundamentally different way. The two different systems have very different upstream regulatory sequences: the sequences upstream of the establishment genes on clade I, III, and IV plasmids are named EGeRS1, and those present on clade II plasmids EGeRS2. The establishment genes of all the clade I, III, and IV plasmids are preceded by a conserved region of 400 to 600 bps that contains multiple inverted repeated sequences. At present, the mechanism responsible for the presumed transient expression of these establishment genes is unknown, but analyses of the region upstream of the *ardC* gene of pLS20 has shown that it does not contain a strong $\sigma^A$-dependent promoter. We also found that the putative conjugative *B. subtilis* plasmid pBS72, which does not belong to the pLS20 family, contains four EGeRS1 sequences. This demonstrates that regulation of the establishment genes by EGeRS1 sequences is not limited to the pLS20 family of plasmids. Moreover, other conjugative plasmids do not contain EGeRS1 or EGeRS2 sequences, indicating that other mechanisms exist to ensure the establishment of the plasmid after transfer into the recipient cell. Indeed, at least one other mechanism exists that is based on the so-called ssDNA promoters, exemplified by the conjugative plasmids F and CollbP-9 of the Gram-negative bacterium *Escherichia coli*. In these cases, ssDNA regions located upstream of several operons contain inverted repeated sequences that, upon hybridisation, generate functional promoters. Hence, these promoters will only be active until the ssDNA has been converted into dsDNA [31–33]. Among the establishment genes on plasmids F and CollbP-9 that are controlled by ssDNA, promoters are anti-restriction genes [34] that are also present on many pLS20 family plasmids (this work). We have considered the possibility that the EGeRS1 sequences might constitute ssDNA promoters. However, this seems highly unlikely for the following reasons. First, the EGeRS sequences are much longer than the ssDNA promoters, and no putative promoter could be identified in the predicted secondary structure generated upon hybridisation of the EGeRS1 sequences. Second, and more importantly, the DNA strand that is nicked by the relaxase is the strand that is transferred into the recipient cell. The DNA strand that is nicked by the relaxase of pLS20 has been determined [35], and this strand corresponds to the non-template strand, i.e., the transferred ssDNA strand cannot function as a template for generation of the establishment gene mRNA.

### 4.2. Establishment Regulons Present on pLS20 Family of Plasmids and Implications of Establishment Regulons in General

Several conclusions can be drawn from the identification of the establishment operons on the pLS20 family of plasmids described here, and also from the published results. For instance, based on their role in the conversion of ss to dsDNA and/or inhibition of host-encoded defence mechanism, it is clear that the establishment genes play a crucial role in the conjugation process, and hence should be considered as part of the conjugation module, together with the genes in the large conjugation operon. This conclusion is supported by the demonstrated importance of *ard* genes in the establishment of conjugative plasmids in the new host [36–39]. Therefore, most genes present on the pLS20 family plasmids are related to the conjugation process, e.g., 69 of the 92 genes (75%) of pLS20 form part of the conjugation module. The work presented here also revealed two interesting differences between the genes of the pLS20 family plasmids that are involved in the first three steps of conjugation (i.e., recipient cell selection and attachment, synthesis of the translocation machinery, and DNA processing), and those corresponding to the fourth step of the conjugation process (i.e., establishment genes). First, there is an overall high level of similarity between the genes involved in the first three steps [11], but the establishment genes involved in step four tend to be more variable. Second, while the genes involved in the first three steps are all located in one single large operon, the establishment genes are distributed in multiple operons, which cluster together on the plasmid.

Based on the arguments described below, we propose that the organisation of establishment genes on a plasmid in a cluster of variable operons has coevolved with the variable clusters of defence mechanisms encoded by bacterial chromosomes. In the last decade or so, it has become clear that bacteria encode (many) more defence mechanisms than the RM and CRISPR–cas systems to protect them against incoming foreign DNA, such as that of phages and conjugative elements [5]. There are striking similarities in the organisation and distribution between the defence genes encoded by bacterial chromosomes and the establishment genes and anti-defence genes present on the pLS20 family of plasmids. Like chromosomal defence genes, establishment genes of the pLS20 family are organised in cluster(s), and there is a big variety of establishment genes harboured by the individual plasmids. Thus, according to the so-called "pan-immune model", a group of related bacterial species together possess a large arsenal of defence mechanisms but individual cells contain only a subset of these defence genes, probably because they pose a high fitness cost to a bacterial cell [5]. The clustered defence mechanisms on bacterial genomes are referred to as defence islands [40]. In analogy, we propose to name the clustered establishment operons as establishment or anti-defence islands. Establishment and anti-defence operons are also clustered on other conjugative plasmids, such as the F-like plasmid that are controlled by ssDNA promoters, and on at least several other conjugative plasmids and mobile elements [41–43], suggesting that island-like organisation of establishment/anti-defence genes/operons is a general feature. Although the origin and mechanism(s) responsible for the formation and plasticity of defence islands are unknown, these regions often also contain mobility genes (e.g., transposases and recombinases, and even conjugation genes), suggesting that horizontal gene transfer events play a role in this [40,44,45]. In addition, or alternatively, regions containing defence clusters may be integration hotspot sites for acquired genes, or they may reflect functional links, including possible coregulation [5,46,47].

At this moment, we do not know why the establishment/anti-defence genes on conjugative elements and other mobile elements are in clusters, but (several of) the arguments proposed to explain the clustered organisation of defence genes on chromosomes may also be accountable for this. Particularly, coregulation may play an important role in clustering of these genes, as we have shown here that the establishment genes present on at least the pLS20 family and the F-like plasmids are regulated "en bloc" by one of the three different mechanisms. In addition, the variability of the establishment/anti-defence genes and their organisation in multiple, rather than one, operons may be directly related to the way the defence genes are organised on bacterial chromosomes. Thus, a given cluster of establishment/anti-defence genes on conjugative elements would suffice to transiently inhibit a subset of defence systems harboured by a recipient cell. Interactions between mobile genetic elements, including conjugative elements, and the host cells would be the driving force for rapid evolutionary alterations in defence and anti-defence genes. One reason why the establishment/anti-defence genes are organised in multiple operons may be that it favours shuffling of the individual operons between plasmids.

In summary, based on the results obtained here, we propose that, besides the anti-restriction genes, several or most of the other establishment genes present on pLS20 family plasmids, and conjugative plasmids in general, function to transiently inhibit known and unknown defence mechanisms encoded by bacterial chromosomes.

**Supplementary Materials:** The following are available online at https://www.mdpi.com/article/10.339 0/microorganisms9122465/s1, Figure S1: Schematic presentation of the regulatory mechanism of the establishment genes on the pLS20-related plasmid p576. Figure S2: Alignment between the *Shigella flexneri* encoded antirestriction protein ArdC (ArdC/pSa) and the putative antirestriction proteins encoded by p576 (ArdC/p576) and pLS20 (ArdC/pLS20). Figure S3: Predicted conserved secondary structure of the EGeRS1 sequences when present in ssDNA or RNA. Figure S4: All seven clade II pLS20 family plasmids encode a homolog of Regp$_{576}$ that in p576 is responsible for regulating the establishment genes. Figure S5: Conserved promoters flanked by the Reg operators in the seven clade II plasmids of the pLS20 family. Table S1: Strains used. Table S2: Plasmids. Table S3: oligonucleotides used. Table S4: Putative establishment genes identified on clade II pLS20 family plasmids. Table S5: Putative establishment genes identified on plasmids pLS20, pBatNRS213, pBamB1895, pBglSRCM103574, pBliYNP2†, pBspNMCC4† and pBS72.

**Author Contributions:** Conceptualization, J.V.-C., L.J.W. and W.J.J.M.; Data curation, J.V.-C., F.F., D.A. and W.J.J.M.; Formal analysis, J.V.-C., A.M.-A., F.F. and D.A.; Funding acquisition W.J.J.M.; Investigation, J.V.-C., F.F., A.M.-A., L.J.W. and W.J.J.M.; Methodology, J.V.-C., A.M.-A., F.F., D.A., L.J.W. and W.J.J.M.; Project administration, W.J.J.M.; Resources, W.J.J.M.; Software, D.A.; Supervision, W.J.J.M.; Validation, D.A.; Visualization, J.V.-C. and W.J.J.M.; Writing—original draft, W.J.J.M.; Writing—review and editing, L.J.W. and W.J.J.M. All authors have read and agreed to the published version of the manuscript.

**Funding:** This research was funded by the MINISTRY OF SCIENCE AND INNOVATION OF THE SPANISH GOVERNMENT grants (bio2016-77883-C2-1-P) and (PID2019-108778GB-C21 (AEI/FEDER, EU)) to W.J.J.M., which also funded J.V-C. and A.M-A. A Wellcome Investigator grant (209500) to Jeff Errington supported L.J.W. Institutional grants from the "Fundación Ramón Areces" and "Banco de Santander" supported the Centro de Biología Molecular "Severo Ochoa".

**Data Availability Statement:** All the data are presented in the paper.

**Acknowledgments:** We acknowledge helpful discussions with members of our laboratories.

**Conflicts of Interest:** The authors declare no conflict of interest.

## References

1. Thomas, C.M.; Nielsen, K.M. Mechanisms of, and barriers to, horizontal gene transfer between bacteria. *Nat. Rev. Microbiol.* **2005**, *3*, 711–721. [CrossRef] [PubMed]
2. Tock, M.R.; Dryden, D.T. The biology of restriction and anti-restriction. *Curr. Opin. Microbiol.* **2005**, *8*, 466–472. [CrossRef] [PubMed]

3.  Oliveira, P.H.; Touchon, M.; Rocha, E.P. The interplay of restriction-modification systems with mobile genetic elements and their prokaryotic hosts. *Nucleic Acids Res.* **2014**, *42*, 10618–10631. [CrossRef] [PubMed]

4.  Mojica, F.J.; Rodriguez-Valera, F. The discovery of CRISPR in archaea and bacteria. *FEBS J.* **2016**, *283*, 3162–3169. [CrossRef] [PubMed]

5.  Bernheim, A.; Sorek, R. The pan-immune system of bacteria: Antiviral defence as a community resource. *Nat. Rev. Microbiol.* **2020**, *18*, 113–119. [CrossRef]

6.  Chilley, P.M.; Wilkins, B.M. Distribution of the ardA family of antirestriction genes on conjugative plasmids. *Microbiology* **1995**, *141 Pt. 9*, 2157–2164. [CrossRef]

7.  Belogurov, A.A.; Delver, E.P.; Rodzevich, O.V. Plasmid pKM101 encodes two nonhomologous antirestriction proteins (ArdA and ArdB) whose expression is controlled by homologous regulatory sequences. *J. Bacteriol.* **1993**, *175*, 4843–4850. [CrossRef] [PubMed]

8.  Belogurov, A.A.; Delver, E.P.; Agafonova, O.V.; Belogurova, N.G.; Lee, L.Y.; Kado, C.I. Antirestriction protein Ard (Type C) encoded by IncW plasmid pSa has a high similarity to the "protein transport" domain of TraC1 primase of promiscuous plasmid RP4. *J. Mol. Biol.* **2000**, *296*, 969–977. [CrossRef] [PubMed]

9.  Wilkins, B.M. Plasmid promiscuity: Meeting the challenge of DNA immigration control. *Environ. Microbiol.* **2002**, *4*, 495–500. [CrossRef]

10. Val-Calvo, J.; Luque-Ortega, J.R.; Crespo, I.; Miguel-Arribas, A.; Abia, D.; Sanchez-Hevia, D.L.; Serrano, E.; Gago-Cordoba, C.; Ares, S.; Alfonso, C.; et al. Novel regulatory mechanism of establishment genes of conjugative plasmids. *Nucleic Acids Res.* **2018**, *46*, 11910–11926. [CrossRef] [PubMed]

11. Val-Calvo, J.; Miguel-Arribas, A.; Abia, D.; Wu, L.J.; Meijer, W.J.J. pLS20 is the archetype of a new family of conjugative plasmids harboured by Bacillus species. *NAR Genom. Bioinform.* **2021**, *3*, lqab096. [CrossRef]

12. Bertani, G. Studies on lysogenesis. I. The mode of phage liberation by lysogenic Escherichia coli. *J. Bacteriol.* **1951**, *62*, 293–300. [CrossRef] [PubMed]

13. Sambrook, J.; Fritsch, E.F.; Maniatis, T. *Molecular Cloning: A Laboratory Manual*; Cold Spring Harbor Laboratory Press: Cold Spring Harbor, NY, USA, 1989.

14. Bron, S.; Meijer, W.J.J.; Holsappel, S.; Haima, P. Plasmid instability and molecular cloning in *Bacillus subtilis*. *Res. Microbiol.* **1991**, *142*, 875–883. [CrossRef]

15. Miguel-Arribas, A.; Val-Calvo, J.; Gago-Cordoba, C.; Izquierdo, J.M.; Abia, D.; Wu, L.J.; Errington, J.; Meijer, W.J.J. A novel bipartite antitermination system widespread in conjugative elements of Gram-positive bacteria. *Nucleic Acids Res.* **2021**. [CrossRef]

16. Gago-Cordoba, C.; Val-Calvo, J.; Miguel-Arribas, A.; Serrano, E.; Singh, P.K.; Abia, D.; Wu, L.J.; Meijer, W.J.J. Surface Exclusion Revisited: Function Related to Differential Expression of the Surface Exclusion System of Bacillus subtilis Plasmid pLS20. *Front. Microbiol.* **2019**, *10*, 1502. [CrossRef] [PubMed]

17. Lorenz, R.; Bernhart, S.H.; Honer Zu Siederdissen, C.; Tafer, H.; Flamm, C.; Stadler, P.F.; Hofacker, I.L. ViennaRNA Package 2.0. *Algorithms Mol. Biol.* **2011**, *6*, 26. [CrossRef]

18. Hofacker, I.L. RNA consensus structure prediction with RNAalifold. *Methods Mol. Biol.* **2007**, *395*, 527–544. [CrossRef]

19. Erb, I.; Gonzalez-Vallinas, J.R.; Bussotti, G.; Blanco, E.; Eyras, E.; Notredame, C. Use of ChIP-Seq data for the design of a multiple promoter-alignment method. *Nucleic Acids Res.* **2012**, *40*, e52. [CrossRef] [PubMed]

20. Robert, X.; Gouet, P. Deciphering key features in protein structures with the new ENDscript server. *Nucleic Acids Res.* **2014**, *42*, W320–W324. [CrossRef] [PubMed]

21. Lai, D.; Proctor, J.R.; Zhu, J.Y.; Meyer, I.M. R-CHIE: A web server and R package for visualizing RNA secondary structures. *Nucleic Acids Res.* **2012**, *40*, e95. [CrossRef]

22. Nguyen, L.T.; Schmidt, H.A.; von Haeseler, A.; Minh, B.Q. IQ-TREE: A fast and effective stochastic algorithm for estimating maximum-likelihood phylogenies. *Mol. Biol. Evol.* **2015**, *32*, 268–274. [CrossRef] [PubMed]

23. Kalyaanamoorthy, S.; Minh, B.Q.; Wong, T.K.F.; von Haeseler, A.; Jermiin, L.S. ModelFinder: Fast model selection for accurate phylogenetic estimates. *Nat. Methods* **2017**, *14*, 587–589. [CrossRef] [PubMed]

24. Singh, P.K.; Ballestero-Beltran, S.; Ramachandran, G.; Meijer, W.J. Complete nucleotide sequence and determination of the replication region of the sporulation inhibiting plasmid p576 from Bacillus pumilus NRS576. *Res. Microbiol.* **2010**, *161*, 772–782. [CrossRef] [PubMed]

25. Meijer, W.J.J.; de Boer, A.; van Tongeren, S.; Venema, G.; Bron, S. Characterization of the replication region of the *Bacillus subtilis* plasmid pLS20: A novel type of replicon. *Nucleic Acids Res.* **1995**, *23*, 3214–3223. [CrossRef]

26. Titok, M.A.; Chapuis, J.; Selezneva, Y.V.; Lagodich, A.V.; Prokulevich, V.A.; Ehrlich, S.D.; Jannière, L. *Bacillus subtilis* soil isolates: Plasmid replicon analysis and construction of a new theta-replicating vector. *Plasmid* **2003**, *49*, 53–62. [CrossRef]

27. Titok, M.; Suski, C.; Dalmais, B.; Ehrlich, S.D.; Janniere, L. The replicative polymerases PolC and DnaE are required for theta replication of the Bacillus subtilis plasmid pBS72. *Microbiology* **2006**, *152*, 1471–1478. [CrossRef] [PubMed]

28. Stanley, S.Y.; Borges, A.L.; Chen, K.H.; Swaney, D.L.; Krogan, N.J.; Bondy-Denomy, J.; Davidson, A.R. Anti-CRISPR-Associated Proteins Are Crucial Repressors of Anti-CRISPR Transcription. *Cell* **2019**, *178*, 1452–1464 e1413. [CrossRef]

29. Hamon, M.A.; Stanley, N.R.; Britton, R.A.; Grossman, A.D.; Lazazzera, B.A. Identification of AbrB-regulated genes involved in biofilm formation by *Bacillus subtilis*. *Mol. Microbiol.* **2004**, *52*, 847–860. [CrossRef] [PubMed]

30. Osuna, B.A.; Karambelkar, S.; Mahendra, C.; Sarbach, A.; Johnson, M.C.; Kilcher, S.; Bondy-Denomy, J. Critical Anti-CRISPR Locus Repression by a Bi-functional Cas9 Inhibitor. *Cell Host Microbe* **2020**, *28*, 23–30 e25. [CrossRef]

31. Bates, S.; Roscoe, R.A.; Althorpe, N.J.; Brammar, W.J.; Wilkins, B.M. Expression of leading region genes on IncI1 plasmid Collb-P9: Genetic evidence for single-stranded DNA transcription. *Microbiology* **1999**, *145 Pt. 10*, 2655–2662. [CrossRef]

32. Masai, H.; Arai, K. Frpo: A novel single-stranded DNA promoter for transcription and for primer RNA synthesis of DNA replication. *Cell* **1997**, *89*, 897–907. [CrossRef]

33. Nasim, M.T.; Eperon, I.C.; Wilkins, B.M.; Brammar, W.J. The activity of a single-stranded promoter of plasmid CollB-P9 depends on its secondary structure. *Mol. Microbiol.* **2004**, *53*, 405–417. [CrossRef] [PubMed]

34. Althorpe, N.J.; Chilley, P.M.; Thomas, A.T.; Brammar, W.J.; Wilkins, B.M. Transient transcriptional activation of the IncI1 plasmid anti-restriction gene (ardA) and SOS inhibition gene (psiB) early in conjugating recipient bacteria. *Mol. Microbiol.* **1999**, *31*, 133–142. [CrossRef] [PubMed]

35. Ramachandran, G.; Miguel-Arribas, A.; Abia, D.; Singh, P.K.; Crespo, I.; Gago-Cordoba, C.; Hao, J.A.; Luque-Ortega, J.R.; Alfonso, C.; Wu, L.J.; et al. Discovery of a new family of relaxases in Firmicutes bacteria. *PLoS Genet.* **2017**, *13*, e1006586. [CrossRef]

36. Goryanin, I.I.; Kudryavtseva, A.A.; Balabanov, V.P.; Biryukova, V.S.; Manukhov, I.V.; Zavilgelsky, G.B. Antirestriction activities of KlcA (RP4) and ArdB (R64) proteins. *FEMS Microbiol. Lett.* **2018**, *365*. [CrossRef]

37. Gonzalez-Montes, L.; Del Campo, I.; Garcillan-Barcia, M.P.; de la Cruz, F.; Moncalian, G. ArdC, a ssDNA-binding protein with a metalloprotease domain, overpasses the recipient hsdRMS restriction system broadening conjugation host range. *PLoS Genet.* **2020**, *16*, e1008750. [CrossRef] [PubMed]

38. Balabanov, V.P.; Kotova, V.Y.; Kholodii, G.Y.; Mindlin, S.Z.; Zavilgelsky, G.B. A novel gene, ardD, determines antirestriction activity of the non-conjugative transposon Tn5053 and is located antisense within the tniA gene. *FEMS Microbiol. Lett.* **2012**, *337*, 55–60. [CrossRef] [PubMed]

39. McMahon, S.A.; Roberts, G.A.; Johnson, K.A.; Cooper, L.P.; Liu, H.; White, J.H.; Carter, L.G.; Sanghvi, B.; Oke, M.; Walkinshaw, M.D.; et al. Extensive DNA mimicry by the ArdA anti-restriction protein and its role in the spread of antibiotic resistance. *Nucleic Acids Res.* **2009**, *37*, 4887–4897. [CrossRef] [PubMed]

40. Makarova, K.S.; Wolf, Y.I.; Snir, S.; Koonin, E.V. Defense islands in bacterial and archaeal genomes and prediction of novel defense systems. *J. Bacteriol.* **2011**, *193*, 6039–6056. [CrossRef] [PubMed]

41. Pinilla-Redondo, R.; Shehreen, S.; Marino, N.D.; Fagerlund, R.D.; Brown, C.M.; Sorensen, S.J.; Fineran, P.C.; Bondy-Denomy, J. Discovery of multiple anti-CRISPRs highlights anti-defense gene clustering in mobile genetic elements. *Nat. Commun.* **2020**, *11*, 5652. [CrossRef] [PubMed]

42. Mahendra, C.; Christie, K.A.; Osuna, B.A.; Pinilla-Redondo, R.; Kleinstiver, B.P.; Bondy-Denomy, J. Broad-spectrum anti-CRISPR proteins facilitate horizontal gene transfer. *Nat. Microbiol.* **2020**, *5*, 620–629. [CrossRef]

43. Roy, D.; Huguet, K.T.; Grenier, F.; Burrus, V. IncC conjugative plasmids and SXT/R391 elements repair double-strand breaks caused by CRISPR-Cas during conjugation. *Nucleic Acids Res.* **2020**, *48*, 8815–8827. [CrossRef]

44. Makarova, K.S.; Wolf, Y.I.; Koonin, E.V. Comparative genomics of defense systems in archaea and bacteria. *Nucleic Acids Res.* **2013**, *41*, 4360–4377. [CrossRef] [PubMed]

45. van Houte, S.; Buckling, A.; Westra, E.R. Evolutionary Ecology of Prokaryotic Immune Mechanisms. *Microbiol. Mol. Biol. Rev.* **2016**, *80*, 745–763. [CrossRef] [PubMed]

46. Koonin, E.V.; Makarova, K.S.; Wolf, Y.I. Evolutionary Genomics of Defense Systems in Archaea and Bacteria. *Annu. Rev. Microbiol.* **2017**, *71*, 233–261. [CrossRef] [PubMed]

47. Oliveira, P.H.; Touchon, M.; Cury, J.; Rocha, E.P.C. The chromosomal organization of horizontal gene transfer in bacteria. *Nat. Commun.* **2017**, *8*, 841. [CrossRef] [PubMed]

MDPI
St. Alban-Anlage 66
4052 Basel
Switzerland
Tel. +41 61 683 77 34
Fax +41 61 302 89 18
www.mdpi.com

*Microorganisms* Editorial Office
E-mail: microorganisms@mdpi.com
www.mdpi.com/journal/microorganisms